Lecture Notes
in Control and Information Sciences   372

Editors: M. Thoma, M. Morari

Jing Zhou, Changyun Wen

Adaptive Backstepping Control of Uncertain Systems

Nonsmooth Nonlinearities, Interactions or Time-Variations

 Springer

Authors

Jing Zhou

Department of Engineering Cybernetics
Faculty of Information Technology,
Mathematics and Electrical Engineering
Norwegian University of Science and Technology
O.S. Bragstads plass 2d
7491 Trondheim, Norway
Email: jing.zhou@itk.ntnu.no

Changyun Wen

School of Electrical and Electronic Engineering
Nanyang Technological University
Nanyang Avenue
639798 Singapore
Email: ecywen@ntu.edu.sg

ISBN 978-3-540-77806-6   e-ISBN 978-3-540-77807-3

DOI 10.1007/978-3-540-77807-3

Lecture Notes in Control and Information Sciences   ISSN 0170-8643

Library of Congress Control Number: 2007943052

Typeset & Cover Design: Scientific Publishing Services Pvt. Ltd., Chennai, India.

Printed in acid-free paper

5 4 3 2 1 0

springer.com

To Feng Zhou, Lingfang Ma, Xiaozhong Shen and Zhile Shen

J. Zhou

To Xiu Zhou, Wen, Wendy, Qingyun and Qinghao

C. Wen

Preface

This book presents new methodologies for the design and analysis of adaptive control systems based on the backstepping approach. Our emphasis is on dynamic uncertain systems with nonsmooth nonlinearities, such as backlash, dead-zone, hysteresis and saturation, or time-varying parameters, or interactions.

The backstepping approach, a recursive Lyapunov-based scheme, was proposed in the beginning of 1990s. With this method the construction of feedback control laws and Lyapunov functions is systematic, following a step-by-step algorithm. Backstepping can be used to relax the matching condition, which blocked the traditional Lyapunov-based design. A major advantage of backstepping is that it has the flexibility to avoid cancellations of useful nonlinearities and achieve regulation and tracking properties. The technique was comprehensively addressed by Krstic, Kanellakopoulos and Kokotovic in [1]. However, there is still no monograph available to address problems such as the handling of non-smooth nonlinearities, time varying parameters and system interactions using this approach.

Nonsmooth nonlinearities such as dead-zone, backlash, hysteresis and saturation are common in industrial control systems, such as mechanical, hydraulic, biomedical, piezoelectric, and physical systems. Such nonlinearities are usually poorly known and may vary with time, and they often limit system performance. An effective control method should be able to accommodate such common practical nonsmooth nonlinearities. In practice, system parameters are changing with time. Parameter time-variations may arise from linear approximations along different motions and may be due to unmodelled dynamics (for instance friction parameters, electric resistances, inertias) and also some other factors such as changes in environmental conditions. For such systems, the control problem is very complicated and becomes even more difficult to deal with when time-varying parameters are unknown. In the control of large scale systems, one usually faces poor knowledge on plant parameters and interactions between subsystems. Decentralized adaptive control strategy is an efficient and effective way for controlling these systems with large amount of uncertainties. Decentralized adaptive controllers are designed independently for local subsystems by using local available signals for feedback. The major challenge is how to compensate the effects of ignored interactions among subsystems.

In this book, we will present research results on adaptive control of such systems with the backstepping based technique, including theoretical success and practical development such as the approaches for stability analysis and the improvement of system tracking and transient performance. These results are given in two parts:

- The first part involves designing and analyzing adaptive backstepping controllers for multi-input multi-output (MIMO) systems, time-varying systems, or larger scale systems. Newly developed strategies are presented. The designed controllers are shown to guarantee all signals bounded in the system and yield good transient and tracking performance.
- In the second part, we will consider systems with four types of nonsmooth nonlinear characteristics, namely backlash, dead-zone, hysteresis and saturation. It will be shown how these four nonsmooth nonlinear characteristics can be adaptively compensated and how desired system performance is achieved, by incorporating the backstepping technique with other methodologies such as the inverse technique. The proposed adaptive control schemes are shown to ensure the stability of the resulting control system. With these schemes, system performances can be precisely characterized as functions of design parameters and thus is tunable in certain sense by designers. Each of these nonsmooth characteristics is considered individually and systematically. The developed adaptive control methodologies are also applied to the control of base isolation mechanism and piezo-positioning mechanism.

This book is helpful to learn and understand the fundamental backstepping schemes for state feedback control and output feedback control. It can be used as a reference book or a textbook on adaptive control with applications for students with some background in feedback control systems. The book is also intended to introduce researchers and practitioners to the area of adaptive control systems involving the treatment on nonsmooth nonlinearities, interactions and time varying parameters. Researchers, graduate students and engineers in the fields of electrical engineering, control, applied mathematics, computer science and others will benefit from this book.

We are grateful to Nanyang Technological University and Norwegian University of Science and Technology for providing plenty of resources for our research work. Jing Zhou would like to acknowledge StatoilHydro for their support.

We would also like to express our deep sense of gratitude to our parents and families who have made us capable enough to write this book. Jing Zhou is very grateful to her parents, Feng Zhou and Lingfang Ma, and her husband, Xiaozhong Shen, for their care, understanding and encouragement. Changyun Wen is greatly indebted to his wife Xiu Zhou and his children Wen Wen, Wendy Wen, Qingyun Wen and Qinghao Wen for their constant support throughout these years.

Finally, we would like to thank the entire team of Springer publications for their cooperation and encouragement in bringing out the work in the form of monograph in such a short span of time.

Norwegian University of Science and Technology, Norway,   Jing Zhou
Nanyang Technological University, Singapore   Changyun Wen
                                                          November 2007

Contents

1 **Introduction** 1
  1.1 Motivation 1
  1.2 Objectives 3
  1.3 Preview of Chapters 4

Part I Adaptive Backstepping Control Design

2 **Adaptive Backstepping Control** 9
  2.1 Backstepping 9
    2.1.1 Integrator Backstepping 9
    2.1.2 Adaptive Backstepping 12
    2.1.3 Adaptive Backstepping with Tuning Functions 15
  2.2 State Feedback Control 17
  2.3 Output Feedback Control 22
    2.3.1 State Estimation Filters 23
    2.3.2 Design Procedure and Stability Analysis 25

3 **Adaptive Control of Time-Varying Nonlinear Systems** 33
  3.1 Background 33
  3.2 System Model and Problem Formulation 34
    3.2.1 Problem Formulation 34
    3.2.2 Preliminary Result 35
  3.3 State Estimation Filters 38
  3.4 Control Design 41
    3.4.1 Design Procedure 41
    3.4.2 Stability Analysis 46
  3.5 An Illustrative Example 48
  3.6 Summary 50

4   Multivariable Adaptive Control 51
    4.1   Introduction 51
    4.2   Problem Formulation 52
    4.3   Preliminary Results.. 54
    4.4   Backstepping Design with SDU Factorization 59
    4.5   Simulation Studies... 63
    4.6   Summary.. 64

5   Decentralized Stabilization of Interconnected Systems 65
    5.1   Introduction 65
    5.2   Problem Formulation 66
    5.3   Local State Estimation Filters 67
    5.4   Design of Adaptive Controllers 68
    5.5   Stability Analysis.. 71
    5.6   An Illustrative Example 74
    5.7   Conclusion 78

Part II Nonsmooth Nonlinearities

6   Nonsmooth Nonlinearities 83
    6.1   Backlash 84
        6.1.1   Valve Control Mechanism 85
        6.1.2   Positioning System 85
        6.1.3   Piezoelectric Actuator 87
    6.2   Dead-Zone... 87
        6.2.1   Upper-Limb Model 87
        6.2.2   Ultrasonic Motor 87
        6.2.3   Servo-Valve....................................... 88
    6.3   Saturation 89
        6.3.1   Active Micro-gravity Isolation System 90
        6.3.2   Power Supply 91
        6.3.3   Fligh Control with Saturating Actuator 91
    6.4   Hysteresis 92
        6.4.1   Magnetic Suspension with Hysteresis................. 95
        6.4.2   Hysteresis Motor 96
        6.4.3   Hysteresis in Brakes 96

**7   Backstepping Control of Systems with Backlash
Nonlinearity** 97
    7.1   Introduction 97
    7.2   State Feedback Control 98
        7.2.1   Problem Formulation 98
        7.2.2   Backstepping Design and Stability Analysis 99
                Control Scheme I.................................. 100
                Control Scheme II 103
        7.2.3   An Illustrative Example 106

7.3   Output Feedback Control............................... 108
      7.3.1 Plant Model 109
      7.3.2 State Estimation Filters 109
      7.3.3 Design of Adaptive Controllers 111
             Design Procedure 111
             Stability Analysis 113
      7.3.4 Extension to Nonlinear Systems 116
      7.3.5 Simulation Studies 119
             Design Example 1: Output Feedback Control 119
             Design Example 2: Valve Control Mechanism 120
7.4   Summary... 122

8   Inverse Control of Systems with Backlash Nonlinearity 125
    8.1   Introduction 125
    8.2   Problem Statement 126
          8.2.1 System Model 126
          8.2.2 Backlash Characteristic 127
    8.3   State Estimation Filters 130
    8.4   Design of Adaptive Controllers 131
    8.5   Simulation Study 136
    8.6   Conclusion 138

9   Stabilization of Interconnected Systems with Backlash
    Nonlinearity 139
    9.1   Introduction 139
    9.2   Problem Formulation 141
    9.3   Local State Estimation Filters 143
    9.4   Design of Adaptive Controllers 145
          9.4.1 Control Scheme I.......................... 145
          9.4.2 Control Scheme II 148
    9.5   Stability Analysis................................ 150
          9.5.1 Control Scheme I.......................... 150
          9.5.2 Control Scheme II 156
    9.6   An Illustrative Example 159
    9.7   Conclusion 164

10 Adaptive Control of Nonlinear Systems with Dead-Zone
    Nonlinearity 165
    10.1 Introduction 165
    10.2 State Feedback Backstepping Control 166
          10.2.1 Problem Statement 166
          10.2.2 Controller Design 168
    10.3 Output Feedback Control Using Backstepping and Inverse
          Technique 170
          10.3.1 System Model 170
          10.3.2 Dead-Zone Characteristic................... 171

10.3.3 State Observer 173
10.3.4 Backstepping Design with Dead-Zone Inverse 174
Design Procedure 174
10.3.5 Stability Analysis 179
10.4 Illustrative Examples 182
10.4.1 Example 1: State Feedback Backstepping Control 182
10.4.2 Example 2: Output Feedback Inverse Control 183
10.4.3 Example 3: Application to Servo-Valve 185
10.5 Summary 188

11 Adaptive Control of Systems with Input Saturation 189
11.1 Introduction 189
11.2 System Description and Problem Statement 190
11.3 Design of Adaptive Controllers 190
11.4 Simulation Study 195
11.5 Conclusion 197

**12 Control of a Hysteretic Structural System in Base
Isolation Scheme** 199
12.1 Introduction 199
12.2 Problem Formulation 200
12.3 Control Design and Main Results 203
12.3.1 Control Scheme I 203
12.3.2 Control Scheme II 207
12.4 Simulation Results 208
12.5 Summary 212

13 Control of a Piezo-Positioning Mechanism with Hysteresis .. 215
13.1 Introduction 215
13.2 System Description 216
13.3 Backstepping Control and Stability Analysis 218
Control Scheme I 218
Control Scheme II 222
13.4 Simulation Results 224
13.5 Conclusion 226

A Appendices 227

References 233

1 Introduction

There have been great efforts to develop high performance control schemes for unknown plants subject to uncertainties. Adaptive control has been proved to be one of the most promising techniques which can be applied to control a wide variety of systems and processes. Adaptive control theory attempts to improve the behavior or performance of physical systems by gathering and exploiting knowledge about the system's operation. Usually this knowledge is encoded as a descriptive mathematical model of the physical plant from which the controller design is derived. Given a mathematical representation, we are interested in designing an adaptive controller to achieve objectives such as stability (convergence of state), output tracking of some reference signals, and transient performance.

1.1 Motivation

Adaptive control has been an important area of active research for over five decades now. Significant development has been seen, including theoretical success and practical development, such as the proof of global stability and the improvement of system tracking and transient performance. One of the reasons for the rapid growth of adaptive control is its ability to control plants with uncertainties during its operation. Adaptive control is a technique of applying some methods to obtain a model of the process and using this model to design a controller. An adaptive controller is designed by combining a parameter estimator, which provides estimates of unknown parameters, with a control law. The parameters of the controller are adjusted during the operation of the plant. In order to obtain desired performances, it also provides adaptation methods to dealt with some uncertainties, such as flow and speed variations, external disturbance and structural uncertainties. One important approach in adaptive control is certainty equivalence based design. Such an approach has been studied extensively and a number of results have been established [2, 3, 4, 5, 6, 7, 8]. Certain schemes have also been proposed to study the robustness issues in the context of both single loop control [9, 10, 11, 12, 13, 14, 15, 16] and decentralized control of multi-loop systems [17, 18, 19, 20, 21, 22, 23, 24, 25]. Problems related

J. Zhou & C. Wen: Adapt. Backstepping Ctrl. of Uncertain Systems, LNCIS 372, pp. 1–5, 2008.
springerlink.com   © Springer-Verlag Berlin Heidelberg 2008

to nonsmooth nonlinearities have also been well addressed in [26, 27, 28, 29]. However, transient performance is difficult to be ensured with this approach.

In the beginning of 1990s, a new approach called "backstepping" was proposed for the design of adaptive controllers. Backstepping is a recursive Lyapunov-based scheme for the class of "strict feedback" systems. In fact, when the controlled plant belongs to the class of systems transformable into the parametric-strict feedback form, this approach guarantees global or regional regulation and tracking properties. An important advantage of the backstepping design method is that it provides a systematic procedure to design stabilizing controllers, following a step-by-step algorithm. With this method the construction of feedback control laws and Lyapunov functions is systematic. Another advantage of backstepping is that it has the flexibility to avoid cancellations of useful nonlinearities and achieve stabilization and tracking. A number of results using this approach has been obtained [30, 31, 32, 33, 34, 35, 36, 37, 38, 39, 40]. However, some important practical problems such as the handling of nonsmooth nonlinearity have not been addressed using this approach.

Most practical systems are multi-input multi-output (MIMO) systems. And usually parameters of practical systems are changing with time. For such systems, the control problem is very complicated due to the coupling among various inputs and outputs. It becomes even more difficult to deal with when there exist unknown parameters in the input or output coupling matrix. In this book, we will address a class of nonlinear MIMO systems with unknown disturbance.

In the control of interconnected systems, decentralized control strategy is an efficient and effective way. In particular, decentralized adaptive control is employed for controlling systems with large amount of uncertainties. Decentralized adaptive controllers are designed independently for local subsystems by using local available signals for feedback. Such decentralized controllers should be robust against the ignored interactions. Due to difficulties to consider the effects of interconnections, extension of single loop results to multi-loop interconnected systems is challenging, which is why the number of available results is still limited. In this book, we present output feedback decentralized stabilizers for a class of interconnected systems with subsystem having arbitrary relative degrees.

Nonsmooth nonlinearities such as dead-zone [41, 42], backlash [28, 43], hysteresis [44, 45] and saturation [46, 47] are common in industrial control systems. Backlash, a dynamic characteristic, exists in mechanical systems, such as in hydraulic actuators. Dead-zone, a static input-output characteristic, often appears in motors, valves and biomedical actuation systems. Hysteresis, another dynamic characteristic, exists in a wide range of physical systems and devices. Saturation is always a potential problem for actuators of control systems as all actuators do saturate at some level.

Such nonlinearities are usually poorly known and may vary with time, and they often limit system performance. Control of systems with nonsmooth nonlinearities is an important area of control system research. A desirable control design approach for such systems should be able to accommodate system uncertainties. The need for effective control methods to deal with nonsmooth systems has motivated growing research activities in adaptive control of systems with such common practical nonsmooth nonlinearities [48, 49]. Various design methods based on different control objectives and system conditions have been developed and verified in theory and practice. Adaptive control schemes have been used to cope with actuator dead zone [50, 51, 52, 53, 54], backlash [28, 44, 55], hysteresis [45, 56, 57, 58] and saturation [46, 59, 60, 61, 62, 63]. Other schemes to handle such nonlinearities have included neural networks control in [42, 64, 65, 66, 67, 68], fuzzy logic control in [41, 69, 70, 71, 72], variable structure control in [43, 52, 73, 74, 75, 76, 77, 78], pole placement control in [46, 47, 60] and recursive least square algorithm in [79].

In this book, adaptive control methods based on backstepping technique incorporated with other methodologies, such as inverse technique, are proposed to handle uncertain dynamic systems containing backlash, hysteresis, dead-zone or saturation in the actuator. Each of these nonsmooth characteristics needs a systematic consideration. In this book it will be shown how nonsmooth nonlinear industrial characteristics can be adaptively compensated and how desired system performance is achieved. The controller designed by using backstepping technique consists of new robust control laws and new estimators to estimate the unknown parameters. Besides showing stability of the system, the transient performance of the tracking error is derived to be an explicit function of design parameters and thus tunable.

1.2 Objectives

The main objectives of this book are as follows:

- The first part of the book is to introduce the backstepping design for state feedback control and output feedback control. New strategy will be proposed for a class of uncertain nonlinear systems with unknown time-varying parameters and unknown sign of high frequency gains in the presence of disturbances. We will also develop a new backstepping control scheme for a class of multi-input multi-output nonlinear systems with respect to parameter uncertainty. In the last chapter of this part, decentralized adaptive stabilization of a class of interconnected systems will be presented.

- The second part of the book is to control systems with four most common component imperfections: backlash, dead-zone, hysteresis and saturation. These nonsmooth nonlinearities have different characteristics, so we will consider each of them individually in each of controller designs. We aim at designing, analyzing and implementing adaptive backstepping control which are able to accommodate such uncertain nonsmooth nonlinearities by introducing new adaptive control schemes to overcome or compensate the effect of these nonlinearities. In the last two chapters of this part, we will present applications in base isolation mechanism and piezo-positioning mechanism.

1.3   Preview of Chapters

This book is divided into thirteen chapters. Each of chapters 3-13 contains a new contribution. We now preview chapters 2-13.

In Chapter 2, we introduce adaptive backstepping tools illustrated with application to a class of nonlinear systems. To grasp the design and analysis procedure easily, we start with simple systems. Recursive design procedures using both state-feedback and output feedback are presented. Approaches of establishing system stability and performances are also given.

In Chapter 3, an adaptive output feedback controller is proposed for single-input single-output uncertain time-varying systems in the presence of unknown bounded disturbances. Both theoretical analysis and simulation results show that the controller designed with the proposed scheme can make the whole adaptive control system stable.

In Chapter 4, we design an adaptive output feedback controller for a class of multiple-input multiple-output systems in the presence of unknown disturbances. In order to reject external disturbances, new filters for state estimation are constructed and an adaptive internal model is employed. The controller design is achieved by using backstepping, tuning functions, SDU factorization and estimation of parameters.

In Chapter 5, adaptive backstepping control is employed to design decentralized adaptive regulators. The interconnected system to be regulated consists of N coupled subsystems having arbitrarily relative degrees. Global stability is established for the closed-loop system and perfect regulation is ensured.

In Chapter 6, we provide certain basic description of nonsmooth nonlinear characteristics: dead-zone, backlash, hysteresis and saturation with some typical examples for illustration.

In Chapter 7, we address adaptive control of uncertain systems preceded by unknown backlash nonlinearity in either state feedback or output feedback control. Detailed design and analysis are given, including structure, stability, and convergence of the algorithms.

In Chapter 8, an adaptive inverse is employed for cancelling the effect of backlash nonlinearity and an adaptive backstepping controller is developed.

In Chapter 9, we provide a solution to the problem on the relaxation of subsystem relative degrees in decentralized control of interconnected systems with each loop preceded by unknown backlash-like hysteresis nonlinearities. It is shown that overall control system is global stable in the sense that all the signals are bounded.

In Chapter 10, adaptive backstepping control technique will be extended to state feedback control and output feedback control of nonlinear systems with unknown dead-zone. The main contribution of this chapter is to develop adaptive schemes for uncertain dynamic nonlinear systems with unknown dead-zone nonlinearity, in the presence of bounded external disturbances. We will give the detailed design procedure and stability analysis.

In Chapter 11, we present a new scheme to design adaptive controllers for uncertain systems in the presence of input saturation. By using backstepping technique, a new robust adaptive control algorithm is developed. Besides showing stability, performance on tracking error is also established.

In Chapter 12, we address a second-order uncertain structural system found in base isolation schemes for seismic active protection of building structures. This system exhibits a hysteretic nonlinear behavior, which is described by the so-called Bouc-Wen model. Numerical results show that the adaptive control laws work satisfactorily in the sense that the response induced by seismic action is significant reduced.

In Chapter 13, we develop robust adaptive backstepping control algorithms for piezo-positioning mechanisms. Due to their materials, nonlinear hysteretic behavior is commonly observed in such mechanisms and can be described by a LuGre model. It is shown that not only global stability is guaranteed, but also both transient and asymptotic tracking performances are quantified as explicit functions of the design parameters so that designers can tune the design parameters in an explicit way to obtain the required closed loop behavior.

Part I

Adaptive Backstepping Control Design

2 Adaptive Backstepping Control

Recursive design in this book is composed of some simple basic steps. They are referred to as "backstepping designs" because they step back toward the control input starting with a scalar equation. This chapter reviews basic backstepping tools for state feedback control and output feedback control. To easily grasp the design and analysis procedures, we start with simple low order systems. Then the ideas are generalized to arbitrarily n order systems.

2.1 Backstepping

2.1.1 Integrator Backstepping

Backstepping is a recursive Lyapunov-based scheme proposed in the beginning of 1990s. The technique was comprehensively addressed by Krstic, Kanellakopoulos and Kokotovic in [1]. The idea of backstepping is to design a controller recursively by considering some of the state variables as "virtual controls" and designing for them intermediate control laws. Backstepping achieves the goals of stabilization and tracking. The proof of these properties is a direct consequence of the recursive procedure, because a Lyapunov function is constructed for the entire system including the parameter estimates.

To give a clear idea of such development, we consider the following third order strict-feedback system.

$$\dot{x}_1 = x_2 + x_1^2$$
$$\dot{x}_2 = x_3 + x_2^2$$
$$\dot{x}_3 = u \tag{2.1}$$

where x_1, x_2 and x_3 are system sates and u is control input. The control objective is to design a state feedback control to asymptotically stabilize the origin.

Step 1. Start with the first equation of (2.1), we define $z_1 = x_1$ and derive the dynamics of the new coordinate

$$\dot{z}_1 = x_2 + x_1^2 \tag{2.2}$$

J. Zhou & C. Wen: Adapt. Backstepping Ctrl. of Uncertain Systems, LNCIS 372, pp. 9–31, 2008.
springerlink.com

We view x_2 as a control variable and define a virtual control law for (2.2), say α_1, and let z_2 be an error variable representing the difference between the actual and virtual controls of (2.2), i.e.,

$$z_2 = x_2 - \alpha_1 \tag{2.3}$$

Thus in terms of the new state variable, we can rewrite (2.2) as

$$\dot{z}_1 = \alpha_1 + x_1^2 + z_2 \tag{2.4}$$

In this step, our objective is to design a virtual control law α_1 which makes $z_1 \to 0$. Consider a control Lyapunov function

$$V_1 = \frac{1}{2} z_1^2 \tag{2.5}$$

The time derivative of which becomes

$$\dot{V}_1 = z_1(\alpha_1 + x_1^2) + z_1 z_2 \tag{2.6}$$

We can now select an appropriate virtual control α_1, which would make the first order system stabilizable.

$$\alpha_1 = -c_1 z_1 - x_1^2 \tag{2.7}$$
$$\dot{\alpha}_1 = -(c_1 + 2x_1)(x_2 + x_1^2) \tag{2.8}$$

where c_1 is a positive constant. Then the time derivative of V_1 becomes

$$\dot{V}_1 = -c_1 z_1^2 + z_1 z_2 \tag{2.9}$$

Clearly, if $z_2 = 0$, then $\dot{V}_1 = -c_1 z_1^2$ and z_1 is guaranteed to converge to zero asymptotically.

Step 2. We derive the error dynamics for $z_2 = x_2 - \alpha_1$.

$$\begin{aligned} \dot{z}_2 &= \dot{x}_2 - \dot{\alpha}_1 \\ &= x_3 + x_2^2 + (c_1 + 2x_1)(x_2 + x_1^2) \end{aligned} \tag{2.10}$$

in which x_3 is viewed as a virtual control input. Define a virtual control law α_2 and let z_3 be an error variable representing the difference between the actual and virtual controls

$$z_3 = x_3 - \alpha_2 \tag{2.11}$$

Then (2.10) becomes

$$\dot{z}_2 = z_3 + \alpha_2 + x_2^2 + (c_1 + 2x_1)(x_2 + x_1^2) \tag{2.12}$$

The control objective is to make $z_2 \to 0$. Choose a control lyapunov function

$$V_2 = V_1 + \frac{1}{2} z_2^2 \tag{2.13}$$

Taking time derivative gives

$$\begin{aligned}
\dot{V}_2 &= \dot{V}_1 + z_2\dot{z}_2 \\
&= -c_1 z_1 + z_1 z_2 + z_2(z_3 + \alpha_2 + x_2^2 + (c_1 + 2x_1)(x_2 + x_1^2)) \\
&= -c_2 z_2^2 + z_2(\alpha_2 + z_1 + x_2^2 + (c_1 + 2x_1)(x_2 + x_1^2) + z_2 z_3 \quad (2.14)
\end{aligned}$$

We can now select an appropriate virtual control α_2 to cancel some terms related to z_1, x_1 and x_2, while the term involving z_3 cannot be removed

$$\alpha_2 = -z_1 - c_2 z_2 - x_2^2 - (c_1 + 2x_1)(x_2 + x_1^2) \quad (2.15)$$

where c_2 is a positive constant. So the time derivative of V_2 becomes

$$\dot{V}_2 = -c_1 z_1^2 - c_2 z_2^2 + z_2 z_3 = -\sum_{i=1}^{2} c_i z_i^2 + z_2 z_3 \quad (2.16)$$

Clearly, if $z_3 = 0$, we have $\dot{V}_2 = -\sum_{i=1}^{2} c_i z_i^2$, and thus both z_1 and z_2 are guaranteed to converge to zero asymptotically.

Step 3. Proceeding to the last equation in (2.1), we derive the error dynamics for $z_3 = x_3 - \alpha_2$.

$$\dot{z}_3 = u - \frac{\partial \alpha_2}{\partial x_1}(x_2 + x_1^2) - \frac{\partial \alpha_2}{\partial x_2}(x_3 + x_2^2) \quad (2.17)$$

In this equation, the actual control input u appears and is at our disposal. Our objective is to design the actual control input u such that z_1, z_2, z_3 converge to zero. Choose a Lyapunov function V_3 as

$$V_3 = V_2 + \frac{1}{2} z_3^2 \quad (2.18)$$

Its time derivative is given by

$$\dot{V}_3 = -\sum_{i=1}^{2} c_i z_i^2 + z_3\left(u + z_2 - \frac{\partial \alpha_2}{\partial x_1}(x_2 + x_1^2) - \frac{\partial \alpha_2}{\partial x_2}(x_3 + x_2^2)\right) \quad (2.19)$$

We are finally in the position to design control u by making $\dot{V}_3 \leq 0$ as follows

$$u = -z_2 - c_3 z_3 + \frac{\partial \alpha_2}{\partial x_1}(x_2 + x_1^2) + \frac{\partial \alpha_2}{\partial x_2}(x_3 + x_2^2) \quad (2.20)$$

where c_3 is a positive constant. Then the derivative of Lyapunov function of V_3 is

$$\dot{V}_3 = -\sum_{i=1}^{3} c_i z_i^2 \quad (2.21)$$

Then the Lasalle Theorem given in Appendix B guarantees the global uniform boundedness of z_1, z_2 and z_3. It follows that $z_1, z_2, z_3 \to 0$ as $t \to \infty$. Since $x_1 = z_1$, x_1 is also bounded and $\lim_{t \to \infty} x_1 = 0$. The boundedness of x_2 follows from boundedness of α_1 in (2.7) and the fact that $x_2 = z_2 + \alpha_1$. Similarly, the boundedness of x_3 then follows from boundedness of α_2 in (2.15) and the fact that and $x_3 = z_3 + \alpha_2$. Combining this with (2.20) we conclude that the control $u(t)$ is also bounded.

With the above example, the idea of backstepping has been illustrated. In the following, we will consider parametric uncertainties and achieve both boundedness of the closed-loop states and asymptotic tracking.

2.1.2   Adaptive Backstepping

In this section, we will consider unknown parameters which appear linearly in system equations. An adaptive controller is designed by combining a parameter estimator, which provides estimates of unknown parameters, with a control law. The parameters of the controller are adjusted during the operation of the plant. In the presence of such parametric uncertainties, the adaptive controller is able to ensure the boundedness of the closed-loop states and asymptotic tracking.

To illustrate the idea of adaptive backstepping, let us first consider the following second order system.

$$\dot{x}_1 = x_2 + \phi_1^T(x_1)\theta$$
$$\dot{x}_2 = u + \phi_2^T(x_1, x_2)\theta \tag{2.22}$$

where $\theta \in R^r$ is an unknown vector constant, and $\phi_1 \in R^r$ and $\phi_2 \in R^r$ are known nonlinear functions. Our problem is to globally stabilize the system and also to achieve the asymptotic tracking of x_r by x_1.

For the development of control laws, the following assumption is made.

Assumption 1: The reference signal x_r and its first second order derivative are piecewise continuous and bounded.

The design procedure is elaborated in the following. Introduce the change of coordinates

$$z_1 = x_1 - x_r \tag{2.23}$$
$$z_2 = x_2 - \alpha_1 - \dot{x}_r \tag{2.24}$$

where α_1 is called virtual control and will be determined in later discussion.

If θ is known, we would apply the static integrator backstepping to design a virtual control law

$$\alpha_1 = -c_1 z_1 - \phi_1^T \theta \tag{2.25}$$

with the control Lyapunov function

$$V = \frac{1}{2}z_1^2 + \frac{1}{2}z_2^2 \tag{2.26}$$

whose derivative is rendered negative definite

$$\dot{V} = z_1 \dot{z}_1 + z_2 \dot{z}_2$$
$$= -c_1 z_1^2 + z_1 z_2 + z_2 (u + \phi_2^T \theta - \frac{\partial \alpha_1}{\partial x_1}(x_2 + \phi_1^T \theta) - \frac{\partial \alpha_1}{\partial x_r}\dot{x}_r - \ddot{x}_r)$$
$$= -c_1 z_1^2 - c_2 z_2^2 \tag{2.27}$$

by choosing the control

$$u = -z_1 - c_2 z_2 - \phi_2^T \theta + \frac{\partial \alpha_1}{\partial x_1}(x_2 + \phi_1^T \theta) + \frac{\partial \alpha_1}{\partial x_r}\dot{x}_r + \ddot{x}_r \tag{2.28}$$

Since θ is unknown, we cannot apply the controller (2.28) with virtual control law (2.25). However, we can use the idea of integrator backstepping and parameter estimation.

Step 1. We start with the first equation of (2.22) by considering x_2 as control variable. The derivative of tracking error z_1 is given as

$$\dot{z}_1 = z_2 + \alpha_1 + \phi_1^T \theta \tag{2.29}$$

Since θ is unknown, this task is fulfilled with an adaptive controller consisting of a control law and an update law to obtain an estimate of θ.

Design the first stabilizing function α_1 and parameter updating law $\dot{\hat{\theta}}_1$ as

$$\alpha_1 = -c_1 z_1 - \phi_1^T \hat{\theta}_1 \tag{2.30}$$
$$\dot{\hat{\theta}}_1 = \Gamma \phi_1 z_1 \tag{2.31}$$

where $\hat{\theta}_1$ is an estimate of θ, c_1 is a positive constant and Γ is a positive definite matrix. Our task in this step is to stabilize (2.29) with respect to the Lyapunov function

$$V_1 = \frac{1}{2}z_1^2 + \frac{1}{2}\tilde{\theta}_1^T \Gamma^{-1} \tilde{\theta}_1 \tag{2.32}$$

where $\tilde{\theta}_1 = \theta - \hat{\theta}_1$. Then the derivative of V_1 is given by

$$\dot{V}_1 = z_1 \dot{z}_1 - \tilde{\theta}_1^T \Gamma^{-1} \dot{\hat{\theta}}_1$$
$$= -c_1 z_1^2 + z_1 z_2 - \tilde{\theta}_1^T \Gamma^{-1}(\dot{\hat{\theta}}_1 - \Gamma \phi_1 z_1)$$
$$= -c_1 z_1^2 + z_1 z_2 \tag{2.33}$$

Step 2. The derivative of z_2 with (2.30) and (2.31) is now expressed as

$$\dot{z}_2 = \dot{x}_2 - \dot{\alpha}_1 - \ddot{x}_r$$
$$= u + \phi_2^T \theta - \frac{\partial \alpha_1}{\partial x_1}(x_2 + \phi_1^T \theta) - \frac{\partial \alpha_1}{\partial \hat{\theta}_1}\dot{\hat{\theta}}_1 - \frac{\partial \alpha_1}{\partial x_r}\dot{x}_r - \ddot{x}_r$$
$$= u - \frac{\partial \alpha_1}{\partial x_1}x_2 + (\phi_2 - \frac{\partial \alpha_1}{\partial x_1}\phi_1)^T \theta - \frac{\partial \alpha_1}{\partial \hat{\theta}_1}\Gamma \phi_1 z_1 - \frac{\partial \alpha_1}{\partial x_r}\dot{x}_r - \ddot{x}_r \tag{2.34}$$

In this equation, the actual control input u appears and is at our disposal. At this point we need to select a control Lyapunov function and design u to make its derivative non-positive. We use a Lyapunov function

$$V_2 = V_1 + \frac{1}{2}z_2^2 \tag{2.35}$$

whose derivative is

$$
\begin{aligned}
\dot{V}_2 &= \dot{V}_1 + z_2 \dot{z}_2 \\
&= -c_1 z_1^2 + z_2 \Big[u + z_1 - \frac{\partial \alpha_1}{\partial x_1} x_2 + (\phi_2 - \frac{\partial \alpha_1}{\partial x_1}\phi_1)^T \theta - \frac{\partial \alpha_1}{\partial \hat{\theta}_1}\Gamma\phi_1 z_1 \\
&\quad - \frac{\partial \alpha_1}{\partial x_r}\dot{x}_r - \ddot{x}_r \Big]
\end{aligned}
\tag{2.36}
$$

The control u is able to cancel the rest six terms in (2.36) to ensure $\dot{V}_2 \le 0$. To deal with the term containing the unknown parameter θ, we try to use the estimate $\hat{\theta}_1$ designed in the first step

$$
\begin{aligned}
u &= -z_1 - c_2 z_2 + \frac{\partial \alpha_1}{\partial x_1} x_2 - (\phi_2 - \frac{\partial \alpha_1}{\partial x_1}\phi_1)^T \hat{\theta}_1 \\
&\quad + \frac{\partial \alpha_1}{\partial \hat{\theta}_1}\Gamma\phi_1 z_1 + \frac{\partial \alpha_1}{\partial x_r}\dot{x}_r + \ddot{x}_r
\end{aligned}
\tag{2.37}
$$

where c_2 is a positive constant. The resulting derivatives of V_2 is given as

$$\dot{V}_2 = -c_1 z_1^2 - c_2 z_2^2 + (\phi_2 - \frac{\partial \alpha_1}{\partial x_1}\phi_1)^T (\theta - \hat{\theta}_1) \tag{2.38}$$

It can be observed that term $(\phi_2 - \frac{\partial \alpha_1}{\partial x_1}\phi_1)^T (\theta - \hat{\theta}_1)$ cannot be cancelled. To eliminate this term, we need to treat θ in the equation (2.34) as a new parameter vector and assign to it a new estimate $\hat{\theta}_2$ by selecting u as

$$
\begin{aligned}
u &= -z_1 - c_2 z_2 + \frac{\partial \alpha_1}{\partial x_1} x_2 - (\phi_2 - \frac{\partial \alpha_1}{\partial x_1}\phi_1)^T \hat{\theta}_2 \\
&\quad + \frac{\partial \alpha_1}{\partial \hat{\theta}_1}\Gamma\phi_1 z_1 + \frac{\partial \alpha_1}{\partial x_r}\dot{x}_r + \ddot{x}_r
\end{aligned}
\tag{2.39}
$$

With this choice, (2.34) becomes

$$\dot{z}_2 = -z_1 - c_2 z_2 + (\phi_2 - \frac{\partial \alpha_1}{\partial x_1}\phi_1)^T (\theta - \hat{\theta}_2) \tag{2.40}$$

Our task in this step is to stabilize the (z_1, z_2) system. The presence of the new parameter estimate $\hat{\theta}_2$ suggests the following form of the Lyapunov function:

$$V_2 = V_1 + \frac{1}{2}z_2^2 + \frac{1}{2}\tilde{\theta}_2^T \Gamma^{-1}\tilde{\theta}_2 \tag{2.41}$$

where $\tilde{\theta}_2 = \theta - \hat{\theta}_2$. Then the derivative of Lapunov function of V_2 is

$$
\begin{aligned}
\dot{V}_2 &= \dot{V}_1 + z_2 \dot{z}_2 - \tilde{\theta}_2^T \Gamma^{-1} \dot{\hat{\theta}}_2 \\
&= -c_1 z_1^2 + z_2(-c_2 z_2 + (\phi_2 - \tilde{\theta}_2^T \frac{\partial \alpha_1}{\partial x_1} \phi_1)) - \tilde{\theta}_2^T \Gamma^{-1} \dot{\hat{\theta}}_2 \\
&= -c_1 z_1^2 - c_2 z_2^2 - \tilde{\theta}_2^T \Gamma^{-1} (\dot{\hat{\theta}}_2 - \Gamma(\phi_2 - \frac{\partial \alpha_1}{\partial x_1} \phi_1) z_2)
\end{aligned}
\tag{2.42}
$$

We choose the update law

$$
\dot{\hat{\theta}}_2 = \Gamma(\phi_2 - \frac{\partial \alpha_1}{\partial x_1} \phi_1) z_2
\tag{2.43}
$$

Then the derivative of V_2 gives

$$
\dot{V}_2 = -c_1 z_1^2 - c_2 z_2^2
\tag{2.44}
$$

By using the Lasalle's Theorem (Appendix B), this Lyapunov function (2.44) guarantees the global uniform boundedness of $z_1, z_2, \hat{\theta}_1, \hat{\theta}_2$ and $z_1, z_2 \to 0$ as $t \to \infty$. It follows that asymptotic tracking is achieved, such that $\lim_{t\to\infty}(x_1 - x_r) = 0$. Since z_1 and x_r are bounded, x_1 is also bounded from $x_1 = z_1 + x_r$. The boundedness of x_2 follows from boundedness of $\dot{x}_r$ and α_1 in (2.30) and the fact that $x_2 = z_2 + \alpha_1 + \dot{x}_r$. Combining this with (2.39) we conclude that the control $u(t)$ is also bounded.

In conclusion, the above adaptive backstepping employs the over parametrization estimation, that is two estimates for the same parameter vector θ in this case. This means that the dynamic order of the controller is not of minimal order. In the next section, a new backstepping design will be presented to avoid such a case happening, which employs the minimal number of parameter estimates.

2.1.3   Adaptive Backstepping with Tuning Functions

To give a clear idea of tuning function design, we consider the same system in (2.22) with the same control objective, namely globally stabilization and also asymptotic tracking of x_r by x_1, and use the same change of coordinates

$$
z_1 = x_1 - x_r
\tag{2.45}
$$
$$
z_2 = x_2 - \alpha_1 - \dot{x}_r
\tag{2.46}
$$

where α_2 is virtual control. The design procedure is elaborated as follows.

Step 1. We start with the first equation of (2.22) by considering x_2 as control variable. The derivative of tracking error z_1 is given as

$$
\dot{z}_1 = z_2 + \alpha_1 + \phi_1^T \theta
\tag{2.47}
$$

Our task in this step is to stabilize (2.47). Choose the control Lyapunov function

$$
V_1 = \frac{1}{2} z_1^2 + \frac{1}{2} \tilde{\theta}^T \Gamma^{-1} \tilde{\theta}
\tag{2.48}
$$

where Γ is a positive definite matrix, $\tilde{\theta} = \theta - \hat{\theta}$. Then the derivative of V_1 is

$$\dot{V}_1 = z_1 \dot{z}_1 - \tilde{\theta}_1^T \Gamma^{-1} \dot{\hat{\theta}}$$
$$= z_1(z_2 + \alpha_1 + \phi_1^T \tilde{\theta}) - \tilde{\theta}^T (\Gamma^{-1} \dot{\hat{\theta}} - \phi_1 z_1) \tag{2.49}$$

We may eliminate $\tilde{\theta}$ by choosing $\dot{\hat{\theta}} = \Gamma \phi_1 z_1$. If x_2 is actual control and let $z_2 = 0$, we choose α_1 to make $\dot{V}_1 \leq 0$.

$$\alpha_1 = -c_1 z_1 - \phi_1^T \hat{\theta} \tag{2.50}$$

where c_1 is a positive constant, $\hat{\theta}$ is an estimate of θ. However, to overcome the over-parametrization problem caused by the appearance of θ as shown in the previous subsection, we do not use this update law to estimate θ. Instead, we define a function τ_1, named tuning function, as follows

$$\tau_1 = \phi_1 z_1 \tag{2.51}$$

The resulting derivative of V_1 is

$$\dot{V}_1 = -c_1 z_1^2 + z_1 z_2 - \tilde{\theta}^T (\Gamma^{-1} \dot{\hat{\theta}} - \tau_1) \tag{2.52}$$

Step 2. We derive the second tracking error for z_2

$$\dot{z}_2 = u - \frac{\partial \alpha_1}{\partial x_1} x_2 + (\phi_2 - \frac{\partial \alpha_1}{\partial x_1}\phi_1)^T \theta - \frac{\partial \alpha_1}{\partial \hat{\theta}} \dot{\hat{\theta}} - \frac{\partial \alpha_1}{\partial x_r} \dot{x}_r - \ddot{x}_r \tag{2.53}$$

In this equation, the actual control input u appears and is at our disposal. The control Lyapunov function is selected as

$$V_2 = V_1 + \frac{1}{2}z_2^2 = \frac{1}{2}z_1^2 + \frac{1}{2}z_2^2 + \frac{1}{2}\tilde{\theta}^T \Gamma^{-1} \tilde{\theta} \tag{2.54}$$

Our task is to make $\dot{V}_2 \leq 0$.

$$\dot{V}_2 = -c_1 z_1^2 + z_2\left(u + z_1 - \frac{\partial \alpha_1}{\partial x_1}x_2 + \hat{\theta}^T(\phi_2 - \frac{\partial \alpha_1}{\partial x_1}\phi_1) - \frac{\partial \alpha_1}{\partial \hat{\theta}}\dot{\hat{\theta}}\right.$$
$$\left. - \frac{\partial \alpha_1}{\partial x_r}\dot{x}_r - \ddot{x}_r\right) + \tilde{\theta}^T\left(\tau_1 + (\phi_2 - \frac{\partial \alpha_1}{\partial x_1}\phi_1)z_2 - \Gamma^{-1}\dot{\hat{\theta}}\right) \tag{2.55}$$

Finally, we can eliminate the $\tilde{\theta}$ term from (2.55) by designing the update law as

$$\dot{\hat{\theta}} = \Gamma \tau_2 \tag{2.56}$$

where τ_2 is called the second tuning function and is selected as

$$\tau_2 = \tau_1 + (\phi_2 - \frac{\partial \alpha_1}{\partial x_1}\phi_1)z_2 \tag{2.57}$$

Then

$$\dot{V}_2 = -c_1 z_1^2 + z_2 \left(u + z_1 - \frac{\partial \alpha_1}{\partial x_1} x_2 + \hat{\theta}^T (\phi_2 - \frac{\partial \alpha_1}{\partial x_1} \phi_1) - \frac{\partial \alpha_1}{\partial \hat{\theta}} \Gamma \tau_2 \right.$$
$$\left. - \frac{\partial \alpha_1}{\partial x_r} \dot{x}_r - \ddot{x}_r \right) \tag{2.58}$$

To stabilize the system (2.53), the actual control input is selected to remove the residual term and make $\dot{V}_2 \leq 0$

$$u = -z_1 - c_2 z_2 + \frac{\partial \alpha_1}{\partial x_1} x_2 - \hat{\theta}^T (\phi_2 - \frac{\partial \alpha_1}{\partial x_1} \phi_1) + \frac{\partial \alpha_1}{\partial \hat{\theta}} \Gamma \tau_2 + \frac{\partial \alpha_1}{\partial x_r} \dot{x}_r + \ddot{x}_r \tag{2.59}$$

where c_2 is a positive constant. The resulting of derivative of V_2 is

$$\dot{V}_2 = -c_1 z_1^2 - c_2 z_2^2 \tag{2.60}$$

This Lyapunov function provides the proof of uniform stability and the proof of asymptotic tracking $x_1(t) - x_r(t) \to 0$.

The controller designed in this section also achieves the goals of stabilization and tracking. By using tuning functions, only one update law is used to estimate unknown parameter θ. This avoids the over-parametrization problem and reduces the dynamic order of the controller to its minimum.

2.2 State Feedback Control

The adaptive backstepping design with tuning functions is now generalized to a class of nonlinear system as in the following parametric strict-feedback form

$$\dot{x}_1 = x_2 + \phi_1^T(x_1)\theta + \psi_1(x_1)$$
$$\dot{x}_2 = x_3 + \phi_2^T(x_1, x_2)\theta + \psi_2(x_1, x_2)$$
$$\vdots \quad \vdots$$
$$\dot{x}_{n-1} = x_n + \phi_{n-1}^T(x_1, \ldots, x_{n-1})\theta + \psi_n(x_1, \ldots, x_{n-1})$$
$$\dot{x}_n = bu + \phi_n^T(x)\theta + \psi_n(x) \tag{2.61}$$

where $x = [x_1, \ldots, x_n]^T \in R^n$, the vector $\theta \in R^r$ is constant and unknown, $\phi_i \in R^r, \psi_1 \in R, \ i = 1, \ldots, n$ are known nonlinear functions, and the high frequency gain b is an unknown constant.

The control objective is to force the output x_1 to asymptotically track the reference signal x_r with the following assumptions.

Assumption 1: The sign of b is known.

Assumption 2: The reference signal x_r and its n order derivatives are piecewise continuous and bounded.

For system (2.61), the number of design steps required is equal to n. At each step, an error variable z_i, a stabilizing function α_i and a tuning function τ_i are generated. Finally, the control u and a parameter estimate $\hat{\theta}$ are developed. Introduce the change of coordinates

$$z_1 = x_1 - x_r \tag{2.62}$$

$$z_i = x_i - \alpha_{i-1} - x_r^{(i-1)}, \quad i = 2, 3, \ldots, n \tag{2.63}$$

where α_i are virtual controllers. The design procedure is elaborated in the following steps.

Step 1. We start with the first equation of (2.61) by considering x_2 as virtual control variable. The derivative of tracking error z_1 is given as

$$\dot{z}_1 = \dot{x}_1 - \dot{x}_r$$
$$= z_2 + \alpha_1 + \phi_1^T \theta + \psi_1 \tag{2.64}$$

Designing the first stabilizing function α_1 as

$$\alpha_1 = -c_1 z_1 - \phi_1^T \hat{\theta} - \psi_1 \tag{2.65}$$

where c_1 is a positive constant and $\hat{\theta}$ is an estimate of θ. Our task in this step is to achieve the tracking task $x_1 \to x_r$ by considering the Lyapunov function

$$V_1 = \frac{1}{2} z_1^2 + \frac{1}{2} \tilde{\theta}^T \Gamma^{-1} \tilde{\theta} \tag{2.66}$$

where Γ is a positive definite matrix and $\tilde{\theta} = \theta - \hat{\theta}$. Then the derivative of V_1 is

$$\dot{V}_1 = z_1 \dot{z}_1 - \tilde{\theta}_1^T \Gamma^{-1} \dot{\hat{\theta}}$$
$$= z_1 (z_2 + \alpha_1 + \phi_1^T \hat{\theta} + \psi_1) - \tilde{\theta}^T (\Gamma^{-1} \dot{\hat{\theta}}_1 - \phi_1 z_1)$$
$$= -c_1 z_1^2 + \tilde{\theta}^T (\tau_1 - \Gamma^{-1} \dot{\hat{\theta}}) + z_1 z_2 \tag{2.67}$$
$$\tau_1 = \phi_1 z_1 \tag{2.68}$$

where τ_1 is the first tuning function.

Step 2. We consider the second equation of (2.61) by considering x_3 as virtual control variable. With (2.63), the z_2 dynamics can be derived

$$\dot{z}_2 = \dot{x}_2 - \dot{\alpha}_1 - \ddot{x}_r$$
$$= x_3 + \phi_2^T \theta + \psi_2 - \frac{\partial \alpha_1}{\partial x_1}(x_2 + \phi_1^T \theta + \psi_1) - \frac{\partial \alpha_1}{\partial \hat{\theta}} \dot{\hat{\theta}} - \frac{\partial \alpha_1}{\partial x_r} \dot{x}_r$$
$$= z_3 + \alpha_2 + \psi_2 - \frac{\partial \alpha_1}{\partial x_1}(x_2 + \psi_1) + (\phi_2 - \frac{\partial \alpha_1}{\partial x_1}\phi_1)^T \theta - \frac{\partial \alpha_1}{\partial \hat{\theta}} \dot{\hat{\theta}} - \frac{\partial \alpha_1}{\partial x_r} \dot{x}_r \tag{2.69}$$

Our task in this step is to stabilize the (z_1, z_2)-system (2.64) and (2.69). The Lyapunov function V_2 is chosen as

$$V_2 = V_1 + \frac{1}{2} z_2^2 \tag{2.70}$$

Now we select

$$\alpha_2 = -z_1 - c_2 z_2 - \psi_2 + \frac{\partial \alpha_1}{\partial x_1}(x_2 + \psi_1) - \hat{\theta}^T(\phi_2 - \frac{\partial \alpha_1}{\partial x_1}\phi_1)$$
$$+ \frac{\partial \alpha_1}{\partial \hat{\theta}}\Gamma\tau_2 + \frac{\partial \alpha_1}{\partial x_r}\dot{x}_r \tag{2.71}$$

$$\tau_2 = \tau_1 + (\phi_2 - \frac{\partial \alpha_1}{\partial x_1}\phi_1)z_2 \tag{2.72}$$

where c_2 is a positive constant and τ_2 is the second tuning function. The resulting derivative of V_2 is

$$\dot{V}_2 = -c_1 z_1^2 + z_2\Big(z_3 + \alpha_2 + z_1 + \psi_2 + \hat{\theta}^T(\phi_2 - \frac{\partial \alpha_1}{\partial x_1}\phi_1) - \frac{\partial \alpha_1}{\partial x_1}(x_2 + \psi_1)$$
$$- \frac{\partial \alpha_1}{\partial x_r}\dot{x}_r - \frac{\partial \alpha_1}{\partial \hat{\theta}}\dot{\hat{\theta}}\Big) + \tilde{\theta}^T\Big(\tau_1 + (\phi_2 - \frac{\partial \alpha_1}{\partial x_1}\phi_1)z_2 - \Gamma^{-1}\dot{\hat{\theta}}\Big)$$
$$= -c_1 z_1^2 - c_2 z_2^2 + z_2 z_3 + z_2\frac{\partial \alpha_1}{\partial \hat{\theta}}(\Gamma\tau_2 - \dot{\hat{\theta}}) + \tilde{\theta}^T(\tau_2 - \Gamma^{-1}\dot{\hat{\theta}}) \tag{2.73}$$

Step 3. Proceeding to the third equation in (2.61) by considering x_4 as a virtual control variable, we obtain

$$\dot{z}_3 = z_4 + \alpha_3 + \psi_3 - \frac{\partial \alpha_2}{\partial x_1}(x_2 + \psi_1) - \frac{\partial \alpha_2}{\partial x_2}(x_3 + \psi_2) - \frac{\partial \alpha_2}{\partial x_r}\dot{x}_r - \frac{\partial \alpha_2}{\partial \dot{x}_r}\ddot{x}_r$$
$$+ (\phi_3 - \frac{\partial \alpha_2}{\partial x_1}\phi_1 - \frac{\partial \alpha_2}{\partial x_2}\phi_2)^T\theta - \frac{\partial \alpha_2}{\partial \hat{\theta}}\dot{\hat{\theta}} \tag{2.74}$$

Now we select

$$\alpha_3 = -z_2 - c_3 z_3 - \psi_3 + \frac{\partial \alpha_2}{\partial x_1}(x_2 + \psi_1) + \frac{\partial \alpha_2}{\partial x_2}(x_3 + \psi_2) + \frac{\partial \alpha_2}{\partial x_r}\dot{x}_r + \frac{\partial \alpha_2}{\partial \dot{x}_r}\ddot{x}_r$$
$$+ (\frac{\partial \alpha_1}{\partial \hat{\theta}}\Gamma z_2 - \hat{\theta}^T)(\phi_3 - \frac{\partial \alpha_2}{\partial x_1}\phi_1 - \frac{\partial \alpha_2}{\partial x_2}\phi_2) + \frac{\partial \alpha_2}{\partial \hat{\theta}}\Gamma\tau_3 \tag{2.75}$$

$$\tau_3 = \tau_2 + (\phi_3 - \frac{\partial \alpha_2}{\partial x_1}\phi_1 - \frac{\partial \alpha_2}{\partial x_2}\phi_2)z_3 \tag{2.76}$$

where c_3 is a positive constant. The Lyapunov function is defined as

$$V_3 = V_2 + \frac{1}{2}z_3^2 \tag{2.77}$$

The derivative of Lapunov function V_3 is

$$\dot{V}_3 = -\sum_{i=1}^{3} c_i z_i^2 + z_3 z_4 + z_2\frac{\partial \alpha_1}{\partial \hat{\theta}}(\Gamma\tau_2 - \dot{\hat{\theta}}) + \tilde{\theta}^T(\tau_2 - \Gamma^{-1}\dot{\hat{\theta}})$$
$$+ z_3(\frac{\partial \alpha_1}{\partial \hat{\theta}}\Gamma z_2 + \tilde{\theta}^T)(\phi_3 - \frac{\partial \alpha_2}{\partial x_1}\phi_1 - \frac{\partial \alpha_2}{\partial x_2}\phi_2) + z_3\frac{\partial \alpha_2}{\partial \hat{\theta}}(\Gamma\tau_3 - \dot{\hat{\theta}})$$

$$= -\sum_{i=1}^{3} c_i z_i^2 + z_3 z_4 + z_2 \frac{\partial \alpha_1}{\partial \hat{\theta}} \left(\Gamma \tau_2 + \Gamma z_3 (\phi_3 - \frac{\partial \alpha_2}{\partial x_1} \phi_1 - \frac{\partial \alpha_2}{\partial x_2} \phi_2) - \dot{\hat{\theta}} \right)$$

$$+ \tilde{\theta}^T \left(\tau_2 + z_3 (\phi_3 - \frac{\partial \alpha_2}{\partial x_1} \phi_1 - \frac{\partial \alpha_2}{\partial x_2} \phi_2) - \Gamma^{-1} \dot{\hat{\theta}} \right) + z_3 \frac{\partial \alpha_2}{\partial \hat{\theta}} (\Gamma \tau_3 - \dot{\hat{\theta}})$$

$$\tag{2.78}$$

Note that

$$\Gamma \tau_2 - \dot{\hat{\theta}} = \Gamma \tau_2 - \Gamma \tau_3 + \Gamma \tau_3 - \dot{\hat{\theta}}$$

$$= -\Gamma z_3 (\phi_3 - \frac{\partial \alpha_2}{\partial x_1} \phi_1 - \frac{\partial \alpha_2}{\partial x_2} \phi_2) + (\Gamma \tau_3 - \dot{\hat{\theta}}) \tag{2.79}$$

Then we have

$$\dot{V}_3 = -\sum_{i=1}^{3} c_i z_i^2 + z_3 z_4 + \left(z_2 \frac{\partial \alpha_1}{\partial \hat{\theta}} + z_3 \frac{\partial \alpha_2}{\partial \hat{\theta}} \right) \left(\Gamma \tau_3 - \dot{\hat{\theta}} \right)$$

$$+ \tilde{\theta}^T \left(\tau_3 - \Gamma^{-1} \dot{\hat{\theta}} \right) \tag{2.80}$$

We can see that the virtual control law α_3 contains the term $\frac{\partial \alpha_1}{\partial \hat{\theta}} \Gamma z_2 (\phi_3 - \frac{\partial \alpha_2}{\partial x_1} \phi_1 - \frac{\partial \alpha_2}{\partial x_2} \phi_2)$. This term is an important term, since it is used to cancel the term $z_2 \frac{\partial \alpha_1}{\partial \hat{\theta}} \Gamma (\tau_2 - \tau_3)$ in the derivative $\dot{V}_3$ of the Lyapunov function by using (2.79).

Step i, $(i = 4, \ldots, n)$. Repeating the procedure in a recursive manner, we derive the i-th tracking error for z_i

$$\dot{z}_i = z_{i+1} + \alpha_i + \psi_i - \sum_{j=1}^{i-1} \frac{\partial \alpha_{i-1}}{\partial x_j} (x_{j+1} + \psi_j) + \theta^T \left(\phi_i - \sum_{j=1}^{i-1} \frac{\partial \alpha_{i-1}}{\partial x_j} \phi_j \right)$$

$$- \frac{\partial \alpha_{i-1}}{\partial \hat{\theta}} \dot{\hat{\theta}} - \sum_{j=1}^{i-1} \frac{\partial \alpha_{i-1}}{\partial x_r^{(j-1)}} x_r^{(j)} \tag{2.81}$$

We select the stabilizing function α_i

$$\alpha_i = -c_i z_i - z_{i-1} - \psi_i + \sum_{j=1}^{i-1} \frac{\partial \alpha_{i-1}}{\partial x_j} (x_{j+1} + \psi_j) - \hat{\theta}^T \left(\phi_i - \sum_{j=1}^{i-1} \frac{\partial \alpha_{i-1}}{\partial x_j} \phi_j \right)$$

$$+ \frac{\partial \alpha_{i-1}}{\partial \hat{\theta}} \Gamma \tau_i + \left(\sum_{j=2}^{i-1} z_j \frac{\partial \alpha_{j-1}}{\partial \hat{\theta}} \right) \Gamma (\phi_i - \sum_{j=1}^{i-1} \frac{\partial \alpha_{i-1}}{\partial x_j} \phi_j) + \sum_{j=1}^{i-1} \frac{\partial \alpha_{i-1}}{\partial x_r^{(j-1)}} x_r^{(j)}$$

$$\tag{2.82}$$

and tuning function

$$\tau_i = \tau_{i-1} + \left(\phi_i - \sum_{j=1}^{i-1} \frac{\partial \alpha_{i-1}}{\partial x_j} \phi_j \right) z_i \tag{2.83}$$

with the Lyapunov function

$$V_i = V_{i-1} + \frac{1}{2} z_i^2 \tag{2.84}$$

Its derivative is given as

$$\dot{V}_i = -\sum_{j=1}^{i} c_j z_j^2 + z_i z_{i+1} + \left(\sum_{j=2}^{i} z_j \frac{\partial \alpha_{j-1}}{\partial \hat{\theta}} \right) (\Gamma \tau_i - \dot{\hat{\theta}}) + \tilde{\theta}^T (\tau_i - \Gamma^{-1} \dot{\hat{\theta}}) \tag{2.85}$$

In the last step n, the actual control input u appears and is at our disposal. We derive the z_n dynamics

$$\dot{z}_n = bu + \psi_n - \sum_{j=1}^{n-1} \frac{\partial \alpha_{i-1}}{\partial x_j} (x_{j+1} + \psi_j) + \theta^T \left(\phi_n - \sum_{j=1}^{n-1} \frac{\partial \alpha_{i-1}}{\partial x_j} \phi_j \right)$$

$$- \frac{\partial \alpha_{n-1}}{\partial \hat{\theta}} \dot{\hat{\theta}} - \sum_{j=1}^{n-1} \frac{\partial \alpha_{i-1}}{\partial x_r^{(j-1)}} x_r^{(j)} - x_r^{(n)} \tag{2.86}$$

We are finally in this position to design control u and update laws $\dot{\hat{\theta}}$ and $\dot{\hat{p}}$ as

$$u = \hat{p} \bar{u} \tag{2.87}$$

$$\bar{u} = \alpha_n + x_r^{(n)} \tag{2.88}$$

$$\dot{\hat{\theta}} = \Gamma \tau_n \tag{2.89}$$

$$\dot{\hat{p}} = -\gamma \text{sign}(b) \bar{u} z_n \tag{2.90}$$

where γ is a positive constant and $\hat{p}$ is an estimate of $p = 1/b$. Note that

$$bu = b\hat{p}\bar{u} = \bar{u} - b\tilde{p}\bar{u} \tag{2.91}$$

where $\tilde{p} = p - \hat{p}$. We choose the Lyapunov function

$$V_n = V_{n-1} + \frac{|b|}{2\gamma} \tilde{p}^2 = \sum_{i=1}^{n} \frac{1}{2} z_i^2 + \frac{1}{2} \tilde{\theta}^T \Gamma^{-1} \tilde{\theta} + \frac{|b|}{2\gamma} \tilde{p}^2 \tag{2.92}$$

where γ is a positive design parameter. Then its derivative is given by

$$\dot{V}_n = -\sum_{i=1}^{n} c_i z_i^2 + \left(\sum_{j=2}^{n} z_j \frac{\partial \alpha_{j-1}}{\partial \hat{\theta}} \right) (\Gamma \tau_n - \dot{\hat{\theta}})$$

$$+ \tilde{\theta}^T (\tau_n - \Gamma^{-1} \dot{\hat{\theta}}) - \frac{|b|}{\gamma} \tilde{p} \left(\dot{\hat{p}} + \gamma \text{sign}(b) \bar{u} z_n \right)$$

$$= -\sum_{i=1}^{n} c_i z_i^2 \leq 0 \tag{2.93}$$

From the Lasalle's Theorem in Appendix B, this Lyapunov function provides the proof of uniform stability, such that $z_1, z_2, \ldots, z_n, \hat{\theta}, \hat{p}$ are bounded and $z_i \to 0, i = 1, \ldots, n$. This further implies that $\lim_{t\to\infty}(x_1 - x_r) = 0$. Since $x_1 = z_1 + x_r$, x_1 is also bounded from the boundedness of z_1 and x_r. The boundedness of x_2 follows from boundedness of $\dot{x}_r$ and α_1 in (2.65) and the fact that $x_2 = z_2 + \alpha_1 + \dot{x}_r$. Similarly, the boundedness of x_i ($i = 3, \ldots, n$) can be ensured from the boundedness of $x_r^{(i-1)}$ and α_i in (2.82) and the fact that $x_i = z_i + \alpha_{i-1} + x_r^{(i-1)}$. Combining this with (2.87) we conclude that the control $u(t)$ is also bounded. Therefore boundedness of all signals and asymptotic tracking are ensured as formally stated in the following Theorem.

Theorem 2.1. *Consider the closed-loop adaptive system (2.61) under Assumptions 1-2, the adaptive controller (2.87), virtual control laws (2.65), (2.71) and (2.82), and updating laws (2.89) and (2.90) guarantee global boundedness of $x(t)$ and $\hat{\theta}, \hat{p}$ and the asymptotic tracking $\lim_{t\to\infty}(x_1 - x_r) = 0$.*

The controller designed in this section achieves the goals of stabilization and tracking. The proof of these properties is a direct consequence of the recursive procedure, because a Lyapunov function is constructed for the entire system including the parameter estimates. The over-parametrization problem is overcomed by using tuning functions. The number of parameter estimates are equal to the number of unknown parameters.

2.3 Output Feedback Control

Now we introduce backstepping design procedures with output feedback for nonlinear systems described in the following form, whose nonlinearities depend only on the output y.

$$\dot{x}_1 = x_2 + \phi_1^T(y)\theta + \psi_1(y)$$

$$\vdots$$

$$\dot{x}_{\rho-1} = x_\rho + \phi_{\rho-1}^T(y)\theta + \psi_{\rho-1}(y)$$
$$\dot{x}_\rho = x_{\rho+1} + \phi_\rho^T(y)\theta + \psi_\rho(y) + b_m u$$

$$\vdots$$

$$\dot{x}_{n-1} = x_n + \phi_{n-1}^T(y)\theta + \psi_{n-1}(y) + b_1 u$$
$$\dot{x}_n = \phi_n^T(y)\theta + \psi_n(y) + b_0 u \tag{2.94}$$
$$y = x_1$$

where $x_1, \ldots, x_n$, y and u are system states, output and input, the vector $\theta \in R^r$ is constant and unknown, $\phi_i(y) \in R^r$, $i = 1, \ldots, n$ are known nonlinear functions, and $b_m, \ldots, b_0$ are unknown constants.

For the development of control laws, the following assumptions are made.

Assumption 1: The sign of b_m is known.

Assumption 2: The relative degree $\rho = n - m$ is known and the system is minimum phase.

Assumption 3: The reference signal y_r and its ρth order derivatives are piecewise continuous, known and bounded.

Our problem is to globally stabilize the system (2.94) and also to achieve the asymptotic tracking of y_r by y.

2.3.1   State Estimation Filters

In order to design the desired adaptive output feedback control law, we rewrite the system (2.94) in the following form

$$\dot{x} = Ax + \Phi(y)\theta + \Psi(y) + \begin{bmatrix} 0 \\ b \end{bmatrix} u \qquad (2.95)$$

where

$$A = \begin{bmatrix} 0 & 1 & 0 & \dots & 0 \\ 0 & 0 & 1 & \dots & 0 \\ \vdots & \vdots & \vdots & \ddots & \vdots \\ 0 & 0 & 0 & \dots & 1 \\ 0 & 0 & 0 & \dots & 0 \end{bmatrix}, \Psi(y) = \begin{bmatrix} \psi_1(y) \\ \vdots \\ \psi_n(y) \end{bmatrix} \qquad (2.96)$$

$$\Phi(y) = \begin{bmatrix} \phi_1^T(y) \\ \vdots \\ \phi_n^T(y) \end{bmatrix}, b = \begin{bmatrix} b_m \\ \vdots \\ b_0 \end{bmatrix} \qquad (2.97)$$

Note that only output y is measured. Thus x is unavailable. We need to design filters to estimate x and generate some signals for controller design. These filters are summarized as

$$\dot{\xi} = A_0 \xi + ky + \Psi(y) \qquad (2.98)$$
$$\dot{\Xi}^T = A_0 \Xi^T + \Phi(y) \qquad (2.99)$$
$$\dot{\lambda} = A_0 \lambda + e_n u \qquad (2.100)$$
$$v_i = A_0^i \lambda, \quad i = 0, 1, \dots, m \qquad (2.101)$$

where $k = [k_1, \dots, k_n]^T$ such that all eigenvalues of $A_0 = A - ke_1^T$ are at some desired stable locations. The state estimates are given by

$$\hat{x}(t) = \xi + \Xi^T \theta + \sum_{i=0}^{m} b_i v_i \qquad (2.102)$$

Note that $\hat{x}$ is unavailable due to the unknown parameters θ and b, so the estimate $\hat{x}$ cannot be used in the later controller design. Instead, it will be used for stability analysis. The derivative of $\hat{x}$ is given as

$$
\dot{\hat{x}}(t) = \dot{\xi} + \dot{\Xi}^T\theta + \sum_{i=0}^{m} b_i \dot{v}_i
$$

$$
= A_0\xi + ky + \Psi(y) + (A_0\Xi^T + \Phi(y))\theta + \sum_{i=0}^{m} b_i A_0^i (A_0\lambda + e_n u)
$$

$$
= A_0(\xi + \Xi^T\theta + \sum_{i=0}^{m} b_i v_i) + ky + \Phi(y)\theta + \Psi(y) + \begin{bmatrix} 0 \\ b \end{bmatrix} u
$$

$$
= A_0\hat{x} + ky + \Phi(y)\theta + \Psi(y) + \begin{bmatrix} 0 \\ b \end{bmatrix} u \tag{2.103}
$$

It can be shown that the state estimation error

$$
\epsilon = x(t) - \hat{x}(t) \tag{2.104}
$$

satisfies

$$
\begin{aligned}
\dot{\epsilon} &= \dot{x}(t) - \dot{\hat{x}}(t) \\
&= Ax - ky - A_0\hat{x} \\
&= (A_0 + ke_1^T)x - ky - A_0\hat{x} \\
&= A_0\epsilon
\end{aligned} \tag{2.105}
$$

Suppose $P \in R^{n \times n}$ is a positive definite matrix, satisfying $PA_0 + A_0^T P \le -I$ and let

$$
V_\epsilon = \epsilon^T P \epsilon \tag{2.106}
$$

It can be shown that

$$
\begin{aligned}
\dot{V}_\epsilon &= \epsilon^T (PA_0 + A_0^T P)\epsilon \\
&\le -\epsilon^T \epsilon
\end{aligned} \tag{2.107}
$$

This Lyapunov function guarantees that $\epsilon \to 0$, which implies $\hat{x}(t) \to x(t)$.

Note that the backstepping design starts with its output y, which is the only available system state allowed to appear in the control law. The dynamic equation of y is expressed as

$$
\begin{aligned}
\dot{y} &= x_2 + \phi_1^T(y)\theta(t) + \psi_1(y) \\
&= b_m v_{m,2} + \xi_2 + \psi_1(y) + \bar{\omega}^T \Theta + \epsilon_2
\end{aligned} \tag{2.108}
$$

where

$$
\Theta = [b_m, \ldots, b_0, \theta^T]^T \tag{2.109}
$$

$$
\omega = [v_{m,2}, v_{m-1,2}, \ldots, v_{0,2}, \Xi_2 + \phi_1^T]^T \tag{2.110}
$$

$$
\bar{\omega} = [0, v_{m-1,2}, \ldots, v_{0,2}, \Xi_2 + \phi_1^T]^T \tag{2.111}
$$

In above equations, $\epsilon_2, v_{i,2}, \xi_2$ and Ξ_2 denote the second entries of ϵ, v_i, ξ and Ξ, respectively, and y, v_i, ξ, Ξ are all available signals.

Combining system (2.108) with our filters (2.98)-(2.101), system (2.94) is represented as

$$\dot{y} = b_m v_{m,2} + \xi_2 + \psi_1(y) + \bar{\omega}^T \Theta + \epsilon_2 \qquad (2.112)$$

$$\dot{v}_{m,i} = v_{m,i+1} - k_i v_{m,1}, \qquad i = 2, 3, \ldots, \rho - 1 \qquad (2.113)$$

$$\dot{v}_{m,\rho} = v_{m,\rho+1} - k_\rho v_{m,1} + u \qquad (2.114)$$

System (2.112)-(2.114) will be our design system, whose states $y, v_{m,2}, \ldots, v_{m,\rho}$ are available. Our task at this stage is to globally stabilize the system and also to achieve the asymptotic tracking of y_r by y.

2.3.2   Design Procedure and Stability Analysis

In this section, we present the adaptive control design using the backstepping technique with tuning functions in ρ steps. Firstly, we take the change of coordinates

$$z_1 = y - y_r \qquad (2.115)$$

$$z_i = v_{m,i} - \alpha_{i-1} - \hat{\varrho} y_r^{(i-1)}, \quad i = 2, 3, \ldots, \rho, \qquad (2.116)$$

where $\hat{\varrho}$ is an estimate of $\varrho = 1/b_m$ and α_{i-1} is the virtual control at each step and will be determined in later discussions.

• *Step* 1: Starting with the equation for the tracking error z_1, we obtain, from (2.112) and (2.115), that

$$\dot{z}_1 = b_m v_{m,2} + \xi_2 + \psi_1(y) + \bar{\omega}^T \Theta + \epsilon_2 - \dot{y}_r \qquad (2.117)$$

By substituting (2.116) for $i = 2$ into (2.117) and using $\tilde{\varrho} = \frac{1}{b_m} - \frac{1}{\hat{b}_m}$, we get

$$\dot{z}_1 = b_m \alpha_1 + \xi_2 + \psi_1(y) + \bar{\omega}^T \Theta + \epsilon_2 - b_m \tilde{\varrho} \dot{y}_r + b_m z_2 \qquad (2.118)$$

By considering $v_{m,2}$ as the first virtual control, we select a virtual control law α_1 as

$$\alpha_1 = \hat{\varrho} \bar{\alpha}_1 \qquad (2.119)$$

$$\bar{\alpha}_1 = -c_1 z_1 - d_1 z_1 - \xi_2 - \psi_1(y) - \bar{\omega}^T \hat{\Theta} \qquad (2.120)$$

where c_1 and d_1 are positive design parameters, and $\hat{\Theta}$ is the estimate of Θ. From (2.118) and (2.119) we have

$$\dot{z}_1 = -c_1 z_1 - d_1 z_1 + \epsilon_2 + \bar{\omega}^T \tilde{\Theta} - b_m(\dot{y}_r + \bar{\alpha}_1)\tilde{\varrho} + b_m z_2$$
$$= -(c_1 + d_1)z_1 + \epsilon_2 + (\omega - \hat{\varrho}(\dot{y}_r + \bar{\alpha}_1)e_1)^T \tilde{\Theta} - b_m(\dot{y}_r + \bar{\alpha}_1)\tilde{\varrho} + \hat{b}_m z_2$$
$$\qquad (2.121)$$

where $\tilde{\Theta} = \Theta - \hat{\Theta}$. Note that

$$b_m\alpha_1 = b_m\hat{\varrho}\bar{\alpha}_1 = \bar{\alpha}_1 - b_m\tilde{\varrho}\bar{\alpha}_1 \tag{2.122}$$

$$\begin{aligned}
\bar{\omega}^T\tilde{\Theta} + b_m z_2 &= \bar{\omega}^T\tilde{\Theta} + \tilde{b}_m z_2 + \hat{b}_m z_2 \\
&= \bar{\omega}^T\tilde{\Theta} + (v_{m,2} - \hat{\varrho}\dot{y}_r - \alpha_1)e_1^T\tilde{\Theta} + \hat{b}_m z_2 \\
&= (\omega - \hat{\varrho}(\dot{y}_r + \bar{\alpha}_1)e_1)^T\tilde{\Theta} + \hat{b}_m z_2 \tag{2.123}
\end{aligned}$$

Define the Lyapunov function V_1 as

$$V_1 = \frac{1}{2}z_1^2 + \frac{1}{2}\tilde{\Theta}^T\Gamma^{-1}\tilde{\Theta} + \frac{|b_m|}{2\gamma}\tilde{\varrho}^2 + \frac{1}{2d_1}\epsilon^T P\epsilon \tag{2.124}$$

where Γ is a positive definite design matrix, γ is a positive design parameter, and P is a definite positive matrix such that $PA_0 + A_0^T P = -I$, $P = P^T > 0$. We examine the derivative of V_1

$$\begin{aligned}
\dot{V}_1 &\leq z_1\dot{z}_1 - \tilde{\Theta}^T\Gamma^{-1}\dot{\hat{\Theta}} - \frac{|b_m|}{\gamma}\tilde{\varrho}\dot{\hat{\varrho}} - \frac{1}{2d_1}\epsilon^T\epsilon \\
&\leq -c_1 z_1^2 + \hat{b}_m z_1 z_2 - \frac{1}{4d_1}\epsilon^T\epsilon - |b_m|\tilde{\varrho}\frac{1}{\gamma}[\gamma\mathrm{sign}(b_m)(\dot{y}_r + \bar{\alpha}_1)z_1 + \dot{\hat{\varrho}}] \\
&\quad + \tilde{\Theta}^T[(\omega - \hat{\varrho}(\dot{y}_r + \bar{\alpha}_1)e_1)z_1 - \Gamma^{-1}\dot{\hat{\Theta}}] - d_1 z_1^2 + z_1\epsilon_2 - \frac{\|\epsilon\|^2}{4d_1} \tag{2.125}
\end{aligned}$$

Now we choose

$$\dot{\hat{\varrho}} = -\gamma\mathrm{sign}(b_m)(\dot{y}_r + \bar{\alpha}_1)z_1 \tag{2.126}$$

Define

$$\tau_1 = (\omega - \hat{\varrho}(\dot{y}_r + \bar{\alpha}_1)e_1)z_1 \tag{2.127}$$

and τ_1 is called the first tuning function. Then the following can be derived by using Young's inequality $ab \leq d_1 a^2 + \frac{1}{4d_1}b^2$, update law (2.126) and (2.127)

$$\dot{V}_1 \leq -c_1 z_1^2 + \hat{b}_m z_1 z_2 - \frac{1}{4d_1}\epsilon^T\epsilon + \tilde{\Theta}^T(\tau_1 - \Gamma^{-1}\dot{\hat{\Theta}}) \tag{2.128}$$

If $z_2 = 0$, we would choose $\dot{\hat{\Theta}} = \Gamma\tau_1$ and the derivative of V_1 would be

$$\dot{V}_1 \leq -c_1 z_1^2 - \frac{1}{4d_1}\epsilon^T\epsilon \leq -c_1 z_1^2 \tag{2.129}$$

which implies that z_1 converges to zero asymptotically. Since $z_2 \neq 0$, we do not use $\dot{\hat{\Theta}} = \Gamma\tau_1$ as an update law for Θ at this step to avoid over-parametrization problem, because Θ will also appear in the following steps.

• *Step* 2: We derive the error dynamics z_2

$$\dot{z}_2 = \dot{v}_{m,2} - \dot{\alpha}_1 - \dot{\hat{\varrho}}\dot{y}_r - \hat{\varrho}\ddot{y}_r$$

$$= v_{m,3} - k_2 v_{m,1} - \frac{\partial \alpha_1}{\partial y}(b_m v_{m,2} + \xi_2 + \psi_1 + \bar{\omega}^T \Theta + \epsilon_2) - \frac{\partial \alpha_1}{\partial y_r}\dot{y}_r$$

$$- \sum_{j=1}^{m+i-1} \frac{\partial \alpha_1}{\partial \lambda_j}(-k_j \lambda_1 + \lambda_{j+1}) - \frac{\partial \alpha_1}{\partial \xi}(A_0 \xi + ky + \Psi(y))$$

$$- \frac{\partial \alpha_1}{\partial \Xi}(A_0 \Xi^T + \Phi(y)) - \frac{\partial \alpha_1}{\partial \hat{\Theta}}\dot{\hat{\Theta}} - \frac{\partial \alpha_1}{\partial \hat{\varrho}}\dot{\hat{\varrho}} - \dot{\hat{\varrho}}\dot{y}_r - \hat{\varrho}\ddot{y}_r$$

$$= v_{m,3} - \hat{\varrho}\ddot{y}_r - \beta_2 - \frac{\partial \alpha_1}{\partial y}(\omega^T \tilde{\Theta} + \epsilon_2)) - \frac{\partial \alpha_1}{\partial \hat{\Theta}}\dot{\hat{\Theta}} \qquad (2.130)$$

where

$$\beta_2 = \frac{\partial \alpha_1}{\partial y}(\xi_2 + \psi_1 + \omega^T \hat{\Theta}) + k_2 v_{m,1} + \frac{\partial \alpha_1}{\partial y_r}\dot{y}_r + (\dot{y}_r + \frac{\partial \alpha_1}{\partial \hat{\varrho}})\dot{\hat{\varrho}}$$

$$+ \sum_{j=1}^{m+i-1} \frac{\partial \alpha_1}{\partial \lambda_j}(-k_j \lambda_1 + \lambda_{j+1}) + \frac{\partial \alpha_1}{\partial \xi}(A_0 \xi + ky + \Psi(y))$$

$$+ \frac{\partial \alpha_1}{\partial \Xi^T}(A_0 \Xi^T + \Phi(y)) \qquad (2.131)$$

By considering $v_{m,3}$ as virtual control input and using $z_3 = v_{m,3} - \alpha_2 - \hat{\varrho}\ddot{y}_r$, we have

$$\dot{z}_2 = z_3 + \alpha_2 - \beta_2 - \frac{\partial \alpha_1}{\partial y}(\omega^T \tilde{\Theta} + \epsilon_2) - \frac{\partial \alpha_1}{\partial \hat{\Theta}}\dot{\hat{\Theta}} \qquad (2.132)$$

With the Lyapunov function

$$V_2 = V_1 + \frac{1}{2}z_2^2 + \frac{1}{2d_2}\epsilon^T P \epsilon \qquad (2.133)$$

We choose the second virtual control law α_2 and tuning function as

$$\alpha_2 = -\hat{b}_m z_1 - \left(c_2 + d_2\left(\frac{\partial \alpha_1}{\partial y}\right)^2\right)z_2 + \beta_2 + \frac{\partial \alpha_1}{\partial \hat{\Theta}}\Gamma \tau_2 \qquad (2.134)$$

$$\tau_2 = \tau_1 - \frac{\partial \alpha_1}{\partial y}\omega z_2 \qquad (2.135)$$

Then

$$\dot{V}_2 = \dot{V}_1 + z_2 \dot{z}_2 - \frac{1}{2d_2}\epsilon^T \epsilon$$

$$\leq -c_1 z_1^2 + \hat{b}_m z_1 z_2 + z_2\left(z_3 + \alpha_2 - \beta_2 - \frac{\partial \alpha_1}{\partial y}(\omega^T \tilde{\Theta} + \epsilon_2) - \frac{\partial \alpha_1}{\partial \hat{\Theta}}\dot{\hat{\Theta}}\right)$$

$$- \frac{1}{2d_2}\epsilon^T \epsilon - \frac{1}{4d_1}\epsilon^T \epsilon + \tilde{\Theta}^T(\tau_1 - \Gamma^{-1}\dot{\hat{\Theta}})$$

$$
\begin{aligned}
&= -c_1 z_1^2 - c_2 z_2^2 + z_2 z_3 - d_2 \Big(\frac{\partial \alpha_1}{\partial y}\Big)^2 z_2^2 - \frac{\partial \alpha_1}{\partial y} \epsilon_2 z_2 - \frac{1}{4 d_2} \epsilon^T \epsilon \\
&\quad - \frac{1}{4 d_2} \epsilon^T \epsilon - \frac{1}{4 d_1} \epsilon^T \epsilon + \tilde{\Theta}^T \big(\tau_1 - \frac{\partial \alpha_1}{\partial y} \omega z_2 - \Gamma^{-1} \dot{\hat{\Theta}}\big) + \frac{\partial \alpha_1}{\partial \hat{\Theta}} (\Gamma \tau_2 - \dot{\hat{\Theta}}) \\
&\leq - \sum_{i=1}^{2} \Big(c_i z_i^2 + \frac{1}{4 d_i} \epsilon^T \epsilon \Big) + z_2 z_3 + \tilde{\Theta}^T (\tau_2 - \Gamma^{-1} \dot{\hat{\Theta}}) + \frac{\partial \alpha_1}{\partial \hat{\Theta}} (\Gamma \tau_2 - \dot{\hat{\Theta}})
\end{aligned}
\tag{2.136}
$$

Following similar arguments as before, we would choose $\dot{\hat{\Theta}} = \Gamma \tau_2$, as this would result in $\dot{V}_2 \leq -c_1 z_1^2 - c_2 z_2^2$ if $z_3 = 0$. But $z_3 \neq 0$ and thus we do not use it as an update law for Θ to overcome the over-parametrization problem.

- *Step i* $(i = 3, \ldots, \rho)$: Choose virtual control laws

$$
\begin{aligned}
\alpha_i &= -z_{i-1} - \big[c_i + d_i \big(\frac{\partial \alpha_{i-1}}{\partial y}\big)^2\big] z_i + \beta_i + \frac{\partial \alpha_{i-1}}{\partial \hat{\Theta}} \Gamma \tau_i \\
&\quad - \Big(\sum_{k=2}^{i-1} z_k \frac{\partial \alpha_{k-1}}{\partial \hat{\Theta}}\Big) \Gamma \frac{\partial \alpha_{i-1}}{\partial y} \omega, \quad i = 3, \ldots, \rho
\end{aligned}
\tag{2.137}
$$

where c_i are positive design parameters and

$$
\tau_i = \tau_{i-1} - \frac{\partial \alpha_{i-1}}{\partial y} \omega z_i
\tag{2.138}
$$

$$
\begin{aligned}
\beta_i &= \frac{\partial \alpha_{i-1}}{\partial y} (\xi_2 + \psi_1 + \omega^T \hat{\Theta}) + k_i v_{m,1} + \sum_{j=1}^{i-1} \frac{\partial \alpha_{i-1}}{\partial y_r^{(j-1)}} y_r^{(j)} + \big(y_r^{(i-1)} + \frac{\partial \alpha_{i-1}}{\partial \hat{\varrho}}\big) \dot{\hat{\varrho}} \\
&\quad + \sum_{j=1}^{m+i-1} \frac{\partial \alpha_{i-1}}{\partial \lambda_j} (-k_j \lambda_1 + \lambda_{j+1}) + \frac{\partial \alpha_{i-1}}{\partial \xi} (A_0 \xi + ky + \Psi(y)) \\
&\quad + \frac{\partial \alpha_{i-1}}{\partial \Xi^T} (A_0 \Xi^T + \Phi(y))
\end{aligned}
\tag{2.139}
$$

In the last step ρ, the adaptive controller and parameter update law are finally given by

$$
u = \alpha_\rho - v_{m,\rho+1} + \hat{\varrho} y_r^{(\rho)}
\tag{2.140}
$$

$$
\dot{\hat{\Theta}} = \Gamma \tau_\rho
\tag{2.141}
$$

We define the final Lyapunov function V_ρ as

$$
V_\rho = \sum_{i=1}^{\rho} \frac{1}{2} z_i^2 + \frac{1}{2} \tilde{\Theta}^T \Gamma^{-1} \tilde{\Theta} + \frac{|b_m|}{2\gamma} \tilde{\varrho}^2 + \sum_{i=1}^{\rho} \frac{1}{2 d_i} \epsilon^T P \epsilon
\tag{2.142}
$$

Note that

$$
\begin{aligned}
\Gamma \tau_{i-1} - \dot{\hat{\Theta}} &= \Gamma \tau_{i-1} - \Gamma \tau_i + \Gamma \tau_i - \dot{\hat{\Theta}} \\
&= \Gamma \frac{\partial \alpha_{i-1}}{\partial y} \omega z_i + (\Gamma \tau_i - \dot{\hat{\Theta}})
\end{aligned}
\tag{2.143}
$$

From (2.137)-(2.141), the derivative of the last Lyapunov function satisfies

$$\dot{V}_\rho = \sum_{i=1}^{\rho} z_i \dot{z}_i - \tilde{\Theta}^T \Gamma^{-1} \dot{\hat{\Theta}} - \frac{|b_m|}{\gamma} \tilde{\varrho} \dot{\hat{\varrho}} - \sum_{i=1}^{\rho} \frac{1}{2d_i} \epsilon^T \epsilon$$

$$\leq -\sum_{i=1}^{\rho} c_i z_i^2 - \sum_{i=1}^{\rho} \frac{1}{4d_i} \epsilon^T \epsilon - \tilde{\Theta}^T \Gamma^{-1} (\dot{\hat{\Theta}} - \Gamma \tau_\rho) + \Big(\sum_{k=2}^{\rho} z_k \frac{\partial \alpha_{k-1}}{\partial \hat{\Theta}} \Big) (\Gamma \tau_\rho - \dot{\hat{\Theta}})$$

$$= -\sum_{i=1}^{\rho} c_i z_i^2 - \sum_{i=1}^{\rho} \frac{1}{4d_i} \epsilon^T \epsilon \qquad (2.144)$$

We have the following stability and performance results based on the designed backstepping controller.

Theorem 2.1. *Consider the system consisting of the parameter estimators given by (2.126) and (2.141), adaptive controllers designed using (2.140) with virtual control laws (2.119), (2.134) and (2.137), the filters (2.98), (2.99) and (2.100), and plant (2.94). The system is stable in the sense that all signals in the closed loop system are globally uniformly bounded. Furthermore*

- *The asymptotic tracking performance is achieved, i.e.,*

$$\lim_{t \to \infty} [y(t) - y_r(t)] = 0 \qquad (2.145)$$

- *The transient tracking error performance is given by*

$$\| y(t) - y_r(t) \|_2 \leq \frac{1}{\sqrt{c_1}} \Big(\frac{1}{2} \tilde{\Theta}(0)^T \Gamma^{-1} \tilde{\Theta}(0) + \frac{|b_m|}{2\gamma} \tilde{\varrho}(0)^2$$

$$+ \frac{1}{2d_0} \| \epsilon(0) \|_P^2 \Big)^{1/2} \qquad (2.146)$$

with $z_i(0) = 0, i = 1, \ldots, \rho$, $d_0 = \Big(\sum_{i=1}^{\rho} \frac{1}{d_i} \Big)^{-1}$ and $\| \epsilon(0) \|_P^2 = \epsilon(0)^T P \epsilon(0)$.

Proof: Due to the piecewise continuity of $y_r(t), \ldots, y_r^{(\rho)}(t)$ and the smoothness of the control law, the parameter updating laws and the filters, the solution of the closed-loop adaptive system exists and is unique. From (2.144), it can be shown that V_ρ is uniformly bounded. Thus $z_i, \hat{\Theta}, \hat{\varrho}$ and ϵ are bounded. Since z_1 and y_r are bounded, y is also bounded. Then from (2.98) and (2.99) we conclude that ξ and Ξ are bounded as A_0 is Hurwitz. From (2.100) and Assumption 2, we have that $\lambda_1, \ldots, \lambda_{m+1}$ are bounded. From the coordinate change (2.116), it gives

$$v_{m,i} = z_i + \hat{\varrho} y_r^{(i-1)} + \alpha_{i-1} \big(y, \xi, \Xi, \hat{\Theta}, \hat{\varrho}, \bar{\lambda}_{m+i-1}, \bar{y}_r^{(i-2)} \big)$$

$$i = 2, 3, \ldots, \rho, \qquad (2.147)$$

where $\bar{\lambda}_k = [\lambda_1, \ldots, \lambda_k]^T$, $\bar{y}_r^{(k)} = [y_r, \ldots, y_r^{(k)}]^T$. For $i = 2$, from the boundedness of $\lambda_{m+1}, z_2, y, \xi, \Xi, \hat{\Theta}, \hat{\varrho}, y_r$ and $\dot{y}_r$, it proves that $v_{m,2}$ is bounded. From (2.101) it follows that λ_{m+2} is bounded. Following the same procedure recursively, we can show that λ is bounded. From (2.102) and the boundedness of $\xi, \Xi, \lambda, \epsilon$, we conclude that x is bounded.

To show the global uniform stability, the boundedness of $m = n - \rho$ dimension states ζ with zero dynamics should be guaranteed. Under a similar transformation as in [1], the states ζ associated with the zero dynamics can be shown to satisfy

$$\dot{\zeta} = A_b\zeta + b_by + T\Phi(y)\theta + T\Psi(y) \tag{2.148}$$

where $\zeta = Tx$, $b_b \in R^m$, the eigenvalues of the $m \times m$ matrix A_b is given as follows

$$A_b = \begin{bmatrix} -b_{m-1}/b_m & & \\ & & I_{m-1} \\ & \vdots & \\ -b_0/b_m & 0 & \dots & 0 \end{bmatrix} \tag{2.149}$$

$$T = [(A_b)^\rho e_1, \dots, A_be_1, I_m]. \tag{2.150}$$

With Assumption 2, we have that A_b is Hurwitz. Hence, there exists matrix P such that

$$PA_b + (A_b)^T P = -2I \tag{2.151}$$

Now we define a Lyapunov function for the zero dynamics of the system as $V_\zeta = \zeta^T P\zeta$. It can be show that

$$\dot{V}_\zeta \le -\zeta^T\zeta + \| P(b_by + T\Phi(y)^T\theta + \Psi(y)) \|^2 \tag{2.152}$$

Because all signals and functions in the second term of (2.152) are bounded, it can be shown that ζ is bounded.

Thus all signals in the closed-loop are globally uniformly bounded. By applying the LaSalle-Yoshizawa theorem to (2.144), it further follows that $z(t) \to 0$ as $t \to \infty$, which implies that $\lim_{t\to\infty}[y(t) - y_r(t)] = 0$.

Now we derive the tracking error bound in term of the L_2 norm. As shown in (2.144), the derivative of V_ρ is

$$\dot{V}_\rho \le -\sum_{i=1}^{\rho} c_iz_i^2 \le -c_1z_1^2 \tag{2.153}$$

Since V_ρ is non-increasing, we have

$$\| z_1 \|_2^2 = \int_0^\infty |z_1(\tau)|^2 d\tau \le \frac{1}{c_1}(V(0) - V(\infty)) \le \frac{1}{c_1}V(0) \tag{2.154}$$

We can set $z_i(0)$ to zero by appropriately initializing the reference trajectory as following

$$y_r(0) = y(0) \tag{2.155}$$

$$y_r^{(i)}(0) = \frac{1}{\hat{\varrho}(0)}\left[v_{m,i+1}(0) - \alpha_i\big(y(0), \xi(0), \varXi(0), \hat{\varTheta}(0), \hat{\varrho}(0), \bar{\lambda}_{m+i}(0), \bar{y}_r^{(i-1)}(0)\big)\right]$$
$$i = 1, \ldots, \rho - 1 \tag{2.156}$$

Thus, by setting $z_i(0) = 0, i = 1, \ldots, n$, we obtain

$$V(0) = \frac{1}{2}\tilde{\varTheta}(0)^T \varGamma^{-1}\tilde{\varTheta}(0) + \frac{|b_m|}{2\gamma}\tilde{\varrho}(0)^2 + \frac{1}{2d_0} \| \epsilon(0) \|_P^2 \tag{2.157}$$

a decreasing function of γ, η and $\varGamma$, independent of c_1. This means that the bound resulting from (2.154) and (2.157) is

$$\| z_1 \|_2 \leq \frac{1}{\sqrt{c_1}}\left(\frac{1}{2}\tilde{\varTheta}(0)^T \varGamma^{-1}\tilde{\varTheta}(0) + \frac{|b_m|}{2\gamma}\tilde{\varrho}(0)^2 + \frac{1}{2d_0} \| \epsilon(0) \|_P^2 \right)^{1/2} \tag{2.158}$$

$\triangle\triangle\triangle$

Remark 2.1. The following conclusions can be obtained:

- The transient performance depends on the initial estimate errors $\tilde{\varTheta}(0)$, $\tilde{\varrho}(0)$ and the explicit design parameters. The closer the initial estimates $\hat{\varTheta}(0)$ and $\hat{\varrho}(0)$ to the true values $\varTheta$ and ϱ, the better the transient performance.
- The bound for $\| y(t) - y_r(t) \|_2$ is an explicit function of design parameters and thus computable. We can decrease the effects of the initial error estimates on the transient performance by increasing the adaptation gains $\varGamma, \gamma, d_0$ or c_1.

3 Adaptive Control of Time-Varying Nonlinear Systems

The task of this chapter is to introduce a new scheme to design adaptive controllers for single-input single-output uncertain time-varying systems in the presence of unknown bounded disturbances. No knowledge is assumed on the sign of the term multiplying the control. The control design is achieved by introducing certain well defined functions, estimating variation rates of parameters and incorporating a Nussbaum gain. To overcome the problem of overparametrization, tuning functions, which are different from the standard ones in [1] due to the use of projection operations, are employed. It is shown that the proposed controller can guarantee the whole system stable.

3.1 Background

Adaptive control has seen significant development. However, only limited number of results are available for nonlinear systems with time-varying parameters and/or without the knowledge on the sign of the term multiplying the control, i.e. high frequency gain in the case of linear systems, in the presence of external disturbances. In this chapter, we shall also call this term as high frequency gain for nonlinear systems for simplicity.

In [80], output feedback control was considered for linear time-varying systems when the sign of high-frequency gain is known. In [34], the problem of adaptive control with unknown sign of high-frequency gain for linear time invariant systems was studied. In [81], Nussbaum gain incorporating with the backstepping technique was used to design adaptive output stabilizer for high order uncertain time invariant nonlinear systems with unknown sign of high-frequency gain in the absence of external disturbances. The nonlinearities considered should satisfy sector conditions. In [35], disturbance decoupling was addressed for nonlinear time invariant systems with known sign of the high frequency gain. The result obtained is critically depending on a function of the system output y and the reference trajectory y_r. In [82], a flat zone was used to handle the problem of nonlinear time invariant systems with unknown sign of high frequency gain in the presence of disturbances. The bound of the disturbance and all the unknown

J. Zhou & C. Wen: Adapt. Backstepping Ctrl. of Uncertain Systems, LNCIS 372, pp. 33–50, 2008.
springerlink.com © Springer-Verlag Berlin Heidelberg 2008

parameters need to be estimated at every step in the backstepping process. This results in the problem of overparametrization and makes the implementation complicated. In [83] state-feedback control was considered for a class of uncertain time-varying nonlinear systems in the presence of disturbances. Due to state feedback, no filter is required for state estimation. Thus the derivatives of the time varying parameters and the disturbance term do not need to be considered in controller design. This also makes the stability analysis greatly simplified. Again, parameters are required to be estimated at every step, which results in the overparametrization problem. In the case of output feedback control of nonlinear time-varying systems in the presence of disturbances, filters are required to estimate system states and the equations of the state estimation error will be used in the design and analysis. In these equations, the external disturbances and derivatives of time-varying parameters will appear and have great impact on the errors. This makes the design and analysis quite difficult, especially when the sign of high frequency gain is unknown and tuning functions are used.

In this chapter, we consider such a case and propose a new control design scheme to solve the problem as in [40]. The nonlinearities considered are not required to satisfy the sector type of conditions like [81]. To handle the disturbances, well defined functions are introduced to eliminate their effects in the Lyapunov functions employed in the recursive design steps. To deal with the time variation problem, an estimator is used to estimate the bound of the variation rates. Furthermore, the overparamterization problem is also solved by using the concept of tuning functions. As projection operation is used, the design of tuning functions are different from existing schemes. With our proposed controller, system stability can be ensured.

3.2 System Model and Problem Formulation

3.2.1 Problem Formulation

Consider the following class of single-input-single-output (SISO) nonlinear time-varying systems in the feedback form

$$\dot{x}_1 = x_2 + \theta_{a1}(t)\psi_{a1}(y) + d_1(t)\phi_{a1}(y) + \psi_{01}(y)$$

$$\vdots$$

$$\dot{x}_{\rho-1} = x_\rho + \theta_{a\rho-1}(t)\psi_{a\rho-1}(y) + d_{\rho-1}(t)\phi_{a\rho-1}(y) + \psi_{0\rho-1}(y)$$

$$\dot{x}_\rho = x_{\rho+1} + \theta_{a\rho}(t)\psi_{a\rho}(y) + d_\rho(t)\phi_{a\rho}(y) + \psi_{0\rho}(y) + b_m(t)u \qquad (3.1)$$

$$\vdots$$

$$\dot{x}_n = \theta_{an}(t)\psi_{an}(y) + d_n(t)\phi_{an}(y) + \psi_{0n}(y) + b_0(t)u$$

$$y = e_1^T x$$

where $x = [x_1, \cdots, x_n]^T \in R^n, u \in R$ and $y \in R$ are system states, input and output, respectively, $b_j(t), j = 0, \ldots, m$ are bounded uncertain time-varying piecewise continuous high-frequency gains, $\theta_{ai}^T(t) \in R^{p_i}$ are uncertain time-varying parameters, $d_i(t)$ denote unknown time-varying bounded disturbances, $\psi_{ai} \in R^{p_i}$, $\psi_{0i} \in R$ and $\phi_{ai} \in R$ are known smooth nonlinear functions, for $i = 1, \ldots, n$. Similar class of systems was analyzed in [84].

For the considered system in (3.1), the following assumptions are imposed.

Assumption 1. The uncertain parameter vector θ is inside a compact set Ω_θ, where $\theta = [b_m(t), \ldots, b_0(t), \theta_{a1}(t), \ldots, \theta_{an}(t)]^T$. In addition, there exists an unknown bounded positive constant q so that $\| \dot{\theta} \| \leq q$. Also q is inside a compact intervals $\Omega_q = [I^-, I^+]$ and $b_m(t) \neq 0, \forall t$.

Assumption 2. The relative degree ρ is fixed and known.

Assumption 3. The reference signal y_r and its $(\rho - 1)$th order derivatives are assumed to be known and bounded.
Assumption 4. The system is minimum phase.

Definition: System is said to be minimum phase if its zero dynamics, subject to appropriate initial conditions and a suitable control producing output identically zero, is stable.

The control objective is to design an adaptive controller for system (3.1) satisfying Assumptions 1-4 such that the closed-loop system is stable and the system output can track a given reference signal $y_r(t)$ as close as possible.

3.2.2   Preliminary Result

In order to cope with the unknown sign of high-frequency gain, the Nussbaum gain technique is employed in this chapter. A function $N(\chi)$ is called a Nussbaum-type function if it has the following properties

$$lim_{s \to \infty} sup \frac{1}{s} \int_0^s N(\chi) d\chi = \infty \tag{3.2}$$

$$lim_{s \to \infty} inf \frac{1}{s} \int_0^s N(\chi) d\chi = -\infty \tag{3.3}$$

In this chapter, the even Nussbaum function $exp(\chi^2) \cos(\frac{\pi}{2}\chi)$ is exploited. The following Lemma will be employed in later analysis.

Lemma 3.1. *Let $V(t)$ and $\chi(t)$ be a smooth function defined on $[0, t_f)$ with $V(t) \geq 0$, $\forall t \in [0, t_f)$, and $N(\chi) = exp(\chi^2) \cos(\frac{\pi}{2}\chi)$ be an even smooth Nussbaum-type function. If the following inequality holds:*

$$V(t) \leq f_0 + e^{-f^*t} \int_0^t g(\tau) N(\chi) \dot{\chi} e^{f^*\tau} d\tau + e^{-f^*t} \int_0^t \dot{\chi}(t) e^{f^*\tau} d\tau \tag{3.4}$$

where constant $f^* > 0$, $g(\tau)$ is a time-varying parameter which takes values in the unknown closed intervals $I_1 = [g^-, g^+]$ with $0 \notin I_1$, and f_0 represents a suitable constant, then $V(t), \chi(t)$ and $\int_0^t g(\tau) N(\chi) \dot{\chi} d\tau$ must be bounded on $[0, t_f)$.

Proof: The boundedness of V can be established based on the Nussbaum gain properties (3.2) and (3.3) via a contradiction argument. We first define

$$V_g(t_i, t_j) = \int_{t_i}^{t_j} [g(\tau)N(\chi) + 1]\dot{\chi}e^{-f^*(t_j - \tau)}d\tau \qquad (3.5)$$

where $g_{max} = \max\{|g^-|, |g^+|\}$ and $g_{min} = \min\{|g^-|, |g^+|\}$. For notation convenience, $V_g(\chi_i, \chi_j) = V_g(\chi(t_i), \chi(t_j)) = V_g(t_i, t_j)$, $t_i \leq t_j$

Using integral inequality $(b - a)f_{min} \leq \int_a^b f(x)dx \leq (b - a)f_{max}$ with $f_{min} = \inf_{a \leq x \leq b} f(x)$ and $f_{max} = \sup_{a \leq x \leq b} f(x)$, and noting the facts that $|g(\tau)| \leq g_{max}$, $0 < e^{-f^*(t - \tau)} \leq 1$ for $\tau \in [0, t]$, we have

$$\begin{aligned}
|V_g(\chi_0, \chi)| &\leq \int_{\chi_0}^{\chi} |g(\tau)N(\chi) + 1|d\chi \\
&\leq (\chi - \chi_0) \sup_{\tau \in [t_0, t], \chi \in [\chi_0, \chi]} |g(\tau)N(\chi) + 1| \\
&\leq (\chi - \chi_0)[g_{max} \sup_{\chi \in [\chi_0, \chi]} |N(\chi)| + 1] \qquad (3.6)
\end{aligned}$$

For $N(\chi) = exp(\chi^2)cos(\pi\chi/2)$, we know that it is positive for $\chi \in (4m-1, 4m+1)$ and negative for $\chi \in (4m + 1, 4m + 3)$ with m an integer.

Then (3.4) is rewritten as

$$0 \leq V(t) \leq f_0 + V_g(\chi_0, \chi) \qquad (3.7)$$

We now show that $\chi(t)$ is bounded on $[0, t_f]$ by seeking a contradiction. Suppose that $\chi(t)$ is unbounded and two cases should be considered: 1) $\chi(t)$ is has no upper bound and 2) $\chi(t)$ has no lower bound.

Case 1): $\chi(t)$ has no upper bound on $[0, t_f]$. In this case, there must exist a monotone increasing variable $\chi_i = \chi(t_i)$ with $\chi_0 = |\chi(t_0)| > 0$, $lim_{i \to \infty}t_i = t_f$, and $lim_{i \to \infty}\chi_i = \infty$.

1. $g(t) > 0$.
   From (3.4), we know for $[\chi_0, \chi_1] = [\chi_0, 4m + 1]$

$$\begin{aligned}
|V_g(\chi_0, \chi_1)| &= |\int_{t_0}^{t_1} [g(\tau)N(\chi) + 1]\dot{\chi}e^{-f^*(t_1 - \tau)}d\tau| \\
&\leq (\chi_1 - \chi_0)[g_{max} \sup_{\chi \in [\chi_0, \chi_1]} |N(\chi)| + 1] \\
&= l_{m1}g_{max}e^{(4m+1)^2} + l_{m1} \qquad (3.8)
\end{aligned}$$

where $l_{m1} = 4m + 1 - \chi_0$. Note that $N(\chi) \leq 0$, $\forall \chi \in [\chi_1, \chi_2] = [4m + 1, 4m + 3]$, we have

$$V_g(\chi_1, \chi_2) \leq \int_{4m+2-c_{m1}}^{4m+2+c_{m1}} [g(\tau)N(\chi) + 1]\dot{\chi}e^{-f^*(t_2 - \tau)}d\tau \qquad (3.9)$$

where $c_{m1} \in (0,1)$, here we select $c_{m1} = \frac{1}{2}$. Using the integral inequality and noting that $g(t) \geq g_{min} > 0$, $e^{-f^*(t_2-\tau)} \geq e^{-f^*(t_2-t_1)} > 0$ for $\tau \in [t_1, t_2]$, we have

$$V_g(\chi_1, \chi_2) \leq 2c_{m1}[g_{min} \inf_{\chi \in [\chi_1, \chi_2]} N(\chi) + 1]e^{-f^*(t_2-t_1)}$$

$$= -g_0 e^{(4m+2-c_{m1})^2} + g_1 \tag{3.10}$$

where $g_0 = 2c_{m1}g_{min}e^{-f^*(t_2-t_1)}cos(\pi c_{m1}/2) > 0$, and $g_1 = 2c_{m1}e^{-f^*(t_2-t_1)} > 0$. Thus from (3.8) and (3.10), we have

$$V_g(\chi_0, \chi_2) = V_g(\chi_0, \chi_1) + V_g(\chi_1, \chi_2)$$

$$\leq e^{(4m+1)^2}\Big[-g_0 e^{\big(2(4m+1)(1-c_{m1})+(1-c_{m1})^2\big)}$$

$$+g_{max}(4m+1-\chi_0) + \frac{4m+1-\chi_0+g_1}{e^{(4m+1)^2}}\Big] \tag{3.11}$$

Note that e^m grows faster than m when $m \to \infty$. From (3.11), we know that $V_g(\chi_0, \chi_2) = V_g(\chi_0, 4m+3) \to -\infty$ as $m \to \infty$.

2. $g(t) < 0$.
Similar to the case $g(t) > 0$, for interval $[\chi_0, \chi_3] = [\chi_0, 4m-1]$, we have

$$|V_g(\chi_0, \chi_3)| = |\int_{t_0}^{t_3} [g(\tau)N(\chi) - 1]\dot{\chi}e^{-f^*(t_3-\tau)}d\tau|$$

$$\leq (\chi_3 - \chi_0)[g_{max} \sup_{\chi \in [\chi_0, \chi_3]} |N(\chi)| + 1]$$

$$= l_{m2}g_{max}e^{(4m-1)^2} + l_{m2} \tag{3.12}$$

where $l_{m2} = 4m - 1 - \chi_0$. Note that $N(\chi) > 0$, $\forall \chi \in [\chi_3, \chi_1] = [4m-1, 4m+1]$, we have

$$V_g(\chi_3, \chi_1) \leq \int_{4m-c_{m1}}^{4m+c_{m1}} [g(\tau)N(\chi) - 1]\dot{\chi}e^{-f^*(t_1-\tau)}d\tau \tag{3.13}$$

Using the integral inequality and noting that $g(t) \leq -g_{min} < 0$, $e^{-f^*(t_1-\tau)} \geq e^{-f^*(t_1-t_3)} > 0$ for $\tau \in [t_3, t_1]$, we have

$$V_g(\chi_3, \chi_1) \leq 2c_{m1}[-g_{min} \inf_{\chi \in [\chi_3, \chi_1]} N(\chi) + 1]e^{-f^*(t_{m1}-t_{m3})}$$

$$= -g_2 e^{(4m-c_{m1})^2} + g_3 \tag{3.14}$$

where $g_2 = 2c_{m1}g_{min}e^{-f^*(t_1-t_3)}cos(\pi c_{m1}/2) > 0$, and $g_3 = 2c_{m1}e^{-f^*(t_1-t_3)} > 0$. Thus from (3.12) and (3.14), we have

$$V_g(\chi_0, \chi_1) = V_g(\chi_0, \chi_3) + V_g(\chi_3, \chi_1)$$

$$\leq e^{(4m-1)^2}[-g_2 e^{[2(4m-1)(1-c_{m1})+(1-c_{m1})^2]}$$

$$+l_{m2}g_{max} + \frac{l_{m2}+g_3}{e^{(4m-1)^2}}] \tag{3.15}$$

From (3.15), we know that $V_g(\chi_0, \chi_1) = V_g(\chi_0, 4m + 1) \to -\infty$ as $m \to \infty$.

From above analysis, we have $V_g(\chi_0, \chi) \to -\infty$ as $m \to \infty$. On the other hand, $V(t) > 0$ for all t. In a conclusion, we can find a subsequence that leads to a contradiction in both $g(t) > 0$ and $g(t) < 0$. Therefore, $\chi(t)$ has upper bound.

Case 2): χ has no lower bound on $[0, t_f]$. Define $\chi = -w$. Accordingly, w has no upper bound. Further noting that $N(.)$ is an even function, (3.4) becomes

$$V(t) \le M - \int_0^t [gN(w) - 1]\dot{w}e^{-f^*(T-\tau)}d\tau$$
$$= M - V_g(w(0), w(t)), \quad \forall t \in [0, t_f) \tag{3.16}$$

Thus, there must exist a monotone increasing variable $\{w_i = w(t_i)\}$ with $w_0 = |w(t_0)| > 0$, $lim_{i\to\infty}t_i = t_f$, and $lim_{i\to\infty}w_i = \infty$. Following the procedure as in Case 1), we can also construct a subsequence that leads to a contradiction. Accordingly, we can claim that w has upper bound on $[0, t_f)$. Since $\chi = -w$, we know that χ has lower bound on $[0, t_f)$.

The above argument is true for all $t_f > 0$. Therefore, χ must be bounded. And also $V(t)$ and $\int_0^t [g(\tau)N(\chi) + 1]\dot{\chi}d\tau$ are bounded on $[0, t_f)$.

3.3   State Estimation Filters

In order to design the desired adaptive control law with output via backstepping procedures, we now transform system (3.1) into the following form

$$\dot{x} = Ax + F(y, u)^T\theta + \Phi_a(y)d(t)^T + \psi_0(y) \tag{3.17}$$

where

$$A = \begin{bmatrix} 0 & 1 & 0 & \dots & 0 \\ 0 & 0 & 1 & \dots & 0 \\ \vdots & \vdots & \vdots & \ddots & \vdots \\ 0 & 0 & 0 & \dots & 1 \\ 0 & 0 & 0 & \dots & 0 \end{bmatrix} \tag{3.18}$$

$$F(y, u)^T = \begin{bmatrix} \begin{bmatrix} 0_{(\rho-1)\times(m+1)} \\ I_{m+1} \end{bmatrix} u, \Psi_a(y) \end{bmatrix} \tag{3.19}$$

$$\Psi_a(y) = \begin{bmatrix} \psi_{a1}^T & 0 & \dots & 0 \\ 0 & \vdots & \ddots & \vdots \\ 0 & 0 & \dots & \psi_{an}^T \end{bmatrix} = \begin{bmatrix} \Psi_{a1}(y) \\ \vdots \\ \Psi_{an}(y) \end{bmatrix} \tag{3.20}$$

$$\Phi_a(y) = \begin{bmatrix} \phi_{a1} & 0 & \dots & 0 \\ 0 & \vdots & \ddots & \vdots \\ 0 & 0 & \dots & \phi_{an} \end{bmatrix} == \begin{bmatrix} \Phi_{a1}^T(y) \\ \vdots \\ \Phi_{an}^T(y) \end{bmatrix} \tag{3.21}$$

$$\theta = [b_m(t), \dots, b_0(t), \theta_{a1}(t), \dots, \theta_{an}(t)]^T \tag{3.22}$$

$$d(t) = [d_1(t), \dots, d_n(t)] \tag{3.23}$$

$$\psi_0(y) = [\psi_{01}(y), \dots, \psi_{0n}(y)]^T \tag{3.24}$$

For state estimation, we employ the filters

$$\dot{\xi} = A_0\xi + ky + \psi_0(y) \tag{3.25}$$

$$\dot{\Omega}^T = A_0\Omega^T + F(y, u)^T \tag{3.26}$$

where

$$k \triangleq [k_1, k_2, \dots, k_n]^T \tag{3.27}$$

$$A_0 = A - ke_1^T \tag{3.28}$$

The vector k in (3.27) is chosen such that the matrix A_0 is strictly stable. Next we lower the dynamic order of the Ω filter by exploiting the structure of $F(y, u)$ in (3.19). We denote the first $m + 1$ columns of Ω^T by $v_m, \dots, v_1, v_0$ and the remaining n columns by Ξ as follows

$$\Omega^T = [v_m, \dots, v_1, v_0, \Xi], \tag{3.29}$$

and show that the equations for the first $n + 1$ columns of Ω^T are governed by

$$\dot{v}_j = A_0 v_j + e_{n-j} u, \quad j = 0, \dots, m \tag{3.30}$$

Due to the special structure of A_0, we have

$$A_0^j e_n = e_{n-j}, \quad j = 0, \dots, n-1 \tag{3.31}$$

The vectors v_j can be obtained from only one input filter

$$\dot{\lambda} = A_0\lambda + e_n u \tag{3.32}$$

Using the algebraic expression, we have

$$v_j = A_0^j \lambda \tag{3.33}$$

We also have

$$\dot{\Xi} = A_0 \Xi + \Psi_a(y) \tag{3.34}$$

We now summarize the reduced-order filters

$$\dot{\xi} = A_0\xi + ky + \psi_0(y) \tag{3.35}$$

$$\dot{\Xi} = A_0 \Xi + \Psi_a(y) \tag{3.36}$$

$$\dot{\lambda} = A_0\lambda + e_n u \tag{3.37}$$

$$v_j = A_0^j \lambda \tag{3.38}$$

With the above filters, a state estimate is given by

$$\hat{x} = \xi + \Omega^T \theta \qquad (3.39)$$

and the estimation errors ϵ is defined as

$$\epsilon = x - \hat{x} \qquad (3.40)$$

We have a static relationship between the state x and the unknown parameter θ

$$x = \xi + \Omega^T \theta + \epsilon \qquad (3.41)$$

Then we have

$$\begin{aligned}
x_2 &= \xi_2 + \Omega_2^T \theta + \epsilon_2 \\
&= \xi_2 + [v_{m,2}, v_{m-1,2}, \dots, v_{0,2}, \Xi_2]\theta + \epsilon_2 \\
&= b_m v_{m,2} + \xi_2 + [0, v_{m-1,2}, \dots, v_{0,2}, \Xi_2]\theta + \epsilon_2
\end{aligned} \qquad (3.42)$$

In above equations, ϵ_2, $v_{i,2}$ and ξ_2 denote the second entries of ϵ, v_i and ξ respectively, ϵ is the estimation error defined in (3.40).

Because the backstepping design starts with its output y, which is the only available system state allowed to appear in the control law, (3.1) is expressed as

$$\begin{aligned}
\dot{y} &= x_2 + \theta_{a1}(t)\psi_{a1}(y) + d_1(t)\phi_{a1}(y) + \psi_{01}(y) \\
&= b_m v_{m,2} + \beta + \bar{\omega}^T \theta + \epsilon_2 + d(t)\Phi_{a1}(y)
\end{aligned} \qquad (3.43)$$

where

$$\beta = \xi_2 + \psi_{01}(y) \qquad (3.44)$$
$$\omega = [v_{m,2}, v_{m-1,2}, \dots, v_{0,2}, \Xi_2 + \Psi_{a1}]^T \qquad (3.45)$$
$$\bar{\omega} = [0, v_{m-1,2}, \dots, v_{0,2}, \Xi_2 + \Psi_{a1}]^T \qquad (3.46)$$

Due to the minimum phase in Assumption 4, system (3.1) is restricted to the first ρ equations. From our designed filters (3.35)-(3.37), the design system is

$$\dot{y} = b_m v_{m,2} + \beta + \bar{\omega}^T \theta + \epsilon_2 + d(t)\Phi_{a1}(y) \qquad (3.47)$$
$$\dot{v}_{m,i} = v_{m,i+1} - k_i v_{m,1}, \qquad i = 2, 3, \dots, \rho - 1 \qquad (3.48)$$
$$\dot{v}_{m,\rho} = v_{m,\rho+1} - k_\rho v_{m,1} + u \qquad (3.49)$$

From the equations (3.17), (3.25), (3.26), (3.39) and (3.40), the estimation error in (3.40) satisfies

$$\dot{\epsilon} = A_0 \epsilon + \Phi_a(y)d(t)^T - \Omega^T \dot{\theta} \qquad (3.50)$$

Remark 1. As the disturbances and derivatives of time-varying parameters appear in (3.50), their effects should be considered in controller design. However for the state-feedback control in [83], no filter is required for state estimation. Their effects may not be necessarily considered in controller design and this makes problem much simpler.

We now divide the error ϵ into two parts, i.e. $\epsilon = \epsilon_a + \epsilon_b$, where ϵ_a satisfies

$$\dot{\epsilon}_a = A_0 \epsilon_a + \Phi_a(y) d(t)^T \tag{3.51}$$

with $\epsilon_a(0) = \epsilon(0)$, and $\epsilon_b = \int_0^t e^{A_0(t-\tau)}(-\Omega^T \dot{\theta}) d\tau$. It can be shown that

$$\begin{aligned}
\| \epsilon_b \| &\le \int_0^t \| e^{A_0(t-\tau)} \| \cdot \| \Omega \| \cdot \| \dot{\theta} \| \, d\tau \\
&\le q \int_0^t \| e^{A_0(t-\tau)} \| \cdot \| \Omega \| \, d\tau \\
&\le q \int_0^t e^{-\lambda_\theta (t-\tau)} k_\theta \| \Omega \| \, d\tau \tag{3.52}
\end{aligned}$$

where λ_θ and k_θ are chosen positive parameters so that

$$k_\theta e^{-\lambda_\theta t} \ge \| e^{A_0 t} \|, \quad \forall t \ge 0 \tag{3.53}$$

Thus ϵ_b satisfies that

$$|\epsilon_b| \le h(t) q \tag{3.54}$$

where $h(t)$ is generated by

$$\dot{h} = -\lambda_\theta h + k_\theta (\| \Omega \|^2 + \frac{1}{4}). \tag{3.55}$$

Suppose $P \in R^{n \times n}$ is a positive definite matrix, satisfying $PA_0 + A_0^T P \le -3I$ and let

$$V_\epsilon = \epsilon_a^T P \epsilon_a \tag{3.56}$$

From equation (3.51), the derivative of V_ϵ is given as

$$\begin{aligned}
\dot{V}_\epsilon &= \dot{\epsilon}_a^T P \epsilon_a + \epsilon_a^T P \dot{\epsilon}_a \\
&= \epsilon_a^T (PA_0 + A_0^T P) \epsilon_a + 2\epsilon_a^T P \Phi_a(y) d(t)^T \\
&\le -2 \| \epsilon_a \|^2 + \| P \Phi_a(y) d(t)^T \|^2 \tag{3.57}
\end{aligned}$$

The problem of this chapter is to design an adaptive controller to make system (3.1) BIBO stable.

3.4 Control Design

3.4.1 Design Procedure

In this section, we present the adaptive control design using the backstepping technique with tuning functions in ρ steps. In order to avoid using the sign of the high frequency gain, we take the change of coordinates

$$z_1 = y - y_r \tag{3.58}$$
$$z_i = v_{m,i} - \alpha_{i-1}, \quad i = 2, 3, \ldots, \rho, \tag{3.59}$$

where α_{i-1} is the virtual control at each step and will be determined in later discussions. Before presenting the detail, a useful function is introduced. Firstly we define $s(x)$ as

$$
s(x) = \begin{cases} x^2 & |x| \geq \delta \\ (\delta^2 - x^2)^\rho + x^2 & |x| < \delta \end{cases} \tag{3.60}
$$

where δ is a positive design parameter. Note that $s(x)$ is $(\rho-1)$th order differentiable and bounded below for $|x| < \delta$. Based on $s(x)$, a function $H(z_1)$ is defined as follows

$$
H(z_1) = \frac{\Phi_a(y)}{s(z_1)} = \begin{cases} \dfrac{\Phi_a(y)}{z_1^2} & |z_1| \geq \delta \\ \dfrac{\Phi_a(y)}{(\delta^2 - z_1^2)^\rho + z_1^2} & |z_1| < \delta \end{cases} \tag{3.61}
$$

Clearly H is well defined and for $|z_1| < \delta$, H is bounded as $s(z_1)$ is bounded below.

Remark 2. In [35], a similar function to (3.61) was used to design controllers for disturbance decoupling. However, the function is undefined at the time instants when $y = y_r$. Thus, the controller presented is undefined at these time instants.

With (3.61) and using Young's inequality $ab \leq \frac{1}{2}a^2 + \frac{1}{2}b^2$, (3.57) gives

$$
\begin{aligned}
\dot{V}_\epsilon &\leq -2 \parallel \epsilon_a \parallel^2 + \frac{1}{2} \parallel P\Phi_a(y) \parallel^4 + \frac{1}{2} \parallel d(t) \parallel^4 \\
&= -2 \parallel \epsilon_a \parallel^2 + \frac{1}{2}s^4 \parallel PH \parallel^4 + \frac{1}{2} \parallel d(t) \parallel^4
\end{aligned} \tag{3.62}
$$

We now illustrate the backstepping design procedures using Nussbaum gain with details given for the first two steps.

Step.1
It follows from (3.47) and (3.58) that

$$
\dot{z}_1 = b_m v_{m,2} + \beta + \bar{\omega}^T \theta + \epsilon_2 + d(t)\Phi_{a1}(y) - \dot{y}_r \tag{3.63}
$$

Without using the sign of b_m, the following virtual control law α_1 is designed

$$
\alpha_1 = N(\chi)\bar{\alpha}_1 e^{-ft} \tag{3.64}
$$

$$
N(\chi) = exp(\chi^2) \cos \frac{\pi}{2}\chi \tag{3.65}
$$

where f is a positive real design parameter, χ is generated by

$$
\dot{\chi} = z_1 \bar{\alpha}_1 \tag{3.66}
$$

and $\bar{\alpha}_1$ is chosen to be

$$
\begin{aligned}
\bar{\alpha}_1 = {}& (c_1 + l_1 + (e_1^T\hat{\theta})^2)z_1 + \beta + \bar{\omega}^T\hat{\theta} - \dot{y}_r + z_1 h^2 \hat{q} \\
& + \frac{1}{4}z_1 \parallel \Phi_{a1}(y) \parallel^2 + \sum_{i=1}^{\rho} \frac{1}{8l_i} z_1 s^3(z_1) \parallel PH \parallel^4
\end{aligned} \tag{3.67}
$$

where c_1 and l_1 are two positive real design parameters, $\hat{\theta}$ and $\hat{q}$ denotes the estimates of θ and q. Notice that

$$b_m v_{m,2} = b_m(z_2 + \alpha_1) = \hat{b}_m z_2 + b_m \alpha_1 + \tilde{b}_m z_2 \tag{3.68}$$

where $\tilde{b}_m = b_m - \hat{b}_m$, $\hat{b}_m$ is the first element of $\hat{\theta}$, i.e. $\hat{b}_m = e_1^T \hat{\theta}$. Then from (3.63) and (3.67) we have

$$\dot{z}_1 - \bar{\alpha}_1 = -(c_1 + l_1 + \hat{b}_m^2)z_1 + (\bar{\omega}^T + z_2 e_1^T)\tilde{\theta} + \epsilon_{a,2} + \epsilon_{b,2} - z_1 h^2 \hat{q} + \hat{b}_m z_2$$
$$+ b_m \alpha_1 + d(t)\Phi_{a1}(y) - \frac{1}{4}z_1 \parallel \Phi_{a1}(y) \parallel^2 - \sum_{i=1}^{\rho} \frac{1}{8l_i} z_1 s^3 \parallel PH \parallel^4 \tag{3.69}$$

where $\tilde{\theta} = \theta - \hat{\theta}$, and $\epsilon_{a,2}$ and $\epsilon_{b,2}$ represent the second entries of ϵ_a and ϵ_b. To proceed, we define the Lyapunov function

$$V_1 = \frac{1}{2}z_1^2 + \frac{1}{2}\tilde{\theta}^T \Gamma^{-1} \tilde{\theta} + \frac{1}{2\gamma}\tilde{q}^2 + \frac{1}{4l_1}V_\epsilon \tag{3.70}$$

where Γ is a positive definite matrix and γ is a positive constant. Then the derivative of V_1 along with (3.62), (3.64) and (3.69) is given by

$$\dot{V}_1 = z_1(\dot{z}_1 - \bar{\alpha}_1) + z_1 \bar{\alpha}_1 + \tilde{\theta}^T \Gamma^{-1}(\dot{\theta} - \dot{\hat{\theta}}) + \frac{1}{\gamma}\tilde{q}\dot{\tilde{q}} + \frac{1}{4l_1}\dot{V}_\epsilon$$
$$\leq -(c_1 + \hat{b}_m^2)z_1^2 + \hat{b}_m z_1 z_2 + \tilde{\theta}^T \Gamma^{-1}(\tau_1 - \dot{\hat{\theta}}) - l_1 z_1^2 + \epsilon_{a,2} z_1 - \frac{1}{2l_1} \parallel \epsilon_a \parallel^2$$
$$+ \epsilon_{b,2} z_1 - \frac{1}{\gamma}\tilde{q}\dot{\hat{q}} - h^2 \hat{q}z_1^2 + d(t)\Phi_{a1}(y)z_1 - \frac{1}{4}z_1^2 \parallel \Phi_{a1}(y) \parallel^2 + b_m \alpha_1 z_1 + \bar{\alpha}_1 z_1$$
$$+ \frac{1}{8l_1}s^4 \parallel PH \parallel^4 - \sum_{i=1}^{\rho} \frac{1}{8l_i} z_1^2 s^3 \parallel PH \parallel^4 + \frac{1}{8l_1} \parallel d(t) \parallel^4 + \tilde{\theta}^T \Gamma^{-1}\dot{\theta} \tag{3.71}$$

where

$$\tau_1 = \Gamma z_1(\bar{\omega} + z_2 e_1) \tag{3.72}$$

Here we know that

$$\epsilon_{b,2} z_1 - h^2 \hat{q}z_1^2 \leq hq|z_1| - h^2 \hat{q}z_1^2 \leq q(h^2 z_1^2 + 1/4) - h^2 \hat{q}z_1^2 = h^2 \tilde{q}z_1^2 + \frac{q}{4}$$

Then we can get

$$\dot{V}_1 \leq (b_m N(\chi)e^{-ft} + 1)\dot{\chi} - c_1 z_1^2 + \tilde{\theta}^T \Gamma^{-1}(\tau_1 - \dot{\hat{\theta}})$$
$$+ \frac{1}{\gamma}\tilde{q}(\iota_1 - \dot{\hat{q}}) - \frac{1}{4l_1} \parallel \epsilon_a \parallel^2 + \frac{1}{4}z_2^2 + M_1 \tag{3.73}$$

where

$$\iota_1 = \gamma h^2 z_1^2 \tag{3.74}$$

$$M_1 = \| d(t) \|^2 + \frac{1}{8l_1} \| d(t) \|^4 - \sum_{i=2}^{\rho} \frac{1}{8l_i} s^4 \| PH \|^4$$

$$+ \tilde{\theta}^T \Gamma^{-1} \dot{\theta} + \frac{1}{4} q + \bar{N} \tag{3.75}$$

$$\bar{N} = \begin{cases} 0 & |z_1| \geq \delta \\ \sum_{i=1}^{\rho} \frac{1}{8l_i} (\delta^2 - z_1^2)^\rho s^3 \| PH \|^4 & |z_1| < \delta \end{cases} \tag{3.76}$$

From (3.61) we know that $\bar{N}$ is bounded.

Step.2

Now, we evaluate the dynamics of the second state z_2. Differentiating (3.59) for $i = 2$ and using (3.48), we have

$$\dot{z}_2 = v_{m,3} - k_2 v_{m,1} - \dot{\alpha}_1 \tag{3.77}$$

Note that α_1 is a function of $y, \hat{\theta}, \hat{q}, \xi, \Xi, \lambda, \chi$ and y_r and following backstepping design by substituting (3.59) with $i = 3$ into (3.77), we get

$$\dot{z}_2 = \alpha_2 - \beta_2 - \frac{\partial \alpha_1}{\partial y} \left(\epsilon_2 + \omega^T \tilde{\theta} + d(t) \Phi_{a1}(y) \right) + z_3$$

$$- \frac{\partial \alpha_1}{\partial y} \omega^T \hat{\theta} - \frac{\partial \alpha_1}{\partial \hat{\theta}} \dot{\hat{\theta}} - \frac{\partial \alpha_1}{\partial \hat{q}} \dot{\hat{q}} \tag{3.78}$$

where

$$\beta_2 \triangleq k_2 v_{m,1} + \frac{\partial \alpha_1}{\partial y} \beta + \frac{\partial \alpha_1}{\partial \Pi} \dot{\Pi} + \sum_{j=1}^{m+1} \frac{\partial \alpha_1}{\partial \lambda_j} (-k_j \lambda_1 + \lambda_{j+1})$$

$$+ \frac{\partial \alpha_1}{\partial y_r} \dot{y}_r + \frac{\partial \alpha_1}{\partial \chi} \dot{\chi} \tag{3.79}$$

where $\Pi = [\xi^T, Vec(\Xi)^T]^T$. Define the Lyapunov function and choose the virtual control for this step as

$$V_2 = V_1 + \frac{1}{2} z_2^2 + \frac{1}{4l_2} V_\epsilon \tag{3.80}$$

$$\alpha_2 = -(c_2 + \frac{1}{4}) z_2 + \frac{\partial \alpha_1}{\partial y} \omega^T \hat{\theta} - z_2 \| \frac{\partial \alpha_1}{\partial \hat{\theta}} \|^2 \| \tau_2 \|^2 - z_2 h^2 \hat{q} \| \frac{\partial \alpha_1}{\partial y} \|^2$$

$$- z_2 \| \frac{\partial \alpha_1}{\partial \hat{q}} \|^2 \iota_2^2 - l_2 \| \frac{\partial \alpha_1}{\partial y} \|^2 z_2 + \beta_2 - \frac{z_2}{4} \| \frac{\partial \alpha_1}{\partial y} \Phi_{a1}(y) \|^2 \tag{3.81}$$

$$\tau_2 = \tau_1 - \Gamma \frac{\partial \alpha_1}{\partial y} \omega z_2 \tag{3.82}$$

$$\iota_2 = \iota_1 + \gamma h^2 \| \frac{\partial \alpha_1}{\partial y} \|^2 z_2^2 \tag{3.83}$$

Using (3.73), (3.80) and (3.81), we have that

$$\dot{V}_2 \le \dot{V}_1 + z_2 \dot{z}_2 + \frac{1}{4l_2}\dot{V}_\epsilon$$

$$\le -\sum_{i=1}^{2} c_i z_i^2 + (b_m N(\chi)e^{-ft} + 1)\dot{\chi} + z_2 z_3 - \sum_{i=1}^{2} \frac{1}{4l_i} \parallel \epsilon_a \parallel^2 + M_2$$

$$+ \tilde{\theta}^T \Gamma^{-1}(\tau_1 - \dot{\hat{\theta}}) - z_2 \frac{\partial \alpha_1}{\partial y}\omega^T \tilde{\theta} + z_2^2 \parallel \frac{\partial \alpha_1}{\partial \hat{\theta}} \parallel^2 \parallel \dot{\hat{\theta}} \parallel^2 - z_2^2 \parallel \frac{\partial \alpha_1}{\partial \hat{\theta}} \parallel^2 \parallel \tau_2 \parallel^2$$

$$+ \frac{1}{\gamma}\tilde{q}(\iota_1 - \dot{\hat{q}}) + h^2 \tilde{q} \parallel \frac{\partial \alpha_1}{\partial y} \parallel^2 z_2^2 + z_2^2 \parallel \frac{\partial \alpha_1}{\partial \hat{q}} \parallel^2 \dot{\hat{q}}^2 - z_2^2 \parallel \frac{\partial \alpha_1}{\partial \hat{q}} \parallel^2 \iota^2$$

$$\le -\sum_{i=1}^{2} c_i z_i^2 + (b_m N(\chi)e^{-ft} + 1)\dot{\chi} + z_2 z_3 + \tilde{\theta}^T \Gamma^{-1}(\tau_2 - \dot{\hat{\theta}})$$

$$+ \frac{1}{\gamma}\tilde{q}(\iota_2 - \dot{\hat{q}}) + z_2^2 (\frac{\partial \alpha_1}{\partial \hat{q}})^2 (\dot{\hat{q}}^2 - \iota_2^2) - \sum_{i=1}^{2} \frac{1}{4l_i} \parallel \epsilon_a \parallel^2$$

$$+ z_2^2 \parallel \frac{\partial \alpha_1}{\partial \hat{\theta}} \parallel^2 (\parallel \dot{\hat{\theta}} \parallel^2 - \parallel \tau_2 \parallel^2) + M_2 \tag{3.84}$$

where

$$M_2 = \sum_{i=1}^{2} \frac{1}{8l_i} \parallel d(t) \parallel^4 + 2 \parallel d(t) \parallel^2 - \sum_{i=3}^{\rho} \frac{1}{8l_i} s^4 \parallel PH \parallel^4$$

$$+ \tilde{\theta}^T \Gamma^{-1}\dot{\theta} + \frac{1}{2} + \frac{1}{2}q + \bar{N} \tag{3.85}$$

Remark 3. Note that M_2 contains $s^4 \parallel PH \parallel^4$ and this term may not be bounded. As seen from our analysis, $\frac{1}{8l_2}s^4 \parallel PH \parallel^4$ disappears in M_2 due to the use of V_ϵ at step 2. If we use V_ϵ at each step, this term will disappear in M_ρ of the last step.

Step.i $(i = 3, \ldots, \rho)$
We define the positive Lyapunov function V_i as

$$V_i = V_{i-1} + \frac{1}{2}z_i^2 + \frac{1}{4l_i}V_\epsilon \tag{3.86}$$

and choose the virtual control law α_i as

$$\alpha_i = -c_i z_i - l_i \parallel \frac{\partial \alpha_{i-1}}{\partial y} \parallel^2 z_i - z_{i-1} + \frac{\partial \alpha_{i-1}}{\partial y}\omega^T \hat{\theta} - \frac{z_i}{4} \parallel \frac{\partial \alpha_{i-1}}{\partial y}\Phi_{a1}(y) \parallel^2$$

$$- z_i \parallel \frac{\partial \alpha_{i-1}}{\partial \hat{\theta}} \parallel^2 \parallel \tau_i \parallel^2 + \left(\sum_{k=2}^{i-1} z_k^2 \parallel \frac{\partial \alpha_{k-1}}{\partial \hat{\theta}} \parallel^2 \right)(\tau_i + \tau_{i-1})^T \Gamma \frac{\partial \alpha_{i-1}}{\partial y}\omega$$

$$+ \beta_i - \left(\sum_{k=2}^{i-1} z_k^2 \parallel \frac{\partial \alpha_{k-1}}{\partial \hat{q}} \parallel^2 \right)(\iota_i + \iota_{i-1})h^2 \parallel \frac{\partial \alpha_{i-1}}{\partial y} \parallel^2 z_i$$

$$-z_i \parallel \frac{\partial \alpha_{i-1}}{\partial \hat{q}} \parallel^2 \iota_i^2 - z_i h^2 \hat{q} \parallel \frac{\partial \alpha_{i-1}}{\partial y} \parallel^2 \tag{3.87}$$

$$\tau_i = \tau_{i-1} - \Gamma \frac{\partial \alpha_{i-1}}{\partial y} \omega z_i \tag{3.88}$$

$$\iota_i = \iota_{i-1} + \gamma h^2 \parallel \frac{\partial \alpha_{i-1}}{\partial y} \parallel^2 z_i^2 \tag{3.89}$$

where

$$\beta_i \triangleq k_i v_{m,1} + \frac{\partial \alpha_{i-1}}{\partial y} \beta + \frac{\partial \alpha_{i-1}}{\partial \Pi} \dot{\Pi} + \frac{\partial \alpha_{i-1}}{\partial y_r} \dot{y}_r$$
$$+ \sum_{j=1}^{m+1} \frac{\partial \alpha_{i-1}}{\partial \lambda_j} (-k_j \lambda_1 + \lambda_{j+1}) + \frac{\partial \alpha_{i-1}}{\partial \chi} \dot{\chi} \tag{3.90}$$

Also note that

$$\parallel \tau_i \parallel^2 = \tau_i^T \tau_i - \tau_{i-1}^T \tau_{i-1} + \tau_{i-1}^T \tau_{i-1}$$
$$= (\tau_i + \tau_{i-1})^T (\tau_i - \tau_{i-1}) + \tau_{i-1}^T \tau_{i-1}$$
$$= -(\tau_i + \tau_{i-1})^T \Gamma \frac{\partial \alpha_{i-1}}{\partial y} \omega z_i + \tau_{i-1}^T \tau_{i-1}$$
$$\iota_i^2 = (\iota_i + \iota_{i-1}) \gamma h^2 \parallel \frac{\partial \alpha_{i-1}}{\partial y} \parallel^2 z_i^2 + \iota_{i-1}^2 \tag{3.91}$$

Then the actual adaptive controller is obtained and given by

$$u(t) = \alpha_\rho - v_{m,\rho+1} \tag{3.92}$$

$$\dot{\hat{\theta}} = Proj(\tau_\rho) \tag{3.93}$$

$$\dot{\hat{q}} = Proj(\iota_\rho) \tag{3.94}$$

where Proj(.) is a smooth projection operation to ensure the estimates belong to compact sets for all time. Such an operation can be found in Appendix C.

Remark 4. Note that the design of tuning functions does not follow the standard tuning function design in [1] as the projection operations are used in the parameter estimators.

3.4.2 Stability Analysis

We construct the final Lyapunov function as

$$V_i = V_{i-1} + \frac{1}{2} z_i^2 + \frac{1}{4l_i} V_\epsilon \tag{3.95}$$

By using the properties $-\tilde{\theta}^T \Gamma^{-1} Proj(\tau) \leq -\tilde{\theta}^T \Gamma^{-1}\tau$ and $Proj(\tau)^T Proj(\tau) \leq \tau^T \tau$ the final Lyapuonv function V_ρ satisfies

$$\dot{V}_\rho \leq -\sum_{k=1}^{\rho} c_k z_k^2 + (b_m N(\chi)e^{-ft} + 1)\dot{\chi} + M_\rho - \sum_{i=1}^{\rho} \frac{1}{4l_i} \parallel \epsilon_a \parallel^2$$

$$+\Big(\sum_{k=2}^{\rho} z_k^2 \parallel \frac{\partial \alpha_{k-1}}{\partial \hat{\theta}} \parallel^2 \Big)(Proj(\tau_\rho)^T Proj(\tau_\rho) - \parallel \tau_\rho \parallel^2)$$

$$+\tilde{\theta}^T \Gamma^{-1}(\tau_\rho - Proj(\tau_\rho)) + \frac{1}{\gamma}\tilde{q}(\iota_\rho - Proj(\iota_\rho))$$

$$+\Big(\sum_{k=2}^{\rho} z_k^2 \big(\frac{\partial \alpha_{k-1}}{\partial \hat{q}}\big)^2\Big)\Big((Proj(\iota_\rho))^2 - \iota_\rho^2\Big)$$

$$\leq -\sum_{k=1}^{\rho} c_k z_k^2 + b_m N(\chi)e^{-ft}\dot{\chi} + \dot{\chi} + M_\rho - \sum_{i=1}^{\rho} \frac{1}{4l_i} \parallel \epsilon_a \parallel^2 \qquad (3.96)$$

where

$$M_\rho = \sum_{i=1}^{\rho} \frac{1}{8l_i} \parallel d(t) \parallel^4 + \rho \parallel d(t) \parallel^2 + \tilde{\theta}^T \Gamma^{-1}\dot{\theta} + \frac{\rho - 1}{2} + \frac{\rho}{4}q + \bar{N} \quad (3.97)$$

Integrating both sides of (3.96) over the interval $[0, t]$ gives

$$\int_0^t \dot{V}_\rho e^{f\tau} d\tau \leq -\int_0^t \sum_{k=1}^{\rho} c_k z_k^2 e^{f\tau} d\tau + \int_0^t b_m N(\chi)\dot{\chi} d\tau + \int_0^t \dot{\chi} e^{f\tau} d\tau$$

$$+\int_0^t M_\rho e^{f\tau} d\tau - \int_0^t \sum_{i=1}^{\rho} \frac{1}{4l_i} \parallel \epsilon_a \parallel^2 e^{f\tau} d\tau \qquad (3.98)$$

Note that $V_\epsilon \leq \parallel P \parallel \parallel \epsilon_a \parallel^2$. Then

$$V_\rho = \sum_{k=1}^{\rho} \frac{1}{2}z_k^2 + \frac{1}{2}\tilde{\theta}^T \Gamma^{-1}\tilde{\theta} + \frac{1}{2\gamma}\tilde{q}^2 + \sum_{i=1}^{\rho} \frac{1}{4l_i} V_\epsilon$$

$$\leq \sum_{k=1}^{\rho} \frac{1}{2}z_k^2 + \frac{1}{2}\tilde{\theta}^T \Gamma^{-1}\tilde{\theta} + \frac{1}{2\gamma}\tilde{q}^2 + \sum_{i=1}^{\rho} \frac{1}{4l_i} \parallel P \parallel \parallel \epsilon_a \parallel^2 \qquad (3.99)$$

This yields

$$0 \leq V_\rho(t) \leq V_\rho(0) + e^{-ft}\int_0^t b_m N(\chi)\dot{\chi} d\tau + \int_0^t \dot{\chi} e^{-f(t-\tau)} d\tau$$

$$+\int_0^t \frac{f}{2}(\tilde{\theta}^T \Gamma^{-1}\tilde{\theta}) + \tilde{q}^2)e^{-f(t-\tau)} d\tau + \int_0^t M_\rho e^{-f(t-\tau)} d\tau$$

$$(3.100)$$

where $f = min\{\frac{1}{\|P\|_2}, 2c_1, 2c_2, \ldots, 2c_\rho,\} > 0$. Due to the utilization of projection operations for $\hat{\theta}$ and $\hat{q}$, the boundedness of $\tilde{\theta}$ and $\tilde{q}$ can be guaranteed. Together the boundedness $d(t)$, q and $\dot{\theta}$, the boundedness of M_ρ and $\int_0^t \frac{f}{2}(\tilde{\theta}^T \Gamma^{-1}\tilde{\theta} + \tilde{q}^2)e^{-f(t-\tau)}d\tau + \int_0^t M_\rho e^{-f(t-\tau)}d\tau$ can be guaranteed. Thus by comparing with (3.4), f_0 is selected as the upper bound of $V_\rho(0) + \int_0^t \frac{f}{2}(\tilde{\theta}^T \Gamma^{-1}\tilde{\theta} + \tilde{q}^2)e^{-f(t-\tau)}d\tau + \int_0^t M_\rho e^{-f(t-\tau)}d\tau$, $g(t) = b_m(t)$. Using Lemma 3.1, we can conclude that $V_\rho(t)$ and $\chi(t)$, hence z_i, $(i = 1, \ldots, \rho)$, θ, q and ϵ_a are bounded. Since z_1 and y_r are bounded, y is also bounded. Because of the boundedness of y, ξ and Ξ in filters (3.35) and (3.36) are bounded as A_0 is Hurwitz. Since the system is minimum phase in Assumption 4 and the boundedness of y, we have $\lambda_1, \ldots, \lambda_{m+1}$ are bounded. Then the coordinate change (3.59) gives

$$v_{m,i} = z_i + \alpha_{i-1}(y, \xi, \Xi, \hat{\theta}, \hat{q}, y_r^{(i-1)}, \lambda_1, \ldots, \lambda_{m+i-1})$$
$$i = 2, 3, \ldots, \rho \tag{3.101}$$

Let $i = 2$, the boundedness of $\lambda_1, \ldots, \lambda_{m+1}$, $z_2, y, \Xi, \xi, \hat{\theta}, \hat{q}, y_r$ and $\dot{y}_r$ proves that $v_{m,2}$ is bounded. Then from (3.38) it follows that λ_{m+2} is bounded. Recursively using the same procedure, we establish that λ and v_j are bounded. From (3.54) and (3.55), the boundedness of ϵ_b is established. Finally, with the boundedness of Ξ, ξ, λ and ϵ, we conclude that x is bounded.

We have thus show that the signals of the closed-loop adaptive system are bounded. To conclude this section, the results established are presented in the following theorem.

Theorem 3.1 *Consider the uncertain time-varying nonlinear system (3.1) satisfying Assumptions 1-4. With the application of the controller (3.92) and the parameter updating laws (3.93) and (3.94), the resulting closed loop system is BIBO stable.*

3.5   An Illustrative Example

For illustration of the proposed scheme, an example is considered. The results of simulation will verify that our adaptive controller makes the system stable. We consider the following second-order system

$$\dot{x}_1 = x_2 + \theta_{a1}(t)x_1^2 + d_1(t)$$
$$\dot{x}_2 = b(t)u + \theta_{a2}(t)x_1 + d_2(t)$$
$$y = x_1 \tag{3.102}$$

where $\theta_{a1}(t) = x_1 e^{-0.5t}$, $\theta_{a2}(t) = 2 + \cos(t)$, $b(t) = 3 + \sin(t)$, $d_1(t) = 0.6\sin(t)$ and $d_2(t) = 0.5\cos^2(t)$, actually these timevarying parameters are not needed to be known in controller design. The objective is to control the system output $y(t)$.to follow a desired trajectory $y_r = \sin(2t) + \sin(t)$. With the application of the filters (3.25) and (3.26), the controller (3.92) and the parameter updating laws (3.93) and (3.94), the resulting closed-loop system is stable.

Since $\rho = 2 = n$, we have $v_\rho = v_n = \lambda$. The filters are designed as

$$\begin{bmatrix} \dot{\xi}_1 \\ \dot{\xi}_2 \end{bmatrix} = \begin{bmatrix} -k_1 & 1 \\ -k_2 & 0 \end{bmatrix} \begin{bmatrix} \xi_1 \\ \xi_2 \end{bmatrix} + \begin{bmatrix} k_1 \\ k_2 \end{bmatrix} y \tag{3.103}$$

$$\begin{bmatrix} \dot{\Xi}_1 \\ \dot{\Xi}_2 \end{bmatrix} = \begin{bmatrix} -k_1 & 1 \\ -k_2 & 0 \end{bmatrix} \begin{bmatrix} \Xi_1 \\ \Xi_2 \end{bmatrix} + \begin{bmatrix} y^2 \\ y \end{bmatrix} \tag{3.104}$$

Since $b(t)$ is unknown, we define $\theta(t) = [b(t), \theta_{a1}(t), \theta_{a1}(t)]$ and the error ϵ satisfies

$$\dot{\epsilon} = A_0 \epsilon + \Phi(y) D(t) \tag{3.105}$$

$$H(y, y_r) = \begin{bmatrix} 1 \\ 0 \end{bmatrix} \tag{3.106}$$

$$P = \begin{bmatrix} \frac{1}{k_1} + \frac{k_2}{k_1} & -1 \\ -1 & \frac{1}{k_1} + \frac{k_2}{k_1} + \frac{1}{k_1 k_2} \end{bmatrix} \tag{3.107}$$

Following the steps presented in the controller design, we have

$$\alpha_1 = -\omega_1^T \hat{\theta} - \dot{y}_r + \frac{1}{2} z_1 \parallel \Phi_{a1}(y) \parallel^2 N(\chi)$$

$$+ \frac{1}{N(\chi)} \frac{1}{16 l_1} z_1^3 \parallel PH(y, y_r) \parallel^4 \tag{3.108}$$

with $\quad \omega_1 = [\bar{\omega}^T, c_1 z_1 + exp(\chi^2) z_1 - \dot{y}_r + \beta]^T$

$$u = -c_2 z_2 + \beta_2 + 2N(\chi) z_1 e^{-ft} - l_2 \left(\frac{\partial \alpha_1}{\partial y} \right)^2 z_2 + \frac{\partial \alpha_1}{\partial y} \omega^T \hat{\theta}$$

$$- \frac{z_2}{4} \parallel \frac{\partial \alpha_1}{\partial y} \Phi_{a1}(y) \parallel^2 - \sum_{i=2}^{\rho} \frac{1}{8 l_i z_2} z_1^4 \parallel PH \parallel^4 \tag{3.109}$$

and the adaptive laws are given as

$$\dot{\hat{\theta}}_1 = Proj\left(- N(\chi) \Gamma_1 \omega_1 z_1 e^{-ft} \right) \tag{3.110}$$

$$\dot{\hat{\theta}} = Proj(\tau_2) \tag{3.111}$$

$$\tau_2 = -\Gamma \frac{\partial \alpha_1}{\partial y} \omega z_2 \tag{3.112}$$

In the simulation, the design parameters were set as $c_1 = c_2 = 2$, $l_1 = l_2 = 1$, $k_1 = 5, k_2 = 4$, $\Gamma = I_3$ and $\gamma = 1$. Figure 3.1 and Figure 3.2 show the system output $y(t)$ with reference signal $y_r(t)$ and the control input $u(t)$. Clearly, simulation results verify the effectiveness of proposed scheme.

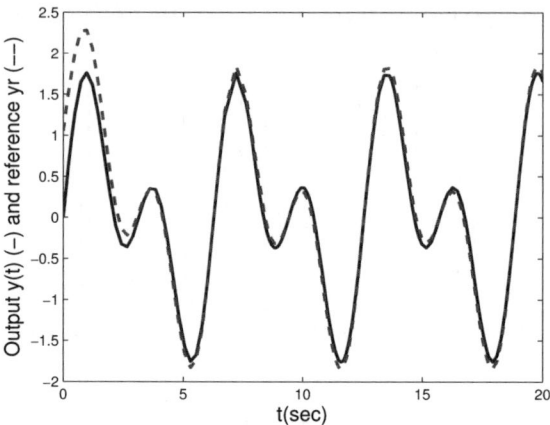

Fig. 3.1. Output y and reference y_r

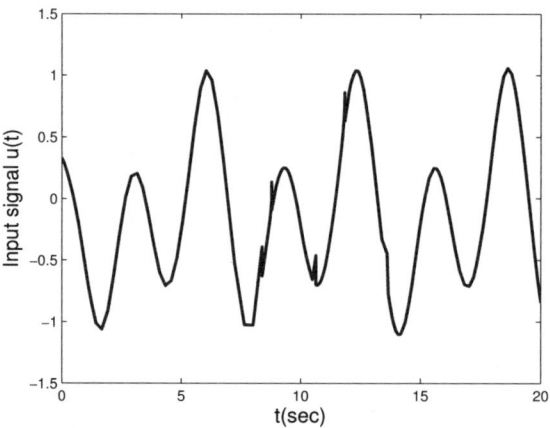

Fig. 3.2. System input u

3.6   Summary

In this chapter, a scheme is proposed to design an adaptive output-feedback controller for uncertain time-varying nonlinear systems with unknown sign of high-frequency gains in the presence of disturbances. No growth conditions on system nonlinearities are imposed. In the design, certain well defined functions are used to cancel the effects of disturbances. Furthermore, the overparamterization problem is also solved by using the concept of tuning functions. It is shown that the controller obtained by the proposed design scheme can make the whole adaptive control system stable.

4 Multivariable Adaptive Control

In this chapter, adaptive output feedback control of a class of multiple-input multiple-output systems is considered in the presence of unknown disturbances. Except the signs of the term multiplying the control are assumed, no other knowledge on the unknown parameters is required. The control design is achieved by using backstepping, tuning functions, SDU (Symmetric Diagonal Unity) factorization and estimation of unknown parameters. It is shown that the proposed controller can guarantee global uniform ultimate boundedness.

4.1 Introduction

In practice, most practical systems considered are multi-input multi-output (MIMO) systems. For such systems, the control problem is very complicated due to the coupling among various inputs and outputs. It becomes even more difficult to deal with when there exist uncertain parameters in the input or output coupling matrix. Due to these difficulties, it is noticed that, in comparison with the vast amount of results on controller design for SISO systems in control area, there are relatively fewer results available for a general class of MIMO systems. Adaptive backstepping control for a class of linear MIMO systems was studied in [85, 86, 87]. In [85] there exists a restrictive assumption about the high frequency gain B_m that a matrix S_m must be known such that $B_m S_m = (B_m S_m)^T > 0$. In [86] a similar restriction is relaxed using the factorization of high frequency gain. Recently a model-reference adaptive control was presented in [87] for MIMO linear systems without external disturbance using factorization of high frequency gain. In [88] a robust adaptive controller was designed for MIMO systems without disturbances by using switch functions. In [89] an output feedback control based on high-order sliding manifold approach was proposed for MIMO plants in the presence of disturbances. The convergence of tracking error is not to zero, but to a small residual set.

In this chapter, a new scheme is developed for a class of MIMO systems in the presence of unknown disturbances. With our scheme, a completely control solution to disturbance rejection is solved. In our design, the signs of the high

J. Zhou & C. Wen: Adapt. Backstepping Ctrl. of Uncertain Systems, LNCIS 372, pp. 51–64, 2008.
springerlink.com

frequency gains are known. To handle the disturbances and unknown parameters, we introduce new filters for states estimation and employ the internal model. As the parameters of the exosystem that generates external disturbances are unknown, an adaptive version of the internal model is proposed. Thus our estimator identifies the unknown parameters in both the system and the exosystem. It is shown that all closed-loop signals are bounded and the tracking error converges to zero.

4.2 Problem Formulation

The objective is to design an adaptive backstepping control scheme to generate the control $u(t)$ for multivariable plant described by

$$y(t) = G(p)\big(u(t) + \bar{d}(t)\big) \tag{4.1}$$

where $p = \frac{d}{dt}$, $u, y \in R^r$, $r > 1$, and $G(p)$ is an $r \times r$ strictly proper rational transfer matrix with unknown parameters, $\bar{d}$ is an unknown bounded disturbance. The similar problem was considered in [90].

A general MIMO plant $G(p)$ in (4.1) can be expressed as

$$G(p) = D^{-1}N(p) = C_g(pI - A_g)^{-1}B_g \tag{4.2}$$
$$D(p) = p^v I_r + A_{v-1}p^{v-1} + \ldots + A_1 p + A_0$$
$$N(p) = B_m p^m + \ldots + B_1 p + B_0$$

$$A_g = \begin{bmatrix} -A_{v-1} & I_r & 0 & \ldots & 0 \\ -A_{v-2} & 0 & I_r & \ldots & 0 \\ \vdots & \vdots & \vdots & \ddots & \vdots \\ -A_1 & 0 & 0 & \ldots & I_r \\ -A_0 & 0 & 0 & \ldots & 0 \end{bmatrix}, \quad B_g = \begin{bmatrix} 0 \\ \vdots \\ 0 \\ B_p \end{bmatrix}, \quad B_p = \begin{bmatrix} B_m \\ \vdots \\ B_1 \\ B_0 \end{bmatrix} \tag{4.3}$$

$$C_g = [I_r \ 0 \ \ldots \ 0 \ 0] \tag{4.4}$$

where v is the observability index of $G(p)$, $rv = n$, I_r is the $r \times r$ identity matrix, and $A_i \in R^{r \times r}$, $i = 0, \ldots, v - 1$, and $B_j \in R^{r \times r}$, $j = 0, \ldots, m$, $m \le v - 1$ are matrices.

Using (4.2), (4.1) can be expressed as the following feedback form

$$\dot{x} = Ax + A_p y + \begin{bmatrix} 0 \\ B_p \end{bmatrix} u + d(t) \tag{4.5}$$
$$y = C_g x$$

where $x \in R^n$ is the system state, $A \in R^{n \times n}$ is the matrix A_g with the first r columns equal to zero, $A_p \in R^{n \times r}$ are the first r columns of A_g and B_P

$\in R^{(m+1)r \times r}$, $d(t) = [0 \ B_p]^T \bar{d}(t) = [d_1^T, \ldots, d_v^T]^T \in R^n$ and $d_i \in R^r, (i = 1, \ldots, v)$.

Suppose that the unknown disturbance is generated from the following exosystem

$$\dot{d}(t) = Sd(t) \qquad (4.6)$$

where S is an unknown $n \times n$ matrix having distinct eigenvalues with zero real parts. The disturbance rejection problem in this chapter is based on the internal model principle in Appendix D.

For system (4.5) there exists an invariant manifold as stated below:

Lemma 4.1. *For the system (4.5) with an exosystem (4.6), there exists* $\pi(d) = \Pi d \in R^n$ *and* $\sigma(d) = \Lambda d \in R^r$ *such that*

$$\Pi S = A\Pi + I + \begin{bmatrix} 0 \\ B_p \end{bmatrix} \Lambda \qquad (4.7)$$

where $\pi_1(d)$ *is the first* r *elements of* $\pi(d)$ *and satisfies* $\pi_1(d) = 0 \in R^r$.

Proof. The existence of Π and Λ in (4.7) follows from the fact that A is a shift matrix. Details are given in Appendix D and also in [91, 92, 93].

With the invariant manifold $\pi(d)$, we define a state transformation as

$$\zeta = x - \pi(d) \qquad (4.8)$$

It can be shown from (4.5), (4.6) and (4.7) that

$$\dot{\zeta} = A\zeta + A_p y + \begin{bmatrix} 0 \\ B_p \end{bmatrix} (u - \sigma(d)) \qquad (4.9)$$

$$y = C_g \zeta$$

Then disturbance rejection problem of (4.5) becomes the stabilization problem of (4.9).

The control objective is that the output $y(t)$ can track a given bounded reference output $y_d(t)$ asymptotically and all closed-loop signals are bounded. Regarding the system and the reference signal, the following assumptions are made:

Assumption 1. The system is minimum phase.

Assumption 2. The leading principle minors of B_m are nonzero and the signs are known.

Assumption 3. The reference output $y_d(t)$ and its ρth order derivatives are assumed to be known, continuous and bounded.

Remark 4.1. Assumption 1 is fundamental in output feedback adaptive control. With Assumption 2, we require an apriori information to achieve simplicity of the resulting multivariable adaptive control scheme and a clear understanding of its properties.

4.3 Preliminary Results

We first perform a factorization of the high frequency gain B_m as stated in the following Lemma.

Lemma 4.2. *Every $r \times r$ real matrix B_m with nonzero leading principal minors $\Delta_1, \Delta_2, \ldots, \Delta_r$, can be factored as*

$$B_m = S_1 D_1 U_1 \tag{4.10}$$

where S_1 is symmetric positive definite, D_1 is diagonal, and U_1 is unity upper triangular.

Proof: Since the leading principal minors of B_m are nonzero, there exists a unique factorization

$$B_m = SDU \tag{4.11}$$

where S and U are unity lower triangular and

$$D = diag\{\Delta_1, \frac{\Delta_2}{\Delta_1}, \ldots, \frac{\Delta_r}{\Delta_{r-1}}\} \tag{4.12}$$

Factoring D as

$$D = D_+ D_1 \tag{4.13}$$

where D_+ is a diagonal matrix with positive entries, we rewrite (4.11) as $B_m = SD_+ S^T S^{-T} D_1 U$, so that (4.10) is satisfied by

$$S_1 = SD_+ S^T \tag{4.14}$$
$$U_1 = D_1^{-1} S^{-T} D_1 U \tag{4.15}$$

Remark 4.2. The factorization $B_m = S_1 D_1 U_1$ is not unique because the positive diagonal matrix D_+ is arbitrary and so D_1 is any diagonal matrix.

To construct a multivariable state observer using $u(t)$ and $y(t)$ for system (4.5), we choose the following matrix

$$K = [k_1 I_r, \ldots, k_v I_r]^T \in R^{n \times r} \tag{4.16}$$

where $k_i > 0, i = 1, \ldots, v$, such that the matrix

$$A_0 = A - KC_g \tag{4.17}$$

is stable. This is sufficient if $s^v + k_1 s^{v-1} + \ldots + k_{v-1} s + k_v$ is a Hurwitz polynomial. Define

$$E_i = e_i \otimes I_r \tag{4.18}$$

where $\otimes$ is the Kronecker product, and e_i is the ith coordinate vector in R^v. We also denote vectors $\xi_v(t), \xi(t), v_j(t)$, as the outputs of the filters

$$\dot{\xi}_v = A_0\xi_v + Ky \tag{4.19}$$

$$\dot{\xi}_i = A_0\xi_i + E_{v-i}y \quad i = 0, 1, \ldots, v-1 \tag{4.20}$$

$$\dot{v}_j = A_0v_j + E_{v-j}u \quad j = 0, 1, \ldots, m \tag{4.21}$$

where $\xi_v = [\xi_{v,1}^T, \ldots, \xi_{v,v}^T]^T$. Since A_0 is Hurwitz and the spectra of A_0 and S are disjoint, there exists $q(d) = Qd$ with $Q \in R^{n \times n}$ such that

$$QS = A_0Q - \begin{bmatrix} 0 \\ B_p \end{bmatrix} \Lambda \tag{4.22}$$

With these filters, we construct a parameterized state observer for the system (4.9) as

$$\hat{\zeta}(t) = \xi_v(t) + q(d) - \sum_{i=0}^{v-1} \bar{A}_i\xi_i(t) - \sum_{j=0}^{m} \bar{B}_jv_j(t) \tag{4.23}$$

where $\bar{A}_i = diag[A_i, \ldots, A_i] \in R^{n \times n}, \bar{B}_i = diag[B_i, \ldots, B_i] \in R^{n \times n}$.

Lemma 4.3. *The state observation error $\epsilon(t) = \zeta - \hat{\zeta}$ satisfies*

$$\dot{\epsilon}(t) = A_0\epsilon \tag{4.24}$$

$$\lim_{t \to \infty} \epsilon(t) = 0 \tag{4.25}$$

exponentially.

Proof. Based on the special structures of $\bar{A}_i$ and A_0, we have

$$\bar{A}_iA_0 = A_0\bar{A}_i, \quad (i = 0, 1, \ldots, v-1) \tag{4.26}$$

$$\bar{B}_jA_0 = A_0\bar{B}_j, \quad (j = 0, 1, \ldots, m) \tag{4.27}$$

From (4.6),(4.9),(4.19-4.27) and from

$$A_py(t) = -\sum_{i=0}^{v-1} \bar{A}_iE_{v-i}y(t) \tag{4.28}$$

$$\begin{bmatrix} 0 \\ B_p \end{bmatrix} u(t) = -\sum_{j=0}^{m} \bar{B}_jE_{v-j}u(t) \tag{4.29}$$

we obtain

$$\dot{\epsilon}(t) = \dot{\zeta} - \dot{\hat{\zeta}} = A_0\epsilon \tag{4.30}$$

where A_0 is a stable matrix. So we get $\lim_{t\to\infty} \epsilon(t) = 0$ exponentially.

Based on the parameterized canonical observer from (4.23), we now design a multivariable adaptive backstepping controller for the plant (4.9).

Let the state variable $\zeta(t)$ be partitioned as

$$\zeta = [\zeta_1^T, \ldots, \zeta_v^T]^T, \quad \zeta_i \in R^r \tag{4.31}$$

Then it follows from (4.4) and (4.9) that

$$y = \zeta_1 \tag{4.32}$$

$$\dot{y} = \zeta_2 - A_{v-1}y \tag{4.33}$$

Let $\epsilon(t), \xi(t)$ and $v_j(t)$ be partitioned as

$$\epsilon = [\epsilon_1^T, \ldots, \epsilon_v^T]^T, \quad \epsilon_i \in R^r$$

$$\xi_i = [\xi_{i,1}^T, \ldots, \xi_{i,v}^T]^T, \quad \xi_{i,k} \in R^r$$

$$v_j = [v_{j,1}^T, \ldots, v_{j,v}^T]^T, \quad v_{j,k} \in R^r \tag{4.34}$$

We also define

$$\Theta_a = [-A_{v-1}, -A_{v-2}, \ldots, -A_1, -A_0],$$

$$\Theta_b = [B_m, B_{m-1}, \ldots, B_1, B_0],$$

$$\xi_{(2)} = [\xi_{v-1,2}^T, \ldots, \xi_{0,2}^T]^T,$$

$$v_{(2)} = [v_{m,2}^T, \ldots, v_{0,2}^T]^T. \tag{4.35}$$

From (4.23) and Lemma 4.3 we have

$$\zeta_2 = \xi_{v,2} + \Theta_a \xi_{(2)} + \Theta_b v_{(2)} + q_2 + \epsilon_2 \tag{4.36}$$

where $\xi_{v,2}, q_2$ and ϵ_2 are the second entries of ξ_v, q and ϵ. Substituting (4.36) into (4.33) yields

$$\dot{y} = \xi_{v,2} + \Theta_a[\xi_{(2)} + E_1 y] + \Theta_b v_{(2)} + q_2 + \epsilon_2$$

$$= \xi_{v,2} + B_m v_{m,2} + \Theta\bar{\omega} + q_2 + \epsilon_2 \tag{4.37}$$

where $E_1 = [T_r, 0_{r\times(v-1)r}]^T$, $\Theta = [\Theta_a, \Theta_b]$, $\bar{\omega} = [\xi_{(2)}^T + (E_1 y)^T, [0_{1,r}, v_{m-1,2}^T, \ldots, v_{0,2}^T]]^T$.

Remark 4.3. The difficulty now is that the term $q(d)$ is not available, because the disturbance $d(t)$ and the matrix Q are unknown. For the adaptive backstepping approach proposed in [1], the state ζ_2 serves as the link between the output and the filter used for the output backstepping. The contribution of $\sigma(d)$ to ζ_2 is reflected by q_2 in ϵ. Following the treatment in [94] and [95], we reparameterize (4.6) for generating $q_2 = Q_2 d$, with Q_2 denoting the second r rows of Q. And we will introduce a new internal model and filter to handle it.

Lemma 4.4. *Given any Hurwitz matrix $F \in R^{v \times v}$ and any vector $G \in R^v$ such that the pair $\{F, G\}, i = 1, \ldots, r$ is controllable, the Sylvester equation*

$$MS - FM = GL \tag{4.38}$$

has a unique solution M, which is non singular.

The existence of a non-singular M is ensured by the facts that S and F have exclusively different eigenvalues and that $\{S, L\}$ is observable. Then we have

$$MSM^{-1} = F + G\psi^T \tag{4.39}$$

where $\psi^T = LM^{-1}$. S is similar to $F + G\psi$. Note that, since $\{F, G\}$ is controllable, and G has just one column, the row vector ψ is precisely the unique solution to the problem of assigning to $F + G\psi$ the poles of S. The system (4.6) is immersed into

$$\dot{\eta} = (F + G\psi^T)\eta \tag{4.40}$$

where $\eta = Md$ and $\psi^T = LM^{-1}$.

From Lemma 4.4, for any known controllable pair $\{F_i, G_i\}, i = 1, \ldots, r$ with $F_i \in R^{v \times v}$ being Hurwitz and $G_i \in R^v$, there exists a $\psi_i \in R^v$ such that

$$\begin{aligned} \dot{\eta}_i &= (F_i + G_i\psi_i^T)\eta_i \\ q_{2,i} &= \psi_i^T\eta_i, \quad i = 1, 2, \ldots, r \end{aligned} \tag{4.41}$$

where $q_{2,i}$ denotes the ith variable of $q_2 = [q_{2,1}, \ldots, q_{2,r}]^T$, the initial value $\eta_i(0)$ dependent on exogenerous variables. We define $\eta = [\eta_1^T, \ldots, \eta_r^T]^T$.

Based on the parametrization (4.41) of the initial model, we design an adaptive internal model as

$$\dot{\delta} = \begin{bmatrix} \dot{\delta}_1 \\ \vdots \\ \dot{\delta}_r \end{bmatrix} = \begin{bmatrix} F_1\delta_1 - G_1\xi_{v,2,1} + F_1G_1y_1 \\ \vdots \\ F_r\delta_r - G_r\xi_{v,2,r} + F_rG_ry_r \end{bmatrix}$$

$$\dot{\hat{q}}_2 = \begin{bmatrix} \dot{\hat{q}}_{2,1} \\ \vdots \\ \dot{\hat{q}}_{2,r} \end{bmatrix} = \begin{bmatrix} \hat{\psi}_1^T\delta_1 \\ \vdots \\ \hat{\psi}_r^T\delta_r \end{bmatrix} \tag{4.42}$$

where $\hat{\psi}_i$ is the estimate of ψ_i, $\xi_{v,2} = [\xi_{v,2,1}, \ldots, \xi_{v,2,r}]^T$, and $y = [y_1, \ldots, y_r]^T$. To further exploit the stability of the internal model, we define the filters

$$\dot{\lambda} = \begin{bmatrix} \dot{\lambda}_1 \\ \vdots \\ \dot{\lambda}_r \end{bmatrix} = \begin{bmatrix} F_1\lambda_1 + G_1\omega^T \\ \vdots \\ F_r\lambda_r + G_r\omega^T \end{bmatrix}$$

$$
\dot{\lambda}_v = \begin{bmatrix} \dot{\lambda}_{v,1} \\ \vdots \\ \dot{\lambda}_{v,r} \end{bmatrix} = \begin{bmatrix} F_1\lambda_{v,1} + F_1G_1v_{m,1}^T \\ \vdots \\ F_r\lambda_{v,r} + F_rG_rv_{m,1}^T \end{bmatrix} \tag{4.43}
$$

where $\omega = [\xi_{(2)}^T + (E_1y)^T, [-(K_1v_{m,1})^T, v_{m-1,2}^T, \ldots, v_{0,2}^T]]^T$.

We define the auxiliary error

$$
e = \begin{bmatrix} e_1 \\ \vdots \\ e_r \end{bmatrix} = \eta - \delta + \begin{bmatrix} \lambda_1\Theta_1^T \\ \vdots \\ \lambda_r\Theta_r^T \end{bmatrix} - \begin{bmatrix} G_1y_1 \\ \vdots \\ G_ry_r \end{bmatrix} + \begin{bmatrix} G_1B_{m,1}v_{m,1} \\ \vdots \\ G_rB_{m,r}v_{m,1} \end{bmatrix} + \begin{bmatrix} \lambda_{v,1}B_{m,1}^T \\ \vdots \\ \lambda_{v,r}B_{m,r}^T \end{bmatrix}
$$

where $\Theta = [\Theta_a, \Theta_b] = \begin{bmatrix} \Theta_1 \\ \vdots \\ \Theta_r \end{bmatrix}$, $B_m = \begin{bmatrix} B_{m,1} \\ \vdots \\ B_{m,r} \end{bmatrix}$.

Remark 4.4. Note that the traditional filters in [1] cannot deal with the unknown disturbance generated from an unknown exosystem. Thus the new adaptive internal model (4.42) and the new auxiliary filters (4.43) are introduced to achieve disturbance rejection.

Lemma 4.5. *The auxiliary error e satisfies*

$$
\dot{e} = = Fe + G\epsilon_2 \tag{4.44}
$$

where $F = diag\{F_1, \ldots, F_r\}, G = diag\{G_1, \ldots, G_r\}$.

Proof. From (4.21), we have

$$
\dot{v}_{m,1} = -K_1v_{m,1} + v_{m,2} \tag{4.45}
$$

From (4.37-4.43), it can be shown that

$$
\dot{e} = \begin{bmatrix} F_1e_1 \\ \vdots \\ F_re_r \end{bmatrix} - \begin{bmatrix} G_1\epsilon_{2,1} \\ \vdots \\ G_r\epsilon_{2,r} \end{bmatrix} \tag{4.46}
$$

With the auxiliary error e, we can express q_2 as

$$
q_2 = \begin{bmatrix} \psi_1^T\eta_1 \\ \vdots \\ \psi_r^T\eta_r \end{bmatrix}
$$

$$= \begin{bmatrix} \psi_1^T e_1 + \psi_1^T \delta_1 + \psi_1^T G_1 y_1 - \Theta_{\psi,1} vec(\lambda_1) - B_{\psi,1}(G_1 \otimes v_{m,1} + vec(\lambda_{v,1})) \\ \vdots \\ \psi_r^T e_r + \psi_r^T \delta_r + \psi_r^T G_r y_r - \Theta_{\psi,r} vec(\lambda_r) - B_{\psi,r}(G_r \otimes v_{m,1} + vec(\lambda_{v,r})) \end{bmatrix}$$

$$= \psi^T e + \psi^T G y + \psi^T \delta - \Theta_\psi \Lambda - B_\psi G \otimes v_{m,1} - B_\psi \Lambda_v \qquad (4.47)$$

where $\Theta_{\psi,i} = \psi_i^T \otimes \Theta_i, B_{\psi,i} = \psi_i^T \otimes B_{m,i}, \Lambda = [vec(\lambda_1)^T, \dots, vec(\lambda_r)^T]^T, \Lambda_v = [vec(\lambda_{v,1})^T, \dots, vec(\lambda_{v,r})^T]^T, i = 1, \dots, r$, with $vec(.)$ denotes the vector obtained by rolling the column vectors of the matrix, and $\psi^T = diag\{\psi_1^T, \dots, \psi_r^T\}$, $\Theta_\psi = diag\{\Theta_{\psi,1}, \dots, \Theta_{\psi,r}\}$ and $B_\psi = diag\{B_{\psi,1}, \dots, B_{\psi,r}\}$. Now all the terms in the right side of (4.47) are products of unknown parameters and unknown filtered signals or exponentially decaying signals.

4.4   Backstepping Design with SDU Factorization

In this section, we design an adaptive controller as in [90] for the plant (4.9), under the given assumptions. With the filters designed in previous section, we will be able to deal with both unknown parameter Θ in the system and the unknown parameter ψ in the exosystem, to design the adaptive control input. From (4.37) and (4.47), we get

$$\dot{y} = \xi_{v,2} + B_m v_{m,2} + \Theta \bar{\omega} + \epsilon_2 + \psi^T e + \psi^T G y + \psi^T \delta$$
$$\quad - \Theta_\psi \Lambda - B_\psi G \otimes v_{m,1} - B_\psi \Lambda_v$$
$$= \xi_{v,2} + \psi^T e + \epsilon_2 + \bar{\Theta} \Omega \qquad (4.48)$$

where $\bar{\Theta} = [\Theta, \psi^T, -\Theta_\psi, -B_\psi]$ and $\Omega = [\omega^T, (Gy + \delta)^T, \Lambda^T, (G \otimes v_{m,1} + \Lambda_v)^T]^T$.

The design procedure is recursive and similar to that for the single-input single-output case

$$z_1 = y - y_d \qquad (4.49)$$
$$z_i = v_{m,i} - \alpha_{i-1}, \quad i = 2, \dots, \rho \qquad (4.50)$$

where y_d is the reference output, α_i is the virtual control and will be designed based on the following procedures.

Step 1. From equations (4.48-4.50), and using Lemma 4.2, we get

$$\dot{z}_1 = \xi_{v,2} + B_m v_{m,2} + \bar{\Theta} \Omega + \psi^T e + \epsilon_2 - \dot{y}_d$$
$$= \xi_{v,2} + S_1 D_1 U_1 v_{m,2} + \bar{\Theta} \Omega + \psi^T e + \epsilon_2 - \dot{y}_d$$
$$= S_1 D_1 z_2 + S_1 D_1 [\alpha_1 + (U_1 - I) v_{m,2}] + \xi_{v,2} + \bar{\Theta} \bar{\Omega} + \psi^T e + \epsilon_2 - \dot{y}_d$$
$$\qquad (4.51)$$

where $\bar{\Omega} = [\bar{\omega}^T, (Gy + \delta)^T, \Lambda^T, (G \otimes v_{m,1} + \Lambda_v)^T]^T$ and $(U_1 - I)$ is strictly upper triangular matrix. In order to remove the zero entries from the above parametrization, we introduce new parameter vectors χ_k and regressor matrix γ_k $(k = 1, \ldots, r)$ as in [86] via the identity

$$(U_1 - I)v_{m,2} = v_{m,2_2} \begin{bmatrix} 1 \\ 0 \\ \vdots \\ 0 \end{bmatrix} [U_{1,2}] + \ldots + v_{m,2_r} \begin{bmatrix} I_{r-1} \\ 0_{1,r-1} \end{bmatrix} \begin{bmatrix} U_{1,r} \\ \vdots \\ U_{r-1,r} \end{bmatrix}$$

$$= \sum_{k=2}^{r} \gamma_k \chi_k \tag{4.52}$$

where $\gamma_k = v_{m,2_k} \begin{bmatrix} I_{k-1} \\ 0_{r-k+1,k-1} \end{bmatrix} \in R^{r\times(k-1)}$, $\chi_1 = 0$, $\chi_k = [U_{1,k} \ U_{2,k} \ \cdots$ $U_{k-1,k}]^t \in R^{k-1}(k = 2, \ldots, r)$ and $v_{m,2_k}$ indicates the kth component of vector $v_{m,2}$. Since $S_1 D_1$ is nonsingular, we can introduce $P = (S_1 D_1)^{-1}$ and new matrix parameters $\theta = P\bar{\Theta}$ and $\Psi = P\psi^T$. Using (4.52) and choosing positive constants C_1 and $\bar{l}_1$, adding and subtracting $(C_1 + \bar{l}_1)z_1$ in equation (4.51) and multiplying both sides by S_1^{-1} we get

$$S_1^{-1}z_1 = -S_1^{-1}C_1 z_1 - S_1^{-1}\bar{l}_1 z_1 + D_1 z_2 + D_1(\alpha_1 + P(C_1 z_1 + \bar{l}_1 z_1 + \xi_{v,2} - \dot{y}_d))$$

$$+ D_1 \sum_{k=2}^{r} \gamma_k \chi_k + D_1 \Theta \bar{\Omega} + D_1 \Psi e + S_1^{-1}\epsilon_2$$

Remark 4.5. The reason for using two positive constants C_1 and $\bar{l}_1$ is to have uniformity with subsequent steps of the backstepping procedure where $\bar{l}_1$ is a coefficient of a damping term countering ϵ_2.

We define the signal

$$\phi = C_1 z_1 + \bar{l}_1 z_1 + \xi_{v,2} - \dot{y}_d \tag{4.53}$$

and introduce $\hat{P}, \hat{\chi}_k, \hat{\theta}$ and $\hat{L}_i(i = 1, \ldots, \rho)$ as estimates of P, χ_k, θ and the upper bound of $\| \Psi \|^2$, respectively. We can choose the first virtual control law α_1 as

$$\alpha_1 = -\hat{P}\phi - \sum_{k=2}^{r}(\gamma_k \hat{\chi}_k) - \hat{\theta}\bar{\Omega} - D_1^T \hat{L}_1 z_1 \tag{4.54}$$

Choose the Lyapunov function as

$$V_1 = \frac{1}{2}Tr(S_1^{-1}z_1 z_1^T) + \frac{1}{2}Tr(\tilde{\theta}\tilde{\theta}^T) + \frac{1}{2}Tr(\tilde{P}\tilde{P}^T)$$

$$+ \frac{1}{\bar{l}_1}\epsilon^T P_\epsilon \epsilon + \frac{1}{2}\tilde{L}_1^2 + \frac{1}{2}e^T P_e e + \frac{1}{2}[\sum_{k=2}^{r} Tr(\tilde{\chi}_k^T \tilde{\chi}_k)] \tag{4.55}$$

where $\tilde{\theta} = \theta - \hat{\theta}, \tilde{P} = P - \hat{P}, \tilde{\chi}_k = \chi_k - \hat{\chi}_k, \tilde{L}_i = L_i - \hat{L}_i, i = 1, \ldots, \rho, P_\epsilon = P_\epsilon^T \in R^{n \times n}$ is the solution of $P_\epsilon A_0 + A_0^T P_\epsilon = -diag(S_1^{-1}, \ldots, S_1^{-1}) - I$ for stable matrix A_0 in (4.25), $P_e = P_e^T \in R^{n \times n}$ is the solution of $P_e F + F^T P_e = -2(\rho + 2)I$.

Using the following update laws for $\hat{\theta}, \hat{P}, \hat{\chi}_k$ and $\hat{L}_1$

$$\dot{\hat{\theta}} = D_1 z_1 \bar{\Omega}^T \tag{4.56}$$

$$\dot{\hat{P}} = D_1 z_1 \phi^T \tag{4.57}$$

$$\dot{\hat{\chi}}_k = \gamma_k^T D_1 z_1, \quad (k = 2, 3, \ldots, r) \tag{4.58}$$

$$\dot{\hat{L}}_1 = Tr(D_1 D_1^T z_1 z_1^T) \tag{4.59}$$

It can be verified that

$$\dot{V}_1 \leq -C_1 Tr(S_1^{-1} z_1 z_1^T) + Tr(D_1 z_2 z_1^T) - \frac{1}{\bar{l}_1} M_1(\epsilon) - \rho e^T e \tag{4.60}$$

where $M_1(\epsilon) = Tr(S_1^{-1} \epsilon_1 \epsilon_1^T + \frac{3}{4} S_1^{-1} \epsilon_2 \epsilon_2^T + \ldots + S_1^{-1} \epsilon_v \epsilon_v^T)$.

Furthermore, we set constant positive reals $\bar{l}_1$ satisfying the following conditions:

$$\frac{1}{\bar{l}_1} \geq \frac{\| P_\epsilon G \|^2}{4} \tag{4.61}$$

Notice that D_1 is a known matrix. By Lemma 4.2, D_1 is any diagonal matrix.

Step i. $(i = 2, \ldots, \rho), \rho = v - m$, we introduce the signal $z_i = v_{m,i} - \alpha_{i-1}$. The corresponding time derivatives can be expressed as

$$\dot{z}_i = v_{m,i+1} + \beta_i - \frac{\partial \alpha_{i-1}}{\partial y}(\bar{\Theta}\Omega - \epsilon_2 - \psi^T e) - \iota_i \tag{4.62}$$

where

$$\frac{\partial \alpha_i}{\partial y} = \begin{bmatrix} \frac{\partial \alpha_{i,1}}{\partial y} \\ \vdots \\ \frac{\partial \alpha_{i,r}}{\partial y} \end{bmatrix} = \begin{bmatrix} \frac{\partial \alpha_{i,1}}{\partial y_1} & \cdots & \frac{\partial \alpha_{i,1}}{\partial y_r} \\ \vdots & \ddots & \vdots \\ \frac{\partial \alpha_{i,r}}{\partial y_1} & \cdots & \frac{\partial \alpha_{i,r}}{\partial y_r} \end{bmatrix} \tag{4.63}$$

$$\iota_i = \left(Tr(\dot{\hat{\Theta}}\frac{\partial \alpha_{i-1,1}}{\partial \hat{\Theta}}), \ldots, Tr(\dot{\hat{\Theta}}\frac{\partial \alpha_{i-1,r}}{\partial \hat{\Theta}})\right)^T \tag{4.64}$$

$$\left(\frac{\partial \alpha_{l,k}}{\partial \hat{\Theta}}\right)_{i,j} = \frac{\partial \alpha_{l,k}}{\partial \hat{\Theta}_{j,i}}, \quad (k = 1, 2, \ldots, r) \tag{4.65}$$

and β_i represents all the known terms except $v_{m,i+1}$. We choose the Lyapunov function as

$$V_2 = V_1 + \frac{1}{2} Tr(z_2 z_2^T) + \frac{1}{l_2} \epsilon^T P_0 \epsilon + \frac{1}{2} Tr(\tilde{\Theta}\Gamma^{-1}\tilde{\Theta}^T) + \frac{1}{2}\tilde{L}_2^2 \tag{4.66}$$

$$V_i = V_{i-1} + \frac{1}{2}Tr(z_i z_i^T) + \frac{1}{l_i}\epsilon^T P_0 \epsilon + \frac{1}{2}\tilde{L}_i^2 \qquad (4.67)$$

where $\tilde{\Theta} = \bar{\Theta} - \hat{\bar{\Theta}}$, Γ is a positive definite matrix, $P_0 = P_0^T > 0$ is the solution of $P_0 A_0 + A_0^T P_0 = -I$ for the stable matric A_0.

The final adaptive control law is designed as

$$u = \alpha_\rho - v_{m,\rho-1} - \sum_{k=2}^{r} u_k \begin{bmatrix} I_{k-1} \\ 0_{r-k+1,k-1} \end{bmatrix} \hat{\chi}_k \qquad (4.68)$$

$$\dot{\hat{\bar{\Theta}}}^T = \tau_\rho \qquad (4.69)$$

$$\alpha_i = -c_i z_i - l_i \left(\frac{\partial \alpha_{i-1}}{\partial y}\right)^T \frac{\partial \alpha_{i-1}}{\partial y} z_i - \beta_i - z_{i-1} + \frac{\partial \alpha_{i-1}}{\partial y}\hat{\Theta}\Omega$$

$$- \sum_{k=2}^{j-1}\{o_{k,j-i}\} + \hat{\iota}_i - \hat{L}_i\left(\frac{\partial \alpha_{i-1}}{\partial y}\right)^T \frac{\partial \alpha_{i-1}}{\partial y} z_i \qquad (4.70)$$

$$\dot{\hat{L}}_i = Tr[\left(\frac{\partial \alpha_{i-1}}{\partial y}\right)^T \frac{\partial \alpha_{i-1}}{\partial y} z_i z_i^T] \qquad (4.71)$$

with

$$\tau_i^T = \tau_{i-1}^T - \Gamma\Omega z_i^T \frac{\partial \alpha_{i-1}}{\partial y} \qquad (4.72)$$

$$\hat{\iota}_i = \left(Tr(\tau_i \frac{\partial \alpha_{i-1,1}}{\partial \hat{\bar{\Theta}}}), \ldots, Tr(\tau_{i-1}\frac{\partial \alpha_{i-1,r}}{\partial \hat{\bar{\Theta}}})\right)^T \qquad (4.73)$$

where $Tr(\frac{\partial \alpha_{j-1,1}}{\partial y})^T z_k \Omega^T \Gamma^T \frac{\partial \alpha_{k-1,i}}{\partial \hat{\bar{\Theta}}})$ is the ith element of $o_{k,j-1} \in R^r$.

Note that the term $\sum_{k=2}^{r}\left(u_k \begin{bmatrix} I_{k-1} \\ 0_{r-k+1,k-1} \end{bmatrix} \hat{\chi}_k\right)$ appears because the signal $v_{m,2_k}$ is used in the regressor matrix γ_k to define the first stabilizing functions α_1 and is passed to the following steps into signal $\beta_j, (j = 2, \ldots, \rho)$ which is formed by the known terms of the derivatives of $\alpha_l, (l = 1, \ldots, j - 1)$.

Upon some algebraic manipulation, we can write $\dot{V}_\rho$ as

$$\dot{V}_\rho \leq \sum_{j=1}^{\rho}\left\{-Tr(c_j z_j z_j^T) - \frac{1}{l_i}M(\epsilon)\right.$$

$$\left.-l_i Tr\left[\left(\frac{\partial \alpha_{j-1}}{\partial y} z_j + \frac{\epsilon_2}{2l_j}\right)\left(\frac{\partial \alpha_{j-1}}{\partial y} z_j + \frac{\epsilon_2}{2l_j}\right)^T\right]\right\} - e^T e \leq 0 \quad (4.74)$$

$$M(\epsilon) = Tr(\epsilon_1 \epsilon_1^T + \frac{3}{4}\epsilon_2 \epsilon_2^T + \ldots + \epsilon_v \epsilon_v^T) \qquad (4.75)$$

where $c_1 = C_1 S_1^{-1}, \frac{1}{l_1} = \frac{1}{l_1}S_1^{-1}$. Since $\dot{V}_\rho \leq 0$, we conclude that the complete system $z_1, \ldots, z_\rho, \hat{\theta}, \hat{P}, \hat{\chi}_k, \hat{L}_k, \hat{\bar{\Theta}}$ are bounded and so the plant output $y(t)$ is bounded. Then it can be shown that all closed-loop signals are bounded and the tracking error $z_1 = y - y_d$ converges to zero.

Theorem 4.1. *Consider the MIMO system (4.5) satisfying Assumptions 1-3. With the application of controller (4.68) and the parameter update laws (4.56)-(4.59), (4.69) and (4.71), all closed-loop signals are bounded and the tracking error converges to zero.*

4.5   Simulation Studies

In this section, we illustrate the above method on a simple MIMO system. Consider the 2×2 plant described by

$$G(p) = \begin{bmatrix} \frac{2(p+1)}{(p-2)^3} & \frac{-(p+1)}{(p-2)^2} \\ \frac{(p+2)}{(p-2)^2(p+1)} & \frac{(p-2)^2}{(p+1)} \end{bmatrix}, \ \dot{\bar{d}}(t) = \begin{bmatrix} 0 & 1 \\ -9\sigma^2 & 0 \end{bmatrix} \bar{d}(t) \qquad (4.76)$$

where $\bar{d}(t)$ is external disturbance, σ is an unknown parameter which ranges between 1 and 4. The only information necessary for the design are the observability index $v = 3$, the order of $N(s)$, $m = 1$, and the signs of the leading principal minors of s. We choose the following design parameters $k_1 = 6, k_2 = 12, k_3 = 8, D_1 = I_2, C_2 = I_2, c_1 = l_1 = l_2 = 1, \Gamma = I_{10}$. The reference signal is given by $y_d = \frac{1}{(p+1)^2}[2\sin(t/2), \ 4\cos(t)]^T$. The plant initial condition is such that $y(0) = [0.4, 0.4]^T$. All other initial conditions are zero. The update laws and the control law are given by (4.56-4.59), (4.69), (4.71) and (4.68). The simulation results presented in the Figure 4.1 and Figure 4.2 show the systems output y and the desired trajectory signal y_d. Clearly, system output y can completely track the trajectory y_d. The result verifies our theoretical findings and show the effectiveness of the control schemes.

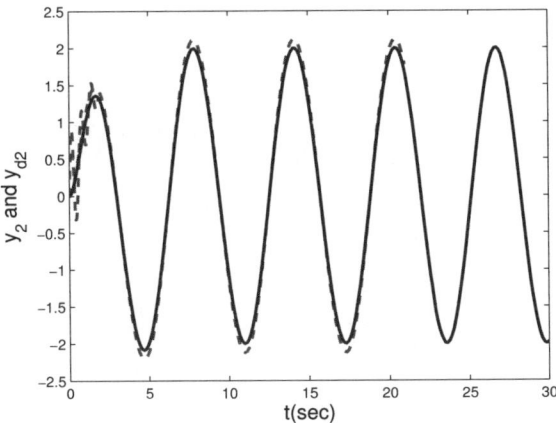

Fig. 4.1. Output y_1 and trajectory y_{d1}

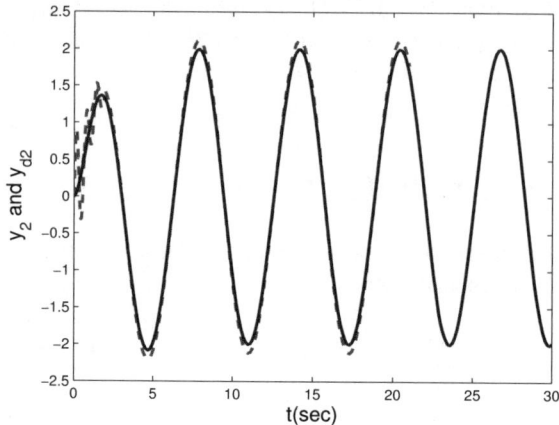

Fig. 4.2. Output y_2 and trajectory y_{d2}

4.6 Summary

In this chapter, a new scheme is proposed to design an adaptive output-feedback controller for uncertain MIMO systems in the presence of disturbances. In order to reject disturbances generated from unknown exosystems, new filters for state estimation are constructed and an adaptive internal model is employed. It is shown that the proposed controller can ensure all the signals in the closed-loop system bounded and the tracking error to converge to zero.

5 Decentralized Stabilization of Interconnected Systems

In this chapter, we propose adaptive backstepping scheme to design decentralized output feedback adaptive controllers. For each subsystem, a general transfer function with arbitrary relative degree is considered. The interactions between subsystems are allowed to satisfy a nonlinear bound with certain structural conditions. The effects due to interactions are taken into consideration in devising local control laws. It is shown that perfect stabilization is ensured and the L_2 norm of the system outputs is also shown to be bounded by a function of design parameters.

5.1 Introduction

In practice, decentralized control, designed independently for local subsystems and using local available signals for feedback, is usually preferred for interconnected systems. In particular, decentralized adaptive control is employed as such systems usually face poor knowledge on the plant parameters and interactions between subsystems. However, only limited number of results are available due to the complexity of the problem, especially difficulties encountered in handling the effects of interconnections, see for examples [20, 22, 24, 25, 31, 32, 33, 96, 97, 98, 99]. In [22], the first result on decentralized adaptive control was reported, but only for subsystems with relative degrees less than or equal to two. In [20], the requirement on subsystem relative degrees was relaxed and unmodelled dynamics were considered at the expense of requiring information exchanging between subsystems. In [96], a structure condition was required and subsystem relative degrees still cannot exceed two. In [99, 100, 101], all conditions related to subsystem degrees, structure conditions and information exchanging between subsystems were relaxed. The relative degree condition was also relaxed by using the concept of high-order tuners in [102] and [103]. In [104], the subsystem relative degree condition with the control scheme in [22] was also relaxed.

Research on decentralized adaptive control using backstpping approach has also received great attentions, due to a number of its advantages such as improving transient performance [1]. In [105], the first decentralized control result

J. Zhou & C. Wen: Adapt. Backstepping Ctrl. of Uncertain Systems, LNCIS 372, pp. 65–79, 2008.

using such a technique was reported without requirement on subsystem relative degrees. More general class of systems with the consideration of unmodelled dynamics was studied in [32] and [33].

In this chapter, we address decentralized adaptive stabilization for a class of interconnected systems. The interactions between subsystems are unknown and allowed to satisfy a high order nonlinear bound. Due to output feedback, local filters are designed to estimate system states. We use the standard backstepping design approach without any modification to design decentralized adaptive controllers. It is shown that the designed controllers can globally stabilize the overall interconnected system asymptotically. This reveals that the standard backstepping controller offers an additional advantage to conventional adaptive controllers in term of its robustness against interactions. In controller design, the term multiplying the control and the system parameters are not assumed to be within known intervals. Besides global stability, the L_2 norms of the system outputs are also shown to be bounded by functions of design parameters. Thus the transient system performance can be tunable by adjusting design parameters.

5.2   Problem Formulation

A system consisting of N interconnected subsystems modelled below is considered.

$$\dot{x}_{oi} = A_{oi}x_{oi} + b_{oi}u_i + \sum_{j=1}^{N} \bar{f}_{ij}(t, x_{oj}) \tag{5.1}$$

$$y_i = c_{oi}^T x_{oi}, \ for \ i = 1, \ldots, N \tag{5.2}$$

where $x_{oi} \in R^{n_i}$, $u_i \in R^1$ and $y_i \in R^1$ are the states, input and output of the ith subsystem, respectively, $\bar{f}_{ij}(t, x_{oj}) \in R^{n_i}$ denotes the nonlinear interactions from the jth subsystem to the ith subsystem for $j \neq i$, or a nonlinear un-modelled part of the ith subsystem for $j = i$. The matrices and vectors in (5.1) and (5.2) have appropriate dimensions, and their elements are constant but unknown.

For each decoupled local system, we make the following assumptions.

Assumption 1: n_i is known;

Assumption 2: The triple $(A_{oi}, \ b_{oi}, \ c_{oi})$ are completely controllable and observable;

Assumption 3: In the transfer function

$$G_i(s) = c_{oi}^T(sI - A_{oi})^{-1}b_{oi} = \frac{N_i(s)}{D_i(s)}$$

$$= \frac{b_i^{m_i}s^{m_i} + \ldots + b_i^1 s + b_i^0}{s^{n_i} + a_i^{n_i-1} + \ldots + a_i^1 s + a_i^0} \tag{5.3}$$

$N_i(s)$ is a Hurwitz polynomial. The sign of $b_i^{m_i}$ and the relative degree $\rho_i(= n_i - m_i)$ of $G_i(s)$ are known;

Assumption 4: The nonlinear interaction terms satisfy

$$\| \bar{f}_{ij}(t, x_{oj}) \| \le \bar{\gamma}_{ij} |y_j \psi_j(y_j)| \tag{5.4}$$

where $\| \cdot \|$ denotes the Euclidean norm, $\bar{\gamma}_{ij}$ are constants denoting the strength of the interaction, and $\psi_j(y_j), j = 1, 2, \ldots, N$ are known nonlinear functions.

Remark 5.1. It is allowed that the interaction $\bar{f}_{ij}$ contains states x_{oj}, as long as it satisfies (5.4).

Remark 5.2. The class of systems considered in [105] and [106] is a special case as their interactions satisfy the Lipschitz condition which implies $\psi_j(y_j) = 1$.

The control objective is to design totally decentralized adaptive controllers for system (5.1) satisfying Assumptions 1-4 such that the closed-loop system is stable.

5.3   Local State Estimation Filters

Clearly, there exists a nonsingular matrix T_i, such that under transformation $x_{oi} = T_i x_i$, (5.1) and (5.2) can be transformed to

$$\dot{x}_i = A_i x_i + a_i y_i + \begin{bmatrix} 0 \\ b_i \end{bmatrix} u_i + f_i \tag{5.5}$$

$$y_i = (e_{n_i}^1)^T x_i, \quad for\ i = 1, \ldots, N \tag{5.6}$$

where

$$A_i = \begin{bmatrix} 0 \\ \vdots & I \\ 0 \ldots 0 \end{bmatrix}, \ a_i = \begin{bmatrix} -a_i^{n_i-1} \\ \vdots \\ -a_i^0 \end{bmatrix}, \ b = \begin{bmatrix} b_i^{m_i} \\ \vdots \\ b_i^0 \end{bmatrix} \tag{5.7}$$

$$f_i = \sum_{j=1}^{N} T_i^{-1} \bar{f}_{ij} \tag{5.8}$$

and e_i^k denotes the kth coordinate vector in R^i.

For state estimation, we use the following filters

$$\dot{\lambda}_i = A_i^0 \lambda_i + e_{n_i}^{n_i} u_i \tag{5.9}$$

$$\dot{\eta}_i = A_i^0 \eta_i + e_{n_i}^{n_i} y_i \tag{5.10}$$

$$\Omega_i^T = [v_i^{m_i}, \ldots, v_i^1, v_i^0, \Xi_i] \tag{5.11}$$

$$v_i^j = (A_i^0)^j \lambda_i, \quad j = 0, \ldots, m_i \tag{5.12}$$

$$\Xi_i = -[(A_i^0)^{n_i-1} \eta_i, \ldots, A_i^0 \eta_i, \eta_i] \tag{5.13}$$

$$\xi_i^{n_i} = -(A_i^0)^{n_i} \eta_i \tag{5.14}$$

where the vector $k_i = [k_i^1, \ldots, k_i^{n_i}]^T$ is chosen so that the matrix $A_i^0 = A_i - k_i(e_{n_i}^1)^T$ is Hurwitz. With these designed filters our state estimate is

$$\hat{x}_i = \xi_i^{n_i} + \Omega_i^T \theta_i \tag{5.15}$$

$$\theta_i^T = [b_i^T, a_i^T] \tag{5.16}$$

and the state estimation error $\epsilon_i = x - \hat{x}$ satisfies

$$\dot{\epsilon}_i = A_i^0 \epsilon_i + f_i \tag{5.17}$$

Let $V_{\epsilon_i} = \epsilon_i^T P_i \epsilon_i$, where P_i is positive definite and satisfies $P_i A_i^0 + (A_i^0)^T P_i = -2I$, $P_i = P_i^T > 0$. Then the derivative of V_{ϵ_i} gives as

$$\dot{V}_{\epsilon_i} = \epsilon_i^T \left[P_i A_i^0 + (A_i^0)^T P_i \right] \epsilon_i + 2\epsilon_i^T P_i f_i$$
$$\leq -\epsilon_i^T \epsilon_i + \| P_i f_i \|^2 \tag{5.18}$$

Because the backstepping design starts with its output y_i, which is the only available system state allowed to appear in the control law, (5.5) is expressed as

$$\dot{y}_i = b_i^{m_i} v_i^{m_i,2} + \xi_i^{n_i,2} + \delta_i^T \theta_i + \epsilon_i^2 + f_i^1 \tag{5.19}$$

where

$$\delta_i = [v_i^{m_i,2}, v_i^{m_i-1,2}, \ldots, v_i^{0,2}, \Xi_i^{(2)} - y_i(e_{n_i}^1)^T]^T \tag{5.20}$$

$$\bar{\delta}_i = [0, v_i^{m_i-1,2}, \ldots, v_i^{0,2}, \Xi_i^{(2)} - y_i(e_{n_i}^1)^T]^T \tag{5.21}$$

and $v_i^{m_i,2}, \epsilon_i^2, \xi_i^{n_i,2}, \Xi_i^{(2)}$ denote the second entries of $v_i^{m_i}, \epsilon_i, \xi_i^{n_i}, \Xi_i$ respectively, f_i^1 and d_i^1 are the first elements of vector f_i and d_i. All states of the local filters in (5.9) and (5.10) are available for feedback.

Due to the Hurwitz polynomial $N_i(s)$ in Assumption 3, system is minimum phase. System (5.5) is restricted to the first ρ_i equations and the design system is given as

$$\dot{y}_i = b_i^{m_i} v_i^{m_i,2} + \xi_i^{n_i,2} + \bar{\delta}_i^T \theta_i + \epsilon_i^2 + f_i^1 \tag{5.22}$$

$$\dot{v}_i^{m_i, q} = v_i^{m_i,q+1} - k_i^q v_i^{m_i,1}, \quad q = 2, \ldots, \rho_i - 1 \tag{5.23}$$

$$\dot{v}_i^{m_i, \rho_i} = v_i^{m_i, \rho_i+1} - k_i^{\rho_i} v_i^{m_i,1} + u_i \tag{5.24}$$

5.4   Design of Adaptive Controllers

The following change of coordinates is made.

$$z_i^1 = y_i \tag{5.25}$$

$$z_i^q = v_i^{m_i,q} - \alpha_i^{q-1}, \quad q = 2, 3, \ldots, \rho_i \tag{5.26}$$

where α_i^{q-1} is the virtual control at the qth step of the ith loop and will be determined in later discussion.

To illustrate the controller design procedures, we now give a brief description on the first step.

• *Step* 1: We start with the equations for the stabilization error z_i^1 obtained from (5.22), (5.25) and (5.26) to get

$$\dot{z}_i^1 = b_i^{m_i}\alpha_i^1 + \xi_i^{n_i,2} + \bar{\delta}_i^T\theta_i + \epsilon_i^2 + f_i^1 + b_i^{m_i}z_i^2 \qquad (5.27)$$

We design the virtual control law α_i^1 as

$$\alpha_i^1 = \hat{p}_i\bar{\alpha}_i^1 \qquad (5.28)$$

$$\bar{\alpha}_i^1 = -2c_i^1 z_i^1 - l_i^1 z_i^1 - l_i^* z_i^1\big(\psi_i(z_i^1)\big)^2 - \xi_i^{n_i,2} - \bar{\delta}_i^T\hat{\theta}_i \qquad (5.29)$$

where c_i^1 and l_i^1 are positive design parameters, $\hat{\theta}_i$ is the estimate of θ_i, $\hat{p}_i$ is an estimate of $p_i = 1/b_i^{m_i}$.

Remark 5.3. The term $l_i^* z_i^1\psi_i^2(z_i^1)$ in (5.29) is designed to compensate the effects of interactions from other subsystems or the un-modelled part of its own subsystem. Note that the scheme in [105] dose not have such a term and thus the result of [105] is not applicable to the systems considered here.

In the following, we have $l_i^1 = 2\bar{l}_i^1$ to make the presentation easier in the stability analysis. Each of $\bar{l}_i^1$ will deal with the terms having ϵ_i^2, or f_i^1 in the evaluation of $\dot{V}_i^1$, respectively.

From (5.27) and (5.28) we have

$$\dot{z}_i^1 = -2c_i^1 z_i^1 - l_i^1 z_i^1 - l_i^* z_i^1\big(\psi_i(z_i^1)\big)^2 + \epsilon_i^2 + \bar{\delta}_i^T\theta_i - b_i^{m_i}\bar{\alpha}_i^1\tilde{p}_i + b_i^{m_i}z_i^2 + f_i^1$$

$$= -2c_i^1 z_i^1 - l_i^1 z_i^1 - l_i^* z_i^1\big(\psi_i(z_i^1)\big)^2 + (\delta_i - \hat{p}_i\bar{\alpha}_i^1 e_{n_i+m_i+1}^1)^T\tilde{\theta}_i + \epsilon_i^2 + f_i^1$$

$$-b_i^{m_i}\bar{\alpha}_i^1\tilde{p}_i + \hat{b}_i^{m_i}z_i^2 \qquad (5.30)$$

where $\tilde{\theta}_i = \theta_i - \hat{\theta}_i$ and using $\tilde{p}_i = \frac{1}{b_i^{m_i}} - \frac{1}{\hat{b}_i^{m_i}}$, we have

$$b_i^{m_i}\alpha_i^1 = b_i^{m_i}\hat{p}_i\bar{\alpha}_i^1 = \bar{\alpha}_i^1 - b_i^{m_i}\tilde{p}_i\bar{\alpha}_i^1 \qquad (5.31)$$

$$\bar{\delta}_i^T\tilde{\theta}_i + b_i^{m_i}z_i^2 = \bar{\delta}_i^T\tilde{\theta} + \tilde{b}_i^{m_i}z_i^2 + \hat{b}_i^{m_i}z_i^2$$

$$= \bar{\delta}_i^T\tilde{\theta}_i + (v_i^{m_i,2} - \alpha_i^1)(e_{n_i+m_i+1}^1)^T\tilde{\theta}_i + \hat{b}_i^{m_i}z_i^2$$

$$= (\delta_i - \hat{p}_i\bar{\alpha}_i^1 e_{n_i+m_i+1}^1)^T\tilde{\theta}_i + \hat{b}_i^{m_i}z_i^2 \qquad (5.32)$$

We consider the Lyapunov function

$$V_i^1 = \frac{1}{2}(z_i^1)^2 + \frac{1}{2}\tilde{\theta}_i^T\Gamma_i^{-1}\tilde{\theta}_i + \frac{|b_i^{m_i}|}{2\gamma_i'}\tilde{p}_i^2 + \frac{1}{2\bar{l}_i^1}V_{\epsilon_i} \qquad (5.33)$$

where Γ_i is a positive definite design matrix and γ_i' is a positive design parameter. We now examine the derivative of V_i^1

$$\dot{V}_i^1 = z_i^1\dot{z}_i^1 - \tilde{\theta}_i^T\Gamma_i^{-1}\dot{\hat{\theta}}_i - \frac{|b_i^{m_i}|}{\gamma_i'}\tilde{p}_i\dot{\hat{p}}_i + \frac{1}{2\bar{l}_i^1}\dot{V}_{\epsilon_i}$$

$$= -2c_i^1(z_i^1)^2 + \hat{b}_i^{m_i} z_i^1 z_i^2 - |b_i^{m_i}| \tilde{e} \frac{1}{\gamma_i'} [\gamma_i' \text{sign}(b_i^{m_i}) \bar{\alpha}_1 z_i^1 + \dot{\hat{p}}_i]$$

$$+ \tilde{\theta}_i^T \Gamma_i^{-1} [\Gamma_i(\delta_i - \hat{p}_i \bar{\alpha}_i^1 e_{n_i+m_i+1}) z_i^1 - \dot{\hat{\theta}}_i] - \bar{l}_i^1 (z_i^1)^2 + z_i^1 f_i^1$$

$$- \bar{l}_i^1 (z_i^1)^2 + z_i^1 \epsilon_i^2 - \frac{1}{2\bar{l}_i^1} \epsilon_i^T \epsilon_i + \frac{1}{2\bar{l}_i^1} \| P_i f_i \|^2 - l_i^*(z_i^1)^2 (\psi_i(z_i^1))^2$$

$$\leq -2c_i^1(z_i^1)^2 + \hat{b}_i^{m_i} z_i^1 z_i^2 - |b_i^{m_i}| \tilde{e} \frac{1}{\gamma_i'} [\gamma_i' \text{sign}(b_i^{m_i}) \bar{\alpha}_1 z_i^1 + \dot{\hat{p}}_i]$$

$$+ \tilde{\theta}_i^T \Gamma_i^{-1} [\Gamma_i(\delta_i - \hat{p}_i \bar{\alpha}_i^1 e_{n_i+m_i+1}) z_i^1 - \dot{\hat{\theta}}_i] - \frac{1}{4\bar{l}_i^1} \epsilon_i^T \epsilon_i$$

$$- l_i^*(z_i^1)^2 (\psi_i(z_i^1))^2 + \frac{1}{2\bar{l}_i^1} \| P_i f_i \|^2 + \frac{1}{4\bar{l}_i^1} \| f_i^1 \|^2 \tag{5.34}$$

where we use $ab \leq a^2 + \frac{1}{4}b^2$. Now we choose

$$\dot{\hat{p}}_i = -\gamma_i' \text{sign}(b_i^{m_i}) \bar{\alpha}_i^1 z_i^1 \tag{5.35}$$

$$\tau_i^1 = (\delta_i - \hat{p}_i \bar{\alpha}_i^1 e_{n_i+m_i+1}) z_i^1 \tag{5.36}$$

Then the following derivation for the derivative of V_i^1 can be carried out by using (5.35)-(5.36)

$$\dot{V}_i^1 \leq -2c_i^1(z_i^1)^2 + \hat{b}_i^{m_i} z_i^1 z_i^2 + \tilde{\theta}_i^T (\tau_i^1 - \Gamma_i^{-1} \dot{\hat{\theta}}_i) - \frac{1}{4\bar{l}_i^1} \epsilon_i^T \epsilon_i$$

$$- l_i^*(z_i^1)^2 (\psi_i(z_i^1))^2 + \frac{1}{2\bar{l}_i^1} \| P_i f_i \|^2 + \frac{1}{4\bar{l}_i^1} \| f_i^1 \|^2 \tag{5.37}$$

• *Step q* $(q = 2, \ldots, \rho_i, i = 1, \ldots, N)$: Choose virtual control laws

$$\alpha_i^2 = -\hat{b}_i^{m_i} z_i^1 - [c_i^2 + l_i^2 (\frac{\partial \alpha_i^1}{\partial y_i})^2] z_i^2 + \bar{B}_i^2 + \frac{\partial \alpha_i^1}{\partial \hat{\theta}_i} \Gamma_i \tau_i^2 \tag{5.38}$$

$$\alpha_i^q = -z_i^{q-1} - [c_i^q + l_i^q (\frac{\partial \alpha_i^{q-1}}{\partial y_i})^2] z_i^q + \bar{B}_i^q + \frac{\partial \alpha_i^{q-1}}{\partial \hat{\theta}_i} \Gamma_i \tau_i^q$$

$$- (\sum_{k=2}^{q-1} z_i^k \frac{\partial \alpha_i^{k-1}}{\partial \hat{\theta}_i}) \Gamma_i \frac{\partial \alpha_i^{q-1}}{\partial y_i} \delta_i \tag{5.39}$$

$$\tau_i^q = \tau_i^{q-1} - \frac{\partial \alpha_i^{q-1}}{\partial y_i} \delta_i z_i^q \tag{5.40}$$

where $c_i^q, l_i^q, q = 3, \ldots, \rho_i$ are positive design parameters, and $\bar{B}_i^q, q = 2, \ldots, \rho_i$ denotes some known terms and its detailed structure can be found in Chapter 2. Then the adaptive controller and parameter update laws are finally given by

$$u_i = \alpha_i^{\rho_i} - v_i^{m_i, \rho_i+1} \tag{5.41}$$

$$\dot{\hat{\theta}}_i = \Gamma_i \tau_i^{\rho_i} \tag{5.42}$$

Remark 5.4. When going through the details of the design procedures, we note that in the equations concerning $\dot{z}_i^q$, $q = 1, 2, \ldots, \rho_i$, just functions f_i^1 from the interactions, and they are always together with ϵ_i^2. This is because only $\dot{y}_i$ from the plant model (5.5) was used in the calculation of $\dot{\alpha}_i^q$ for steps $q = 2, \ldots, \rho_i$.

5.5 Stability Analysis

In this section, the stability of the overall closed-loop system consisting of the interconnected plants and decentralized controllers will be established.

Firstly, Define $z_i(t) = [z_i^1, z_i^2, \ldots, z_i^{\rho_i}]^T$. A mathematical model for each local closed-loop control system is derived from (5.30) and the rest of the design steps $2, \ldots, \rho_i$.

$$\dot{z}_i = A_{z_i} z_i + W_{\epsilon i}(\epsilon_i^2 + f_i^1) + W_{\theta i}^T \tilde{\theta}_i - b_i^{m_i} \bar{\alpha}_i^1 \tilde{p}_i e_{\rho_i}^1 - l_i^* z_i^1 \big(\psi_i(z_i^1)\big)^2 e_{\rho_i}^1 \quad (5.43)$$

where A_{z_i} is a matrix as the following.

$$A_{z_i} = \begin{bmatrix} -2c_i^1 - l_i^1 & \hat{b}_i^{m_i} & 0 & \cdots & 0 \\ -\hat{b}_i^{m_i} & -c_i^2 - l_i^2\big(\frac{\partial\alpha_i^1}{\partial y_i}\big)^2 & 1 + \sigma_i^{2,3} & \cdots & \sigma_i^{2,\rho_i} \\ 0 & -1 - \sigma_i^{2,3} & -c_i^3 - l_i^3\big(\frac{\partial\alpha_i^2}{\partial y_i}\big)^2 & \cdots & \sigma_i^{3,\rho_i} \\ \vdots & \vdots & \vdots & \vdots & \vdots \\ 0 & -\sigma_i^{2,\rho_i} & -\sigma_i^{3,\rho_i} & \cdots & -c_i^{\rho_i} - l_i^{\rho_i}\big(\frac{\partial\alpha_i^{\rho_i-1}}{\partial y_i}\big)^2 \end{bmatrix}$$

$$\quad (5.44)$$

$$W_{\epsilon i} = \begin{bmatrix} 1 \\ -\frac{\partial\alpha_i^1}{\partial y_i} \\ \vdots \\ -\frac{\partial\alpha_i^{\rho_i-1}}{\partial y_i} \end{bmatrix}, \quad W_{\theta i}^T = W_{\epsilon i}\delta_i^T - \hat{p}_i\bar{\alpha}_i^1 e_{\rho_i}^1 e_{\rho_i}^{1}{}^T \quad (5.45)$$

where the terms $\sigma_i^{k,q}$ are due to the terms $\frac{\partial\alpha_i^{k-1}}{\partial\hat{\theta}_i}\Gamma_i(\tau_i^q - \tau_i^{q-1})$ in the z_i^q equation.

To show the system stability, the variables of the filters in (5.10) and the zero dynamics of subsystems should be included in the Lyapunov function. The variables ζ_i associated with the zero dynamics of the ith subsystem can be shown to satisfy

$$\dot{\zeta}_i = A_i^{b_i}\zeta_i + \bar{b}_i z_i^1 + \bar{f}_i \quad (5.46)$$

where the eigenvalues of the $m_i \times m_i$ matrix $A_i^{b_i}$ are the zeros of the Hurwitz polynomial $N_i(s)$, $\bar{b}_i \in R^{m_i}$ and $\bar{f}_i \in R^{m_i}$ denoting the effects of the transformed interactions.

Now we define a Lyapunov function of the overall decentralized adaptive control system as

$$V = \sum_{i=1}^{N} V_i \tag{5.47}$$

where

$$V_i = \sum_{q=1}^{\rho_i} \left(\frac{1}{2}(z_i^q)^2 + \frac{1}{2\overline{l}_i^q} V_{\epsilon_i} \right) + \frac{1}{2}\tilde{\theta}_i^T \Gamma_i^{-1} \tilde{\theta}_i + \frac{|b_i^{m_i}|}{2\gamma_i'}(\tilde{p}_i)^2$$

$$+ \frac{1}{2l_i^{\eta_i}}\eta_i^T P_i \eta_i + \frac{1}{2l_i^{\zeta_i}}\zeta_i^T P_i^{b_i} \zeta_i \tag{5.48}$$

where $P_i^{b_i}$ is positive definite and satisfies $P_i^{b_i} A_i^{b_i} + (A_i^{b_i})^T P_i^{b_i} = -2I$, $P_i^{b_i} = (P_i^{b_i})^T > 0$, and $l_i^{\zeta_i}$ and $l_i^{\eta_i}$ are constants satisfying

$$l_i^{\eta_i} \geq \frac{\| P_i e_i^{n_i} \|^2}{c_i^1}$$

$$l_i^{\zeta_i} \geq \frac{2 \| P_i^{b_i} \overline{b}_i \|^2}{c_i^1}$$

The derivative of the V_i satisfies

$$\dot{V}_i \leq -\sum_{q=1}^{\rho_i} \left[c_i^q (z_i^q)^2 - \frac{1}{4\overline{l}_i^q}\epsilon_i^T \epsilon_i \right] - \frac{1}{2l_i^{\eta_i}}\eta_i^T \eta_i - \frac{1}{4l_i^{\zeta_i}}\zeta_i^T \zeta_i$$

$$+ \tilde{\theta}_i^T (\tau_i^{\rho_i} - \Gamma_i^{-1}\dot{\hat{\theta}}_i) - |b_i^{m_i}|\tilde{p}_i \frac{1}{\gamma_i'}[\gamma_i' sgn(b_i^{m_i})\bar{\alpha}_i^1 z_i^1 + \dot{\hat{p}}_i]$$

$$- l_i^*(z_i^1)^2 (\psi_i(z_i^1))^2 + \sum_{q=1}^{\rho_i} \frac{1}{4\overline{l}_i^q}(2 \| P_i f_i \|^2 + \| f_i \|^2) + \frac{1}{2l_i^{\zeta_i}} \| P_i^{b_i} \overline{f}_i \|^2$$

$$\leq -\sum_{q=1}^{\rho_i} \left[c_i^q (z_i^q)^2 - \frac{1}{4\overline{l}_i^q}\epsilon_i^T \epsilon_i \right] - \frac{1}{2l_i^{\eta_i}}\eta_i^T \eta_i - \frac{1}{4l_i^{\zeta_i}}\zeta_i^T \zeta_i$$

$$- l_i^*(z_i^1)^2 (\psi_i(z_i^1))^2 + \sum_{q=1}^{\rho_i} \frac{1}{4\overline{l}_i^q}(2 \| P_i f_i \|^2 + \| f_i \|^2) + \frac{1}{2l_i^{\zeta_i}} \| P_i^{b_i} \overline{f}_i \|^2$$

$$\tag{5.49}$$

From Assumption 4, we can show that

$$\sum_{q=1}^{\rho_i} \frac{1}{4\overline{l}_i^q}(2 \| P_i f_i \|^2 + \| f_i \|^2) + \frac{1}{2l_i^{\zeta_i}} \| P_i^{b_i} \overline{f}_i \|^2 \leq \sum_{j=1}^{N} \gamma_{ij}|z_j^1 \psi_j(z_j^1)|^2 \tag{5.50}$$

where $\gamma_{ij} = O(\bar{\gamma}_{ij}^2)$ indicating the coupling strength from the jth subsystem to the ith subsystem depending on $\overline{l}_i^q, l_i^{\zeta_i}, \| P_i \|, \| P_i^{b_i} \|$. $O(\bar{\gamma}_{ij}^2)$ denotes that γ_{ij}

and $O(\bar{\gamma}_{ij}^2)$ are in the same order mathematically. Clearly there exist γ_{ij}^* such that for all $\gamma_{ij} \leq \gamma_{ij}^*$,

$$l_i^* \geq \sum_{j=1}^{N} \gamma_{ji} \tag{5.51}$$

$$if \qquad l_i^* \geq \sum_{j=1}^{N} \gamma_{ji}^* \tag{5.52}$$

Now taking the summation of the first term in the second line of (5.49) into account and using (5.50) and (5.51), we get

$$\sum_{i=1}^{N} -[l_i^*(z_i^1)^2(\psi_i(z_i^1))^2 - \sum_{k=1}^{\rho_i}\frac{1}{4\bar{l}_i^k}(2 \parallel P_i f_i \parallel^2 + \parallel f_i \parallel^2) - \frac{1}{2l_i^{\zeta_i}} \parallel P_i^{b_i}\bar{f}_i \parallel^2]$$
$$\leq \sum_{i=1}^{N} -[l_i^* - \sum_{j=1}^{N}\gamma_{ji}]|z_i^1\psi_i(z_i^1)|^2 \leq 0 \tag{5.53}$$

Then from (5.47)-(5.49), the derivative of the V satisfies

$$\dot{V} \leq \sum_{i=1}^{N}\left[-\sum_{q=1}^{\rho_i}(c_i^q(z_i^q)^2 - \frac{1}{4\bar{l}_i^q}\epsilon_i^T\epsilon_i) - \frac{1}{2l_i^{\eta_i}}\eta_i^T\eta_i - \frac{1}{4l_i^{\zeta_i}}\zeta_i^T\zeta_i\right] \tag{5.54}$$

This shows that $z_i^1, z_i^2, \ldots, z_i^{\rho_i}, \epsilon_i, \zeta_i, \lambda_i, \eta_i, \tilde{\theta}_i, \tilde{p}_i$ and x_i are bounded. Therefore boundedness of all signals in the system is ensured as formally stated in the following theorem.

Theorem 5.1. *Consider the closed-loop adaptive system consisting of the plant (5.1) under Assumptions 1-4, the controller (5.41), the estimators (5.35), (5.42), and the filters (5.9) and (5.10). There exist γ_{ij}^* such that for all $\gamma_{ij} \leq \gamma_{ij}^*$, $i, j = 1, \ldots, N$, all the states of the system asymptotically approach to zero and the bound $\parallel y_i \parallel_2$ is given by*

$$\parallel y_i \parallel_2 \leq \frac{1}{2\sqrt{c_i^1}}(\sum_{i=1}^{N} y_i(0)+ \parallel \tilde{\theta}_i(0) \parallel_{\Gamma_i^{-1}}^2 + \frac{|b_i^{m_i}|}{\gamma_i'}|\tilde{p}_i(0)|^2 + d_i^0 \parallel \epsilon_i(0) \parallel_{P_i}^2)^{1/2} \tag{5.55}$$

by setting $z_i^q(0) = 0$, $q = 2,\ldots,\rho_i$, $i = 1,\ldots,N$ and definign $d_i^0 = \sum_{q=1}^{\rho_i}\frac{1}{2\bar{l}_i^q}$, $\parallel \epsilon_i \parallel_{P_i}^2 = \epsilon_i^T(0)P_i\epsilon_i(0)$, and $\parallel \tilde{\theta}_i(0) \parallel_{\Gamma_i^{-1}}^2 = \tilde{\theta}_i^T(0)\Gamma_i^{-1}\tilde{\theta}_i(0)$.

Proof: From (5.54), We have

$$\dot{V} \leq -c_i^0 \parallel z_i \parallel_2^2 \leq 0 \tag{5.56}$$

where $c_i^0 = min\{c_i^q\}, q = 1,\ldots,\rho_i$, $i = 1,\ldots,N$. This proves that the uniform stability and the uniform boundedness of $z_i^q, \hat{p}_i, \hat{\theta}_i, \epsilon_i, \zeta_i, \eta_i, v_i^j, x_i$ and u_i. It can be shown that both $\dot{V}$ and $\ddot{V}$ are bounded as well as $\dot{V}$ is integrable over $[0, \infty]$. Therefore, $\dot{V}$ tends to zero and thus the system states x_i converge to zero from (5.56).

Since V is non-increasing, we have

$$\dot{V} \leq -c_i^1 \parallel z_i^1 \parallel^2 \tag{5.57}$$

$$\parallel y_i(t) \parallel_2^2 = \int_0^\infty \parallel z_i^1(t) \parallel^2 dt$$

$$\leq \frac{1}{c_i^1}\left(V(0) - V(\infty)\right) \leq \frac{1}{c_i^1}V(0) \tag{5.58}$$

Note that the initial values $z_i^q(0)$ depends on $c_i^1, \gamma_i', \Gamma_i$. We can set $z_{i,q}(0), q = 2, \ldots, \rho_i$ to zero by suitably initializing our designed filters and initialize $\tilde{\eta}_i(0) = \zeta_i(0) = 0$. We have

$$V(0) = \sum_{i=1}^N \frac{1}{2}(y_i(0))^2 + d_i^0 \parallel \epsilon_i(0) \parallel_{P_i}^2 + \parallel \tilde{\theta}_i(0) \parallel_{\Gamma_i^{-1}}^2 + \frac{|b_i^{m_i}|}{\gamma_i'}|\tilde{p}_i(0)|^2 \tag{5.59}$$

Thus the bounds for $y_i(t)$ can be obtained clearly.

Remark 5.5. The L_2 norm of the system states is shown to be bounded by a function of design parameters. This implies that the transient system performance in terms of L_2 bounds can be adjusted by choosing suitable design parameters.

Remark 5.6. The condition that $\gamma_{ij} \leq \gamma_{ij}^*$ now has the following two implications:

(1) If we know $\bar{\gamma}_{ij}$, then we can get an estimate of its bound γ_{ij}^* and design l_i^* according to (5.51). This means that the coupling strength of the interconnection between subsystems can be allowed arbitrarily strong.

(2) If we do not know $\bar{\gamma}_{ij}$, then the designed local controllers are able to stabilize any interconnected system with coupling strength satisfying (5.52). This implication is similar to the interpretations of the results in [22], [24], [31] and [32], where sufficiently weak interactions are allowed.

5.6   An Illustrative Example

We consider the following interconnected system with two second order subsystems.

$$\dot{x}_{i1} = x_{i2} + a_{i1}y_i + f_{i1}$$
$$\dot{x}_{i2} = b_i u_i + a_{i2}y_i + f_{i2}$$
$$y_i = x_{i1}, i = 1, 2 \tag{5.60}$$

where $b_1 = b_2 = 1$, $a_{11} = 0, a_{12} = 1$, $a_{21} = 0, a_{21} = 2$, the nonlinear interaction terms $f_{11} = y_2, f_{12} = 2y_2$, $f_{11} = 0.5y_1, f_{12} = y_1$. The objective is to stabilize system (5.60). The controller (5.41) and the estimator (5.35) and (5.42) are implemented, where $\hat{p}_i$ and $\hat{\theta}_i$ are estimates of $p_i = 1/b_i$ and $\theta_i = a_{ij}, i, j = 1, 2$, respectively. The initials are set as $y_1(0) = 0.4, y_2(0) = 0.4$.

The design parameters are chosen as $k_i = [4, 4]^T, i = 1, 2, \gamma_1' = \gamma_2' = 0.1, \Gamma_1 = \Gamma_2 = I_3, c_1^1 = c_1^2 = c_2^1 = c_2^2 = 5, l_1^1 = l_1^2 = l_2^1 = l_2^2 = 2, l_1^* = l_2^* = 5$. The simulation results presented in Figures 5.1-5.4 show the system outputs y_1, y_2 and inputs u_1, u_2. Clearly, the system is stabilized and the outputs of both subsystems converge to zero. This verifies that the proposed scheme is effective in handling interactions.

1. **Effects of Parameters c_i^1**
   To see the effects of changing design parameters c_i^1 as indicated in Theorem, we fix $\gamma_1 = \gamma_2 = 0.4, \Gamma_1 = \Gamma_2 = 0.1, c_1^2 = c_2^2 = 1, l_1^1 = l_1^2 = l_2^1 = l_2^2 = 1, l_1^* = l_2^* = 1$. The outputs of the two subsystem system outputs y_1 and y_2 are compared in Figures 5.5-5.6 when c_i^1 is chosen as 0.1, 5, respectively for $i = 1, 2$. Obviously, the L_2 norms of the outputs decrease as c_i^1 for $i = 1, 2$ increase. The corresponding control u_1, u_2 are illustrated in Figures 5.7-5.8.

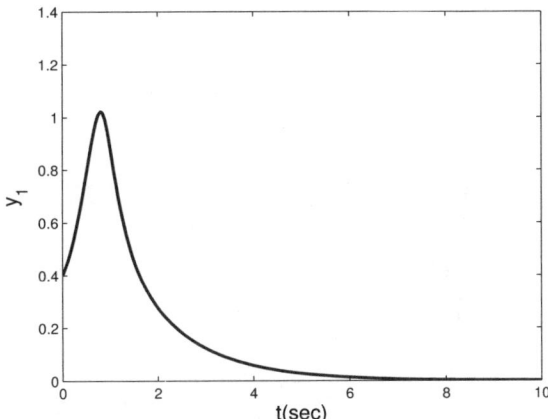

Fig. 5.1. Output y_1

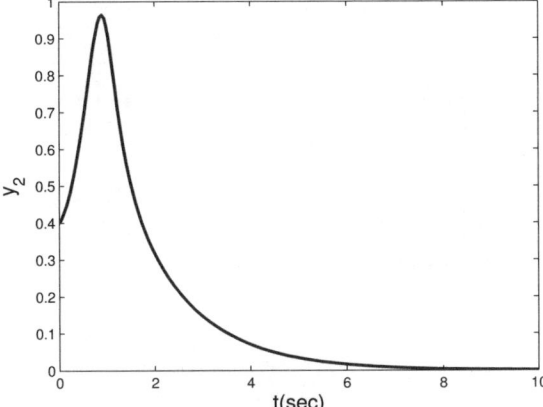

Fig. 5.2. Output y_2

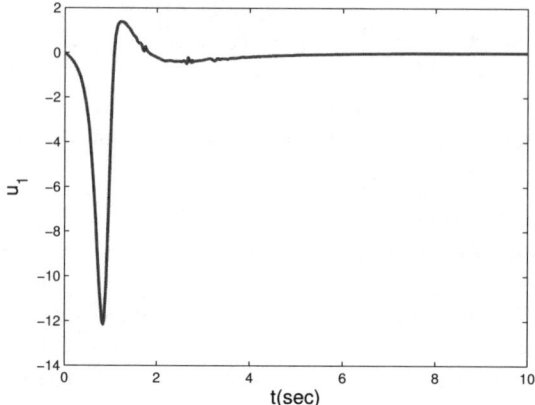

Fig. 5.3. Input u_1

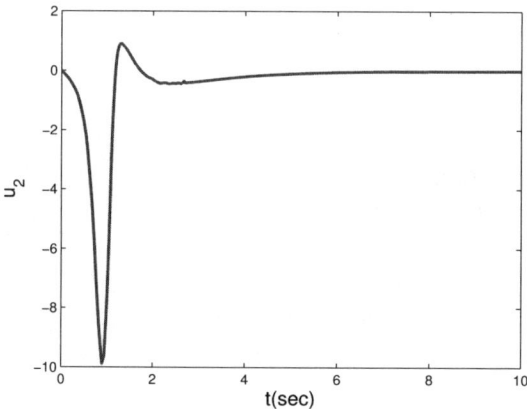

Fig. 5.4. Input u_2

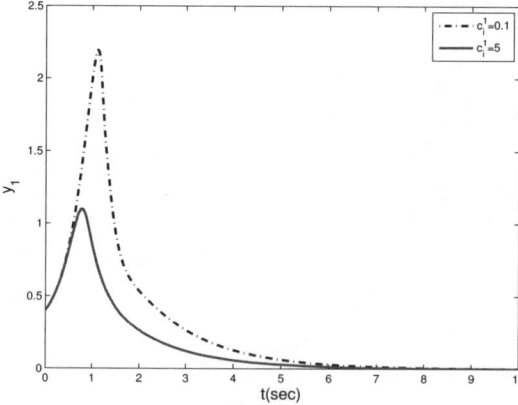

Fig. 5.5. Output y_1 with $c_i^1 = 0.1, 5$

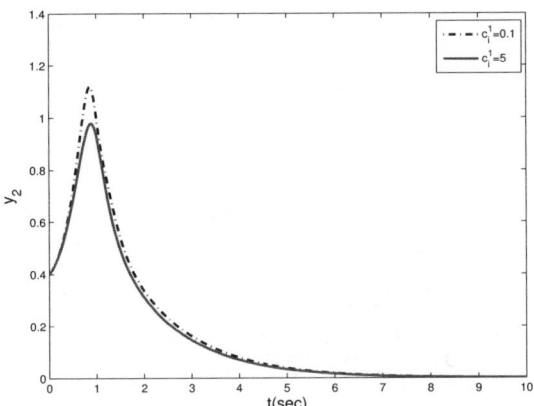

Fig. 5.6. Output y_2 with $c_i^1 = 0.1, 5$

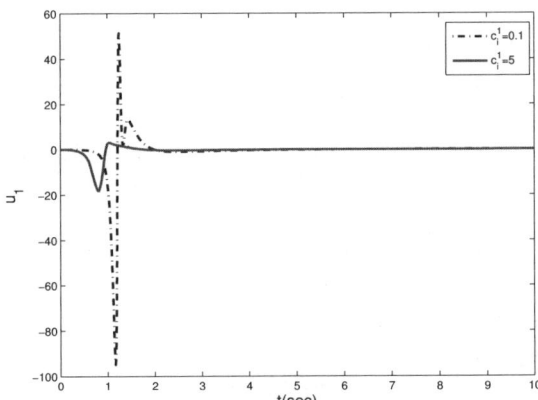

Fig. 5.7. Input u_1 with $c_i^1 = 0.1, 5$

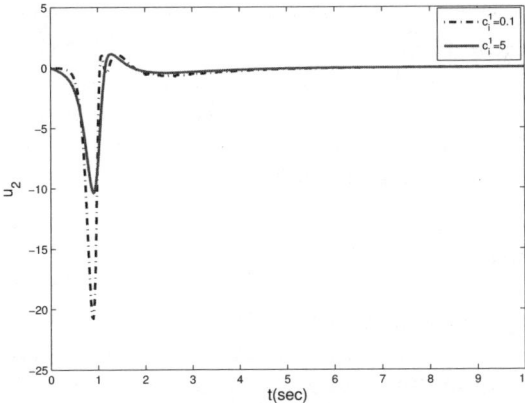

Fig. 5.8. Input u_2 with $c_i^1 = 0.1, 5$

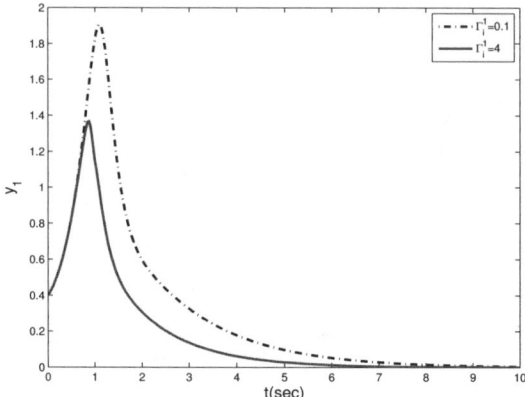

Fig. 5.9. Output y_1 with $\Gamma_i = 0.1, 4$

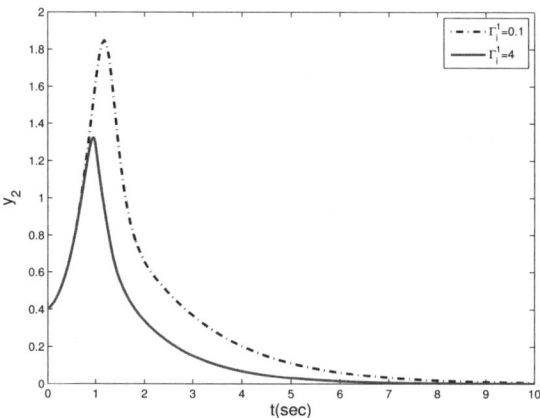

Fig. 5.10. Output y_2 with $\Gamma_i = 0.1, 4$

2. **Effects of Parameters Γ_i**

We now fix $c_1^1 = c_1^2 = c_1^2 = c_2^2 = 2$, $\gamma_1 = \gamma_2 = 0.1$, $l_1^1 = l_1^2 = l_2^1 = l_2^2 = 2$, $l_1^* = l_2^* = 1$. The subsystem outputs y_1, y_2 are compared in Figures 5.9-5.10 when $\Gamma_i = 0.1, 4$ respectively for $i = 1, 2$. Clearly, the transient tracking performances are found significantly improved by increasing Γ_i.

5.7 Conclusion

In this chapter, decentralized adaptive output feedback stabilization of a class of interconnected systems is considered. Each local adaptive controller is designed based on a general transfer function of the local subsystem with arbitrary relative degree by developing an adaptive backstepping control scheme. The effects of interactions are considered in the design. The nonlinear interactions between subsystems are allowed to satisfy higher-order nonlinear bounds. It is shown that the designed local adaptive controllers stabilize the overall interconnected systems. Perfect stabilization is ensured and the L_2 norm of the system outputs is also shown to be bounded by a function of design parameters. The strengths can be allowed arbitrary strong if their upper bounds are available in this case. Simulation results illustrate the effectiveness of our proposed scheme.

Part II

Nonsmooth Nonlinearities

6 Nonsmooth Nonlinearities

When dealing with real control problems, the designer is inevitably led to face the difficulties tied to the presence of real physical components, which often contain nonsmooth nonlinearities. In particular, actuators used in practice almost always contain static (e.g., dead-zone) or dynamic (e.g., backlash, hysteresis) nonlinearities, whose parameters are unknown and may vary with time. Dead-zone, backlash, hysteresis and saturation nonlinearities exist in mechanical, hydraulic, magnetic, and other types of system components. Nonsmooth nonlinearities are among the key factors limiting both static and dynamic performance of feedback control systems. As a matter of fact, these nonlinearities are particularly harmful and usually lead to a relevant deterioration of system performance. These nonsmooth nonlinear characteristics are often neglected in control system design. Nevertheless, certain design methods based on different control objectives and system conditions have been developed and verified in theory and practice. Some of them are reviewed below.

The development of control techniques to mitigate effects of unknown nonsmooth nonlinearities has been studied for decades and has attracted a lot of attentions in engineering and science [45, 53, 107, 108, 109, 110]. A number of techniques are available in literature to compensate these nonlinearities in the actuator. Starting from the pioneering work in [50], the idea of employing an adaptive inverse of the nonlinearity itself to cancel its effect has been widely used to cope with actuator dead zone [41, 51, 53, 73], backlash [43, 44, 55], or hysteresis [45, 48, 56] with unknown parameters. These schemes assumed that the system parameters must be inside known compact sets. Sometimes it is difficult to obtain its inverse. Intelligent control using neural networks (NN) is presented in [42, 64, 65, 66, 67, 68], while fuzzy logic is used in [41, 69, 70, 71]. The system states and uncertain weights must be within known compact sets. With this, the error resulted from using NN or fuzzy logic to approximate system functions will be bounded with known bounds. Variable structure control has been used as well in [43, 52, 74, 75, 76] and a describing function-based model is adopted for the input nonlinearities. Although the sliding motion is essential in variable structure control, it is an undesirable phenomenon from the adaptive control point of view. In practice, sliding motions cause chattering and also may lead to a theoretical loss of uniqueness of solutions. Model reference approaches have recently been proposed to handle such nonlinearity, see for examples to cancel the

J. Zhou & C. Wen: Adapt. Backstepping Ctrl. of Uncertain Systems, LNCIS 372, pp. 83–96, 2008.
springerlink.com   © Springer-Verlag Berlin Heidelberg 2008

effects of dead-zone in [41], backlash in [28], hysteresis in [56], saturation [59, 63] and actuator failure in [27]. Systematic design procedures for saturation have been developed for pole placement control of discrete time plants [46, 47, 60], where all the poles and zeros of the plant are strictly inside the unit circle or the plant has only one pole at $z = 1$ and the others are within the unit circle. Very recently, the fusion of relay feedback control with robust nominal model following control has been used and experimentally tested to handle actuator dead-zone nonlinearities [111]. A recursive least square (RLS) algorithm avoiding nonlinearity inversion holding for dead zones in sensors is described in [79]. In [46, 47, 59, 60, 63, 79, 111], the plant linearity assumption is still required.

In the second part of this book, the controller designed consists of new robust control laws and new estimators to estimate the unknown parameters. Besides showing stability of the system, the transient performance of the tracking error is derived to be an explicit function of design parameters and thus can be tuned. It will be shown how nonsmooth nonlinear characteristics can be adaptively compensated and how desired system performance is achieved in the presence of such nonlinearities. In our pragmatic approach, we have chosen general class of models with sufficient number of adjustable parameters which provide significant flexibility in matching real situations. This flexibility will be exploited for our backstepping control schemes with such nonsmooth nonlinearities. The considered plant is supposed to be preceded by the actuating device $u = f(v)$ as in Figure 6.1, u being the plant input not available for control.

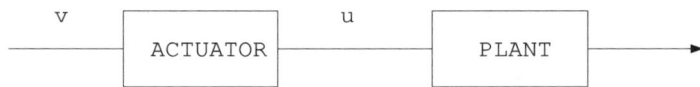

Fig. 6.1. Block scheme of a plant driven by the actuator

This chapter introduces basic description of nonsmooth nonlinear characteristics, such as dead-zone, backlash, hysteresis and saturation. These nonsmooth nonlinear characteristics are illustrated by a few examples in this chapter.

6.1 Backlash

Backlash is a dynamic input-output relationship. It exists in a wide range of physical systems and devices, such as biology optics, electro-magnetism, mechanical actuators, electronic relay circuits and other areas. The analytical expression of the backlash characteristic is

$$\dot{u}(t) = \begin{cases} m\dot{v}(t) & \textit{if } \dot{v}(t) \geq 0 \textit{ and } u(t) = m(v(t) - c_r), \textit{ or} \\ & \quad \textit{if } \dot{v}(t) \leq 0 \textit{ and } u(t) = m(v(t) - c_l) \\ 0 & \textit{otherwise} \end{cases} \tag{6.1}$$

where $m > 0$, $c_l < c_r$ are constant parameters. The motion on any inner segment is characterized by $\dot{u}(t) = 0$. A widely accepted characteristic of backlash is shown in Figure 6.2 where v is the input, u is the output, and c_r and c_l are the right and left "crossing".

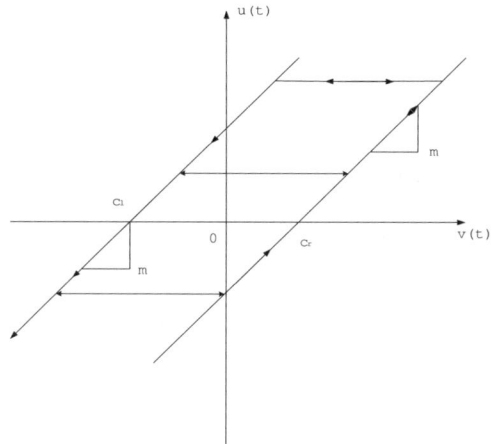

Fig. 6.2. Backlash hysteresis

Another expression of the backlash characteristic is described by using a differential equation as

$$\frac{du}{dt} = \alpha \left| \frac{dv}{dt} \right| (cv - u) + B_1 \frac{dv}{dt} \qquad (6.2)$$

where α, c and B_1 are constants, $c > 0$ is the slope of the lines satisfying $c > B_1$. Figure 6.3 shows that the dynamic equation (6.2) can be used to model a class of backlash nonlinearities, where the parameters $\alpha = 1, c = 3.1635$, and $B_1 = 0.345$, the input signal $v(t) = 6.5sin(2.3t)$ and the initial condition $u(0) = 0$. The simple backlash model appears in numerous studies of a wide variety of phenomena. We now briefly describe three typical examples.

6.1.1   Valve Control Mechanism

An input backlash example as in [49] is shown in Figure 6.4, where the backlash is in the valve control mechanism and $G(s) = k/s$ is the transfer function relating the liquid level h with the difference between the controlled inflow u and the uncontrolled outflow d.

6.1.2   Positioning System

An output backlash example is a simple servo for positioning of a low inertia object in [49], as shown in Figure 6.5. In this case, the transfer function is from

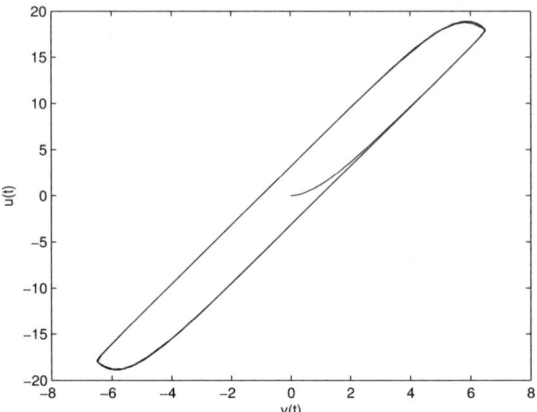

Fig. 6.3. Backlash hysteresis

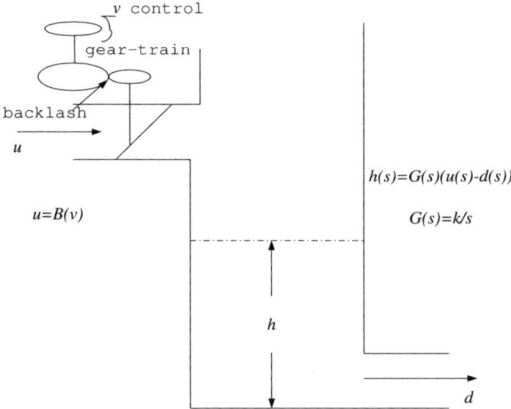

Fig. 6.4. Backlash in the valve control mechanism of a liquid tank

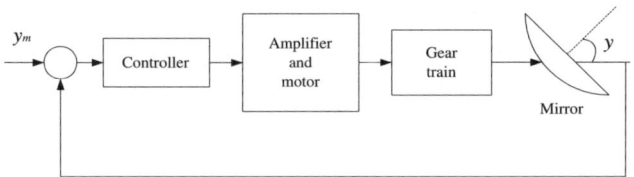

Fig. 6.5. Output backlash in a positioning system

the amplifier/motor unit. Effects of gear-train backlash in such classical servo mechanism have been extensively studied.

6.1.3   Piezoelectric Actuator

A piezoelectric actuator [112] is an electrically controllable positioning element which functions on the basis of the piezoelectric effect. A major limitation of piezoelectric actuator is the rate-independent hysteresis exhibited between voltage and displacement as shown in equation (6.2) and Figure 6.3, which severely limits system performance such as giving rise to undesirable inaccuracy or oscillations, even leading to instability.

6.2   Dead-Zone

Dead-zone is a static input-output relationship which for a range of input values gives no output. Once the output appears, the slope between the input and the output is constant. The analytical expression of the dead-zone characteristic is

$$
u(t) = \begin{cases} m_r(v(t) - b_r) & v(t) \geq b_r \\ 0 & b_l < v(t) < b_r \\ m_l(v(t) - b_l) & v(t) \leq b_l \end{cases} \tag{6.3}
$$

A graphical representation of the dead-zone is shown in Figure 6.6, where v is the input and u is the output. In general, neither the break-points $b_r \geq 0, b_l \leq 0$ nor the slopes $m_r, m_l > 0$ are equal. There is no loss of generality in assuming that the zero input point is inside the dead-zone because this can always be achieved with a redefinition of the input v.

The simple dead-zone model appears in numerous studies of a wide variety of phenomena, not limited to man-made systems. We briefly describe three typical examples, starting with a bioengineering application.

6.2.1   Upper-Limb Model

In functional neuromuscular stimulation a controlled electrical stimulus v is applied to inactive nerve in an attempt to replace upper motor neuron control which may be lost through cerebral stroke, brain injury, tumor, or spinal cessation. In [113] this approach has been applied to stimulation of the upper limb, concentrating on elbow flexion/extension. Two dead-zone models are employed to represent the biceps and triceps nonlinear "gains" appearing at the input of limb dynamics block in Figure 6.7. A similar model was employed in [49], [107] and [114] to adaptively control the knee joint of paraplegics.

6.2.2   Ultrasonic Motor

Ultrasonic motor(USM) is a new type motor as in [115], which is driven by the ultrasonic vibration force of piezoelectric elements. This motor has a nonlinear speed characteristics, which vary with drive conditions. In position control system, the motor shows a variable dead-zone in the control input (phase difference

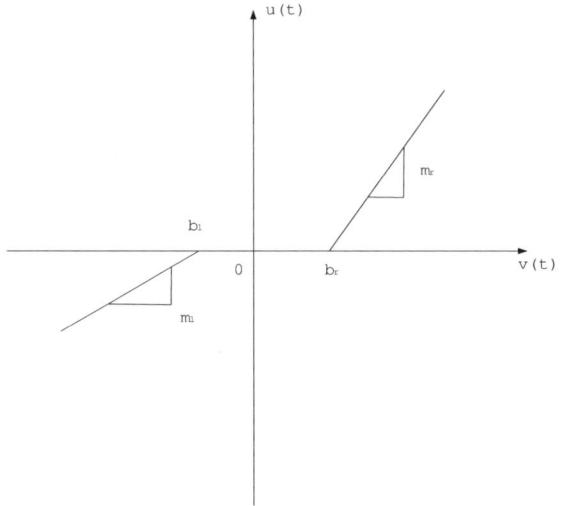

Fig. 6.6. Dead-zone

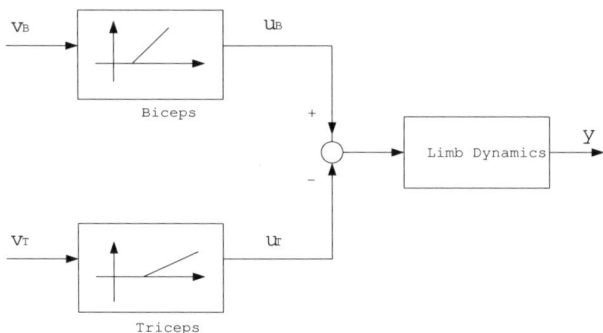

Fig. 6.7. Dead-zones in upper-limb model

of applied voltages) against load torque. The block diagram of USM is shown in Figure 6.8. This USM is typical travelling-wave type USM and consists of a stator and rotor made by elastic body, piezoelectric elements.

6.2.3   Servo-Valve

A common example from industrial applications is servo-valve in Figure 6.9. Its spool occludes the orifice with some overlap so that for a range of spool positions v there is no fluid flow u. This overlap prevents leakage losses which increase with wear and tear. Considering the spool position as the input v, and the load position y as output, the hydraulic system in Figure 6.9 is represented in Figure 6.10 as a dead-zone block. It is located as the input of linear dynamics

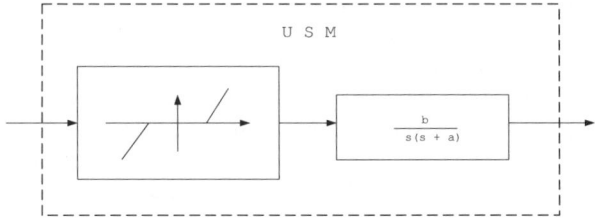

Fig. 6.8. Block diagram of USM

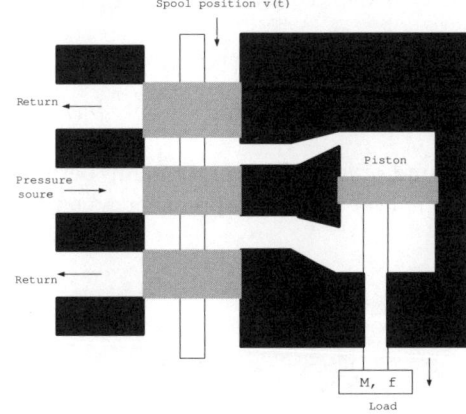

Fig. 6.9. Dead-zone in servo-valve

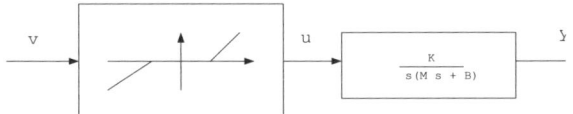

Fig. 6.10. Block diagram of the servo-valve

with transfer function $G(s) = \frac{K}{Ms^2+Bs}$, where $K = \frac{Ak_x}{k_p}$, $B = f + \frac{A^2}{k_p}$, $k_x = \frac{\partial g}{\partial x}$, $k_p = \frac{\partial g}{\partial P}$, $g = g(x, P) = $ flow, $A = $ area of piston, $P = $ pressure, and $f = $ viscous friction.

6.3 Saturation

It is known that all real dynamic systems are subject to hard limits on input. This is due to inherent physical constraints of the dynamical system and constraints in the controller actuators. Dynamical systems with hard limit constraints on the amplitude of control input, such as finite voltage of electrical motors and finite capacity of a pump, are the most common cases, where the hard limit constraint is modelled by a saturation nonlinearity. Saturation is always a potential problem for actuators of control systems as all actuators do saturate at some level. Actuator saturation affects the transient performance and even leads to system instability. Ignoring their existence may lead to severe performance deterioration and even instability in some cases. Thus, the impact of these constraints upon the closed-loop feedback control system needs to be addressed. In well-designed plants the operational requirements have been taken into consideration, and in addition the performance specs that the plant will be expected to meet are in line with the applicable physical constraints.

Saturation nonlinearity is defined as follows and shown in Figure 6.11

$$u = sat(v(t)) = \begin{cases} \text{sign}(v(t))u_M & |v(t)| \geq u_M \\ v(t) & |v(t)| < u_M \end{cases} \tag{6.4}$$

where u_M is a known bound of $u(t)$. The relationship between the applied control $u(t)$ and the control input $v(t)$ has a sharp corner when $|v(t)| = u_M$.

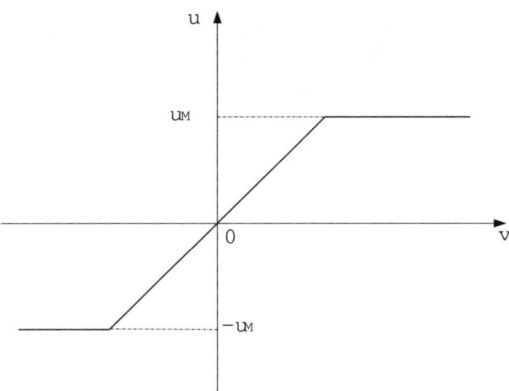

Fig. 6.11. Saturation

Input saturation constitutes a class of most encountered nonlinearities in control design. The control variables of all real-world systems are constrained or limited due to the physical nature of the actuator.

6.3.1 Active Micro-gravity Isolation System

An active micro-gravity isolation system [116] has saturation which limits the actuator force. A schematic of system is shown in Figure 6.12. The goal of the control design is to achieve a level of isolation between the base acceleration x_{off} and the inertial acceleration x_{on} of the isolated platform. The isolated platform must operate in a limited rattle space. Hence, an additional design constraint is that the relative displacement $x_{on} - x_{off}$ does not exceed the 0.5 inch rattle space limit in order to prevent the platform from bumping into its hard stops.

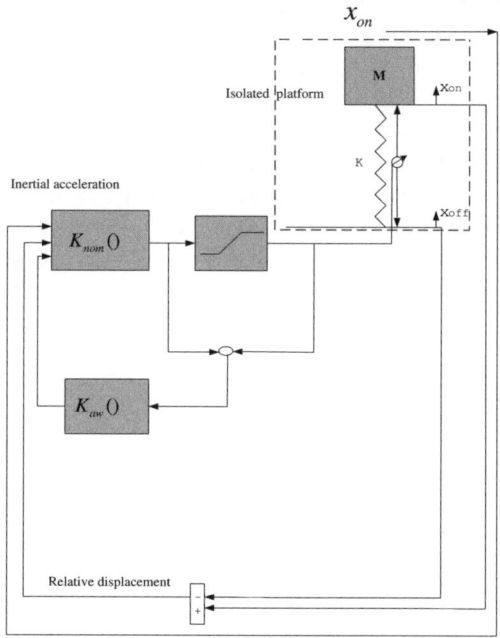

Fig. 6.12. Schematic Isolation system

6.3.2   Power Supply

The control of the current, position and shape of an elongated cross-section tokamak plasma [117] is complicated by the instability of the plasma vertical position. Due to the size and therefore the cost of ITER, there will naturally be smaller margins in the Poloidal Field coil power supplies implying that the feedback will experience actuator saturation during large transients due to a variety of plasma disturbances. Current saturation is relatively good due to the integrating nature of the tokamak, resulting in a reasonable time horizon for strategically handling this problem. On the other hand, voltage saturation is produced by the feedback controller itself, with no intrinsic delay.

6.3.3   Fligh Control with Saturating Actuator

In flight control the sizing and placement of control surfaces on an aircraft are determined by the performance requirements as in [118]. Obviously, realistic performance specs must be stipulated. Furthermore, the available control authority must be properly allocated among the tasks at hand. In well-designed plants, the saturation constraints are generally of minimal impact, and industry has fared well in plant design and closed-loop feedback control in Figure 6.13. There are, however, situations where actuator saturation can become a problem in operational flight control systems. For example, dogfights and aerial demonstrations at the boundary of the aircraft's operational envelope may require high-amplitude slewing maneuvers at the extreme edge of an aircraft's capabilities. In the quest for high performance, and when these systems are "pushed to their limits", it is reasonable to expect that actuator saturations may in fact occur, and the consideration of these nonlinear effects in the design phase might indeed reduce the degree of conservativeness of an flight control system and thus enhance the system performance. Additionally, there is the quest for reconfigurable flight control, which is driven by the need to accommodate failed control surfaces; saturation of the actuators may realistically become a problem in the event of a control surface failure or when battle damage is sustained, and performance is to be recovered.

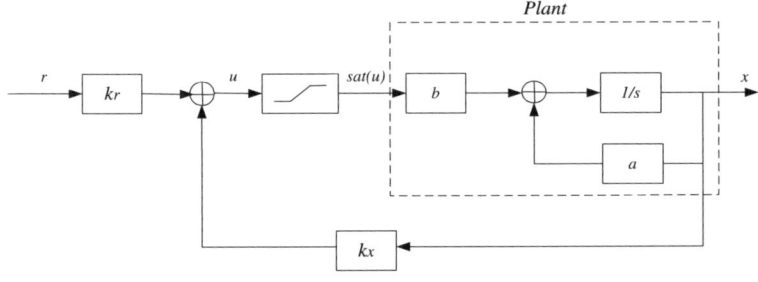

Fig. 6.13. Closed-loop flight control systems

Physical dynamic systems are commonly modelled as linear or nonlinear systems, whereas in practice, all physical systems are subject to hard constraints, e.g., and in particular in control systems, actuator displacement and rate saturations. Thus, the impact of these constraints upon the closed-loop feedback control system needs to be addressed.

6.4 Hysteresis

A hysteresis characteristic can be tuned by eight parameters: four slopes m_r, m_l, m_t, m_b and four crossing parameters c_r, c_l, c_t, c_b, where the subscripts l, r, t, b respectively indicate "left", "right", "top", "bottom" sides of the hysteresis loop. The hysteresis is a dynamic nonlinearity and is described by two half-lines, two line segments, and the quadrilateral formed by those half-lines and segments. The two half-lines and two line segments are described by:

$$u(t) = m_t v(t) + c_t, \ v(t) > v_1 = \frac{c_t + m_l c_l}{m_l - m_t} \tag{6.5}$$

$$u(t) = m_b v(t) + c_b, \ v(t) < v_2 = \frac{c_b + m_r c_r}{m_r - m_b} \tag{6.6}$$

$$u(t) = m_r(v(t) - c_r), \ v_2 < v(t) < v_3 = \frac{c_t + m_r c_r}{m_r - m_t}, \dot{v}(t) > 0, \dot{u}(t) > 0 \tag{6.7}$$

$$u(t) = m_l(v(t) - c_l), \ \frac{c_b + m_l c_l}{m_l - m_b} = v_4 < v(t) < v_1, \ \dot{v}(t) < 0, \dot{u}(t) < 0 \tag{6.8}$$

where v_1, v_2, v_3, v_4 are the values of $v(t)$ at the upper-left, lower-right, upper-right, and lower-left corners of the quadrilateral. The motion on any inner segment is characterized by $\dot{u}(t) = 0$ even if $v(t)$ increases or deceases.

The hysteresis phenomena occur inside the loop formed by the half-lines (6.5-6.6) and the segments (6.7-6.8). Inside the hysteresis loop, the relationship between $u(t)$ and $v(t)$ is

$$u(t) = \begin{cases} m_t v(t) + c_d(t) \text{ for } \dot{v}(t) < 0 \\ m_b v(t) + c_u(t) \text{ for } \dot{v}(t) > 0 \end{cases} \tag{6.9}$$

where $c_d(t) \in (c_t, c_1), c_u(t) \in (c_2, c_b)$ are piecewise constant functions which depend on the point where $\dot{v}(t)$ changes its sign and on the past trajectories of $(v(t), u(t))$, with

$$c_1 = \begin{cases} (m_b - m_t)\frac{c_b + m_l c_l}{m_l - m_b} + c_b \text{ for } m_t < m_b \\ (m_b - m_t)\frac{c_b + m_r c_r}{m_r - m_b} + c_b \text{ for } m_t > m_b \\ c_b \qquad\qquad\qquad \text{ for } m_t = m_b \end{cases} \tag{6.10}$$

$$c_2 = \begin{cases} (m_t - m_b)\frac{c_t + m_r c_r}{m_r - m_t} + c_t \text{ for } m_t > m_b \\ (m_t - m_b)\frac{c_t + m_l c_l}{m_l - m_t} + c_t \text{ for } m_t < m_b \\ c_t \qquad\qquad\qquad \text{ for } m_t = m_b \end{cases} \tag{6.11}$$

The relationship (6.9) holds for a part of one of the half-lines: when $m_t > m_b$, on the half-line (6.5) with $v_1 < v(t) < v_3$, $u(t) = m_t v(t) + c_t$ for $\dot{v}(t) < 0$; when $m_t < m_b$, on the half-line (6.6) with $v_4 < v(t) < v_2$, $u(t) = m_b v(t) + c_b$ for $\dot{v}(t) > 0$.

The signs of $\dot{u}(t)$ and $\dot{v}(t)$ are not restricted on other parts of these two half-lines: $u(t) = m_t v(t) + c_t, v(t) \geq v_3$; $u(t) = m_b v(t) + c_b, v(t) \leq v_4$; and $u(t) = m_t v(t) + c_t, \ v_1 < v(t) < v_3$ when $m_t < m_b$ or $u(t) = m_b v(t) + c_b$, $v_4 < v(t) < v_2$ when $m_t > m_b$.

The model of the hysteresis and its two typical minor loops are shown in Figure 6.14.

The motion of $u(t)$ and $v(t)$ on the half-lines (6.5)-(6.6) and the segments (6.7)-(6.8) and inside the hysteresis loop can be mathematically described as

$$\dot{u}(t) = \begin{cases} m_t \dot{v}(t) & \text{if } v(t) \geq v_3 \text{ and } u(t) = m_t v(t) + c_t, \\ & \text{or if } v_4 < v(t) < v_3, \dot{v}(t) < 0, u(t) = m_t v(t) + c_d, \\ & u(t) \neq m_l(v(t) - c_l) \text{ and } u(t) \neq m_b v(t) + c_b, \\ & \text{or if } v_4 < v(t) < v_3, \ \dot{v}(t) < 0, \\ & u(t) = m_b v(t) + c_b \text{ and } m_t < m_b \\ & \text{or if } v_4 < v(t) < v_3, \ \dot{v}(t) > 0, \\ & u(t) = m_t v(t) + c_t \text{ and } m_t < m_b \\ m_b \dot{v}(t) & \text{if } v(t) \leq v_4 \text{ and } u(t) = m_b v(t) + c_b, \\ & \text{or if } v_4 < v(t) < v_3, \ \dot{v}(t) > 0, u(t) = m_b v(t) + c_u \\ & u(t) \neq m_r(v(t) - c_r) \text{ and } u(t) \neq m_t v(t) + c_t, \\ & \text{or if } v_4 < v(t) < v_3, \ \dot{v}(t) > 0, \\ & u(t) = m_t v(t) + c_t \text{ and } m_t > m_b \\ & \text{or if } v_4 < v(t) < v_3, \ \dot{v}(t) < 0, \\ & u(t) = m_b v(t) + c_b \text{ and } m_t > m_b \\ m_r \dot{v}(t) & \text{if } v_4 < v(t) < v_3, \ \dot{v}(t) > 0 \text{ and } u(t) = m_r(v(t) - c_r) \\ m_l \dot{v}(t) & \text{if } v_4 < v(t) < v_3, \ \dot{v}(t) < 0 \text{ and } u(t) = m_l(v(t) - c_l) \\ 0 & \text{if } \dot{v}(t) = 0 \end{cases} \quad (6.12)$$

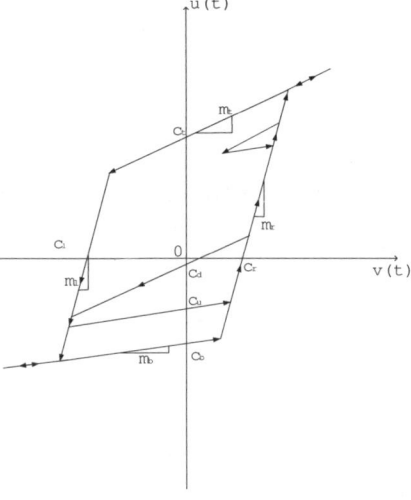

Fig. 6.14. Hysteresis model

6.4.1 Magnetic Suspension with Hysteresis

Typical examples of control systems with input hysteresis are magnetic suspensions and bearings. An oversimplified schematic representation of a magnetic suspension systems is shown in Figure 6.15. The position of an iron ball is detected by a light source L and a photocell P and compared with a desired reference r. The error signal $y - r$ is sent to a controller which generates the control signal - the electromagnet current I.

The magnetic force acting upon the iron ball is a nonlinear function of the ball position y and the magnetic flux ϕ. The remaining nonlinearity is the ferromagnetic hysteresis characteristic $\phi(I)$. To cast this system, we consider the current I as the signal $v(t)$ and the magnetic force F acting upon the iron ball as the signal $u(t)$. To hold the ball at some desired position $y = r$ the required force is u_s. The amount of the current v needed to generate this force depends on the operating point on the hysteresis characteristic.

The input-output model of the magnetic suspension system from the current I to the ball position y can be represented by the block diagram in Figure 6.16 where under certain simplifying assumptions the transfer function $G(s)$ has two poles: $G(s) = k/(s-p_1)(s-p_2)$. One of the poles, say p_1, is necessarily unstable, $p_1 > 0$, because the magnetic force decays with the distance and the gravitation force is constant.

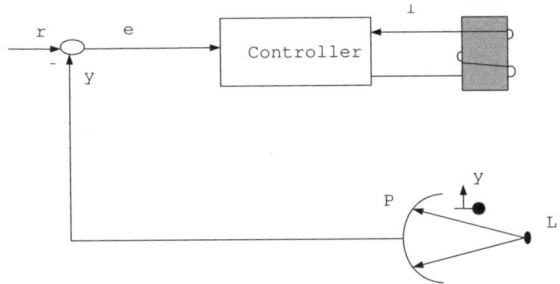

Fig. 6.15. A magnetic suspension with solenoid hysteresis

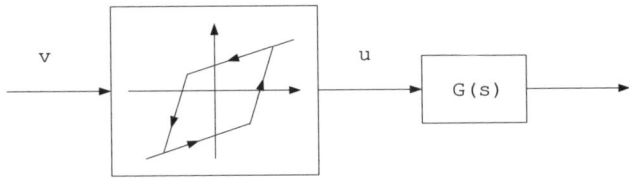

Fig. 6.16. Plant with input hysteresis

6.4.2   Hysteresis Motor

Hysteresis motor [119] is a self-starting synchronous motor that uses the hysteresis characteristic of the semi-hard magnetic materials. It consists of polyphase stator and rotor which contains hysteresis ring. Most of cases, semi-hard magnetic material is used for the hysteresis ring. It needs to determine the adequate thickness of the hysteresis ring in the hysteresis motor and the motor torque is calculated by the area of hysteresis loop determined by the field intensity in the ring. The hysteresis ring is affected by the rotational hysteresis caused by the stator windings and the direction of the magnetization of each element of the ring is different from that of the magnetic field or magnetic flux density. That is to say, the thicker the hysteresis ring becomes, the larger the rotational hysteresis increases and to make matters worse, the output of the thicker ring motor becomes less than that of thin rotor motor.

6.4.3   Hysteresis in Brakes

Disk brakes are becoming important actuators in advance automotive control systems which improves safety, drivability, and the overall performance of passenger cars and trucks. A common air disk (ADB-1560) for trucks has an input-output force hysteresis characteristic. In a feedback control loop, this large hysteresis would limit the achievable dynamic performance. Therefore, for high dynamic performance the effect of hysteresis must be compensated.

7 Backstepping Control of Systems with Backlash Nonlinearity

In this chapter, we consider uncertain dynamic systems preceded by unknown backlash nonlinearity. By using backstepping technique, new schemes for both state feedback and output feedback are proposed. Besides showing global stability of the system, the transient performance in terms of L_2 norm of the tracking error is derived to be an explicit function of design parameters. For output feedback control we develop a new scheme for a class of uncertain linear systems preceded by unknown backlash nonlinearities. The controller designed by using backstepping technique consists of a new robust control law and a new estimator to estimate the unknown parameters. The result is also extended to nonlinear systems.

7.1 Introduction

The development of control techniques to mitigate effects of unknown backlash has been studied for decades. Much of this interest is a consequence of its importance in present application. Interest in studying dynamic systems with backlash is motivated by their role as nonlinearities for which traditional control methods are insufficient and so requiring development of new approaches. Several adaptive control schemes have recently been proposed, see for examples [43, 44, 48, 55, 56, 120]. In [44], [55] and [56], an adaptive inverse cascaded with the plant was employed to cancel the effects of nonlinearity. In [43] a dynamic backlash model is defined to pattern a backlash rather than constructing an inverse model to mitigate the effects of the backlash. However in [43], the term multiplying the control and the uncertain parameters of the system must be within known intervals and the 'disturbance-like' term must be bounded by a known bound. Projection was used to handle the 'disturbance-like' term and unknown parameters. System stability was established and the tracking error was shown to converge to a residual. In [120], a state feedback backstepping design was developed to deal with the backlash nonlinearity, where the effect of backlash was treated as a bounded disturbance and an estimate was used to estimate its bound.

J. Zhou & C. Wen: Adapt. Backstepping Ctrl. of Uncertain Systems, LNCIS 372, pp. 97–123, 2008.
springerlink.com

7.2 State Feedback Control

In this section, we develop two simple backstepping adaptive control schemes for the same class of nonlinear systems as in [43] and [120]. In the first scheme, a sign function is involved and this can ensure perfect tracking. To avoid possible chattering caused by the sign function, we propose an alternative smooth control law and the tracking error is still ensured to approach a prescribed bound in this case. In our design, the term multiplying the control and the system parameters are not assumed to be within known intervals. The bound of the 'disturbance-like' term is not required. To handle such a term, an estimator is used to estimate its bound. Besides showing global stability of the system, transient performance in terms of L_2 norm of the tracking error is derived to be an explicit function of design parameters and thus our scheme allows designers to obtain the closed loop behavior by tuning design parameters in an explicit way.

7.2.1 Problem Formulation

The class of systems is modelled as follows:

$$x^{(n)}(t) + \sum_{i=1}^{r} a_i Y_i\big(x(t), \dot{x}(t), \dots, x^{(n-1)}(t)\big) = bu(w) + \bar{d}(t) \qquad (7.1)$$

where Y_i are known continuous linear or nonlinear functions, $\bar{d}(t)$ denotes bounded external disturbances, parameters a_i are unknown constants and control gain b is an unknown constant, w is the control input, $u(w)$ denotes backlash type of nonlinearities described as the following.

$$\frac{du}{dt} = \alpha \Big| \frac{dw}{dt} \Big| (cw - u) + B_1 \frac{dw}{dt} \qquad (7.2)$$

where α, c and B_1 are constants, $c > 0$ is the slope of the lines satisfying $c > B_1$. This equation can be solved explicitly for w piecewise monotone

$$u(t) = cw(t) + d_1(w) \qquad (7.3)$$

$$d_1(w) = [u_0 - cw_0]e^{-\alpha(w-w_0)\mathrm{sign}(\dot{w})} + e^{-\alpha w\,\mathrm{sign}(\dot{w})} \int_{w_0}^{w} [B_1 - c]e^{\alpha \xi\,\mathrm{sign}(\dot{w})} d\xi \qquad (7.4)$$

for $\dot{w}$ constant and $u(w_0) = w_0$. The solution indicates that dynamic equation (7.2) can be used to model a class of backlash nonlinearities. Analyzing the solution (7.3), we see that it is composed of a line with the slop c, together with a term $d_1(w)$. For $d_1(w)$, it is bounded clearly. It can be shown that if $u(w; w_0, u_0)$ is the solution of (7.4) with initial values (w_0, u_0), then, if $\dot{w} > 0(\dot{w} < 0)$ and $w \to +\infty(-\infty)$, one has

$$lim_{w\to\infty} d_1(w) = lim_{w\to\infty}[u(w; w_0, u_0) - f(w)] = -\frac{c - B_1}{\alpha} \qquad (7.5)$$

$$lim_{w\to-\infty} d_1(w) = lim_{w\to-\infty}[u(w; w_0, u_0) - f(w)] = \frac{c - B_1}{\alpha} \qquad (7.6)$$

It should be noted that the above convergence is exponential at the rate of α. And we get $d_1(w)$ is bounded.

From the solution structure (7.3) of model (7.2), (7.1) becomes

$$x^{(n)}(t) + \sum_{i=1}^{r} a_i Y_i\big(x(t), \dot{x}(t), \ldots, x^{(n-1)}(t)\big) = \theta w(t) + d(t) \qquad (7.7)$$

where $\theta = bc$ and $d(t) = bd_1(w(t)) + \bar{d}(t)$. The effect of $d(t)$ is due to both external disturbances and $bd_1(w(t))$. We call $d(t)$ a 'disturbance-like' term for simplicity of presentation.

Now equation (7.7) is rewritten in the following form

$$\dot{x}_1 = x_2$$
$$\vdots$$
$$\dot{x}_{n-1} = x_n$$
$$\dot{x}_n = -\sum_{i=1}^{r} a_i Y_i\big(x_1(t), x_2(t), \ldots, x_{(n-1)}(t)\big) + \theta w(t) + d(t)$$
$$= a^T Y + \theta w(t) + d(t) \qquad (7.8)$$

where $x_1 = x, x_2 = \dot{x}, \ldots, x_n = x^{(n-1)}$, $a = [-a_1, -a_2, \ldots, -a_r]^T$ and $Y = [Y_1, Y_2, \ldots, Y_r]^T$.

For the development of control laws, the following assumptions are made.

Assumption 1. The uncertain parameters b and c are such that $\theta > 0$.

Assumption 2. The desired trajectory $y_r(t)$ and its $(n-1)$th order derivatives are known and bounded.

The control objectives are to design backstepping adaptive control laws such that

- The closed loop is globally stable in the sense that all the signals in the loop are uniformly ultimately bounded;
- The tracking error $x(t) - y_r(t)$ is adjustable during the transient period by an explicit choice of design parameters and $\lim_{t\to\infty} x(t) - y_r(t) = 0$ or $\lim_{t\to\infty} |x(t) - y_r(t)| - \delta_1 = 0$ for an arbitrary specified bound δ_1.

Remark 7.1. Compared with [43], the uncertain parameters θ and a_i are not assumed inside known intervals. The bound D for $d(t)$ is not assumed to be known and it will be estimated by our adaptive controllers. Also the control objectives are not only to ensure global stability, but also transient performance.

7.2.2 Backstepping Design and Stability Analysis

Before presenting the adaptive control design using the backstepping technique to achieve the desired control objectives, the following change of coordinates is made.

$$z_1 = x_1 - y_r \qquad (7.9)$$

$$z_i = x_i - y_r^{(i-1)} - \alpha_{i-1}, \qquad i = 2, 3, \ldots, n \tag{7.10}$$

where α_{i-1} is the virtual control at the *ith step* and will be determined in later discussion. In the following, two control schemes are proposed.

Control Scheme I

To illustrate the backstepping procedures, only the last step of the design, i.e. *step n* below, is elaborated in details.

- *Step* 1: For $i = 2$, it follows from (7.8) to (7.10) that

$$\dot{z}_1 = z_2 + \alpha_1 \tag{7.11}$$

We design the virtual control law α_1 as

$$\alpha_1 = -c_1 z_1 \tag{7.12}$$

where c_1 is a positive design parameter. From (7.11) and (7.12) we have

$$z_1 \dot{z}_1 = -c_1 z_1^2 + z_1 z_2 \tag{7.13}$$

- *Step* i $(i = 2, \ldots, n-1)$: Choose

$$\alpha_i = -c_i z_i - z_{i-1} + \dot{\alpha}_{i-1}(x_1, \ldots, x_{i-1}, y_r, \ldots, y_r^{(i-1)}) \tag{7.14}$$

where $c_i, i = 2, \ldots, n-1$ are positive design parameters. From (7.10) and (7.14) we obtain

$$z_i \dot{z}_i = -z_{i-1} z_i - c_i z_i^2 + z_i z_{i+1} \tag{7.15}$$

- *Step* n: From (7.8) and (7.10) we obtain

$$\dot{z}_n = \theta w(t) + a^T Y + d(t) - y_r^{(n)} - \dot{\alpha}_{n-1} \tag{7.16}$$

Then the adaptive control law is designed as follows

$$w = \hat{\vartheta}\bar{w} \tag{7.17}$$

$$\bar{w} = -c_n z_n - z_{n-1} - \hat{a}^T Y - \text{sign}(z_n)\hat{D} + y_r^{(n)} + \dot{\alpha}_{n-1} \tag{7.18}$$

$$\dot{\hat{\vartheta}} = -\gamma \bar{w} z_n \tag{7.19}$$

$$\dot{\hat{a}} = \Gamma Y z_n \tag{7.20}$$

$$\dot{\hat{D}} = \eta |z_n| \tag{7.21}$$

where c_n, γ and η are three positive design parameters, Γ is a positive definite matrix, $\hat{\vartheta}$, $\hat{a}$ and $\hat{D}$ are estimates of $\vartheta = 1/\theta$, a and D. Let $\tilde{\vartheta} = \vartheta - \hat{\vartheta}$, $\tilde{a} = a - \hat{a}$ and $\tilde{D} = D - \hat{D}$. Note that $\theta w(t)$ in (7.16) can be expressed as

$$\theta w = \theta \hat{\vartheta}\bar{w} = \bar{w} - \theta \tilde{\vartheta}\bar{w} \tag{7.22}$$

From (7.16),(7.18) and (7.22) we obtain

$$\dot{z}_n = -c_n z_n - z_{n-1} + \tilde{a}^T Y - \text{sign}(z_n)\hat{D} + d(t) - \theta \tilde{\vartheta} \bar{w} \qquad (7.23)$$

We define Lyapunov function as

$$V = \sum_{i=1}^{n} \frac{1}{2} z_i^2 + \frac{1}{2} \tilde{a}^T \Gamma^{-1} \tilde{a} + \frac{\theta}{2\gamma} \tilde{\vartheta}^2 + \frac{1}{2\eta} \tilde{D}^2 \qquad (7.24)$$

Then the derivative of V along with (7.8) and (7.17) to (7.21) is given by

$$\dot{V} = \sum_{i=1}^{n} z_i \dot{z}_i + \tilde{a}^T \Gamma^{-1} \dot{\tilde{a}} + \frac{\theta}{\gamma} \tilde{\vartheta} \dot{\tilde{\vartheta}} + \frac{1}{\eta} \tilde{D} \dot{\tilde{D}}$$

$$= -\sum_{i=1}^{n} c_i z_i^2 + \tilde{a}^T Y z_n - |z_n|\hat{D} + d(t)z_n - \theta \tilde{\vartheta} \bar{w} z_n - \tilde{a} \Gamma^{-1} \dot{\hat{a}} - \frac{\theta}{\gamma} \tilde{\vartheta} \dot{\hat{\vartheta}} - \frac{1}{\eta} \tilde{D} \dot{\hat{D}}$$

$$\leq -\sum_{i=1}^{n} c_i z_i^2 + \tilde{a}^T \Gamma^{-1}(\Gamma Y z_n - \dot{\hat{a}}) - \frac{\theta}{\gamma} \tilde{\vartheta}(\gamma \bar{w} z_n + \dot{\hat{\vartheta}}) + \frac{1}{\eta} \tilde{D}(\eta |z_n| - \dot{\hat{D}})$$

$$= -\sum_{i=1}^{n} c_i z_i^2 \qquad (7.25)$$

where we have used (7.13),(7.15),(7.23) and $z_n d(t) \leq |z_n|D$ to obtain (7.25).

We have the following stability and performance results based on this scheme.

Theorem 7.1. *Consider the uncertain nonlinear system (7.1) satisfying Assumptions 1-2. With the application of controller (7.17) and the parameter update laws (7.19)-(7.21), the following statements hold:*

- *The resulting closed loop system is globally stable.*
- *The asymptotic tracking is achieved, i.e.,*

$$\lim_{t \to \infty} [x(t) - y_r(t)] = 0 \qquad (7.26)$$

- *The transient tracking error performance is given by*

$$\| x(t) - y_r(t) \|_2 \leq \frac{1}{\sqrt{c_1}} \left(\frac{1}{2} \tilde{a}(0)^T \Gamma^{-1} \tilde{a}(0) + \frac{\theta}{2\gamma} \tilde{\vartheta}(0)^2 + \frac{1}{2\eta} \tilde{D}(0)^2 \right)^{1/2} \qquad (7.27)$$

Proof: From (7.25) we established that V is non increasing. Hence, $z_i, i = 1, \ldots, n$, $\hat{\vartheta}, \hat{a}, \hat{D}$ are bounded. By applying the LaSalle-Yoshizawa theorem in Appendix B to (7.25), it further follows that $z_i(t) \to 0, i = 1, \ldots, n$ as $t \to \infty$, which implies that $\lim_{t \to \infty}[x(t) - y_r(t)] = 0$.

Then we have

$$\| z_1 \|_2^2 = \int_0^\infty |z_1(\tau)|^2 d\tau \leq \frac{1}{c_1}(V(0) - V(\infty)) \leq \frac{1}{c_1} V(0) \qquad (7.28)$$

Thus, by setting $z_i(0) = 0, i = 1, \ldots, n$, we obtain

$$V(0) = \frac{1}{2}\tilde{a}(0)^T \Gamma^{-1} \tilde{a}(0) + \frac{\theta}{2\gamma}\tilde{\vartheta}(0)^2 + \frac{1}{2\eta}\tilde{D}(0)^2, \qquad (7.29)$$

a decreasing function of γ, η and Γ, independent of c_1. This means that the bound resulting from (7.28) and (7.29) is

$$\| z_1 \|_2 \leq \frac{1}{\sqrt{c_1}}\left(\frac{1}{2}\tilde{a}(0)^T \Gamma^{-1} \tilde{a}(0) + \frac{\theta}{2\gamma}\tilde{\vartheta}(0)^2 + \frac{1}{2\eta}\tilde{D}(0)^2\right)^{1/2} \qquad (7.30)$$

$$\triangle\triangle\triangle$$

Remark 7.2. From Theorem 7.1 the following conclusions can be obtained:

- The transient performance depends on the initial estimate errors $\tilde{\vartheta}(0)$, $\tilde{a}(0)$, $\tilde{D}(0)$ and the explicit design parameters. The closer the initial estimates $\hat{\vartheta}(0), \hat{a}(0)$ and $\hat{D}(0)$ to the true values ϑ, a and D, the better the transient performance.
- The bound for $\| x(t) - y_r(t) \|_2$ is an explicit function of design parameters and thus computable. We can decrease the effects of the initial error estimates on the transient performance by increasing the adaptation gains γ, η and Γ.
- To improve the tracking error performance we can also increase the gain c_1. However, increasing c_1 will influence other performance such as $\| \dot{x} - \dot{y}_r \|_2$ as shown below.

Since $\dot{V} \leq 0$, immediately from (7.24) we know

$$V(t) = \sum_{i=1}^{n}\frac{1}{2}z_i^2 + \frac{1}{2}\tilde{a}^T \Gamma^{-1}\tilde{a} + \frac{\theta}{2\gamma}\tilde{\vartheta}^2 + \frac{1}{2\eta}\tilde{D}^2 \leq V(0) \qquad (7.31)$$

Then

$$\| z_i \|_\infty \leq \sqrt{2V(0)}, \quad i = 1, \ldots, n \qquad (7.32)$$

$$\| \tilde{a} \|_\infty \leq \sqrt{\bar{\lambda}(\Gamma)}\sqrt{2V(0)} \qquad (7.33)$$

From equations (7.10) for $i = 2$ and (7.12), we get

$$\| \dot{x} - \dot{y}_r \|_2 = \| z_2 - c_1 z_1 \|_2$$
$$\leq \| z_2 \|_2 + c_1 \| z_1 \|_2 \qquad (7.34)$$

Similar to the proof of (7.30), we can get $\| z_2 \|_2 \leq \frac{1}{\sqrt{c_2}}\sqrt{V(0)}$ and thus

$$\| \dot{x} - \dot{y}_r \|_2 \leq \left(\frac{1}{\sqrt{c_2}} + \sqrt{c_1}\right)\sqrt{V(0)} \qquad (7.35)$$

From equation (7.35) we can see that increasing c_1 also increase the error $\| \dot{x} - \dot{y}_r \|_2$. This suggests to fix the gain c_1 to some acceptable value and adjust the other gains such as γ, η and Γ.

Control Scheme II

In the previous scheme, a discontinuous function $sgn(z_n)$ is involved in the control and this may cause chattering. To avoid this, we now propose an alternative smooth control scheme.

Firstly we define a function $sg_i(z_i)$ as follows

$$sg_i(z_i) = \begin{cases} \dfrac{z_i}{|z_i|} & |z_i| \geq \delta_i \\[2mm] \dfrac{z_i^{(2q+1)}}{(\delta_i^2 - z_i^2)^{n-i+2} + |z_i|^{(2q+1)}} & |z_i| < \delta_i \end{cases} \tag{7.36}$$

where $\delta_i (i = 1, \ldots, n)$ is a positive design parameter and $q = round\{(n - i + 2)/2\}$, where $round\{x\}$ means the element of x to the nearest integer. Clearly $2q + 1 \geq (n - i + 2)$.

Remark 7.3. Note that $sg_i(z_i)$ is $(n - i + 2)th$ order differentiable, so that this function can be used in the recursive backstepping control design, which required the function continuous differentiable. The function is used in the control scheme to remove the effect of disturbance and avoid chattering problem caused by discontinuous function.

We also design a function $f_i(z_i)$ as

$$f_i(z_i) = \begin{cases} 1 & |z_i| \geq \delta_i \\ 0 & |z_i| < \delta_i \end{cases} \tag{7.37}$$

Then we can get

$$sg_i(z_i) f_i(z_i) = \begin{cases} 1 & z_i \geq \delta_i \\ 0 & |z_i| < \delta_i \\ -1 & z_i \leq \delta_i \end{cases} \tag{7.38}$$

To ensure the resultant functions are differentiable, we replace z_i^2 by $(|z_i| - \delta_i)^{n-i+2} sg_i(z_i)$ in the Lyapunov functions for $i = 1, \ldots, n$ in Scheme I and we also replace z_i by $(|z_i| - \delta_i)^{n-i+1} sg_i$ in the design procedure as detailed below.

• *Step 1:* we design virtual control law α_1 as

$$\alpha_1 = -(c_1 + \frac{1}{4})(|z_1| - \delta_1)^n sg_1(z_1) - (\delta_2 + 1)sg_1(z_1) \tag{7.39}$$

where c_1 is a positive design parameter. We choose Lyapunov function V_1 as

$$V_1 = \frac{1}{n+1}(|z_1| - \delta_1)^{n+1} f_1 \tag{7.40}$$

Then the derivative of V_1 is

$$\dot{V}_1 = (|z_1| - \delta_1)^n f_1 sg_1(z_1) \dot{z}_1$$
$$\leq -(c_1 + \frac{1}{4})(|z_1| - \delta_1)^{2n} f_1 + (|z_1| - \delta_1)^n (|z_2| - \delta_2 - 1) f_1 \qquad (7.41)$$

where (7.39) has been used.

• *Step 2:* we design virtual control law α_2 as

$$\alpha_2 = -(c_2 + \frac{5}{4})(|z_2| - \delta_2)^{n-1} sg_2(z_2) + \dot{\alpha}_1 - (\delta_3 + 1) sg_2(z_2) \qquad (7.42)$$

where c_2 is positive design parameter.

We design Lyapunov function V_2 as

$$V_2 = \frac{1}{n}(|z_2| - \delta_2)^n f_2 + V_1 \qquad (7.43)$$

Then the derivative of V_2 is

$$\dot{V}_2 \leq -\sum_{i=1}^{2} c_i(|z_i| - \delta_i)^{2(n-i+1)} f_i + M_2 + (|z_2| - \delta_2)^{n-1}(|z_3| - \delta_3 - 1) f_2 \qquad (7.44)$$

where $M_2 = -\frac{1}{4}(|z_1| - \delta_1)^{2n} f_1 + (|z_1| - \delta_1)^n (|z_2| - \delta_2 - 1) f_1 - (|z_2| - \delta_2)^{2(n-1)} f_2$. Now we show that $M_2 < 0$. It is clear that $M_2 \leq 0$ for $|z_2| < \delta_2 + 1$. For $|z_2| \geq \delta_2 + 1$

$$M_2 \leq -\frac{1}{4}(|z_1| - \delta_1)^{2n} f_1 + \frac{1}{4}(|z_1| - \delta_1)^{2n} f_1^2$$
$$+(|z_2| - \delta_2 - 1)^2 - (|z_2| - \delta_2)^{2(n-1)}$$
$$< (|z_2| - \delta_2)^2 - (|z_2| - \delta_2)^{2(n-1)}$$
$$= (|z_2| - \delta_2)^2(1 - (|z_2| - \delta_2)^{2(n-2)})$$
$$\leq 0 \qquad (7.45)$$

Then (7.44) is written as

$$\dot{V}_2 \leq -\sum_{i=1}^{2} c_i(|z_i| - \delta_i)^{2(n-i+1)} f_i + (|z_2| - \delta_2)^{n-1}(|z_3| - \delta_3 - 1) f_2 \quad (7.46)$$

• *Step i $(i = 3, \ldots, n-1)$:* Choose

$$\alpha_i = -(c_i + \frac{5}{4})(|z_i| - \delta_i)^{n-i+1} sg_i(z_i) + \dot{\alpha}_{i-1} - (\delta_{i+1} + 1) sg_i(z_i) \qquad (7.47)$$

where c_i is positive design parameter.

• *Step n:* The control law and parameter update laws are designed as follows

$$w = \hat{\vartheta}\bar{w} \qquad (7.48)$$

$$\bar{w} = -(c_n + 1)(|z_n| - \delta_n)sg_n(z_n) - \hat{a}^T Y - sg_n\hat{D} + y_r^{(n)} + \dot{\alpha}_{n-1} \quad (7.49)$$

$$\dot{\hat{\vartheta}} = -\gamma\bar{w}(|z_n| - \delta_n)f_n sg_n(z_n) \quad (7.50)$$

$$\dot{\hat{a}} = \Gamma Y(|z_n| - \delta_n)f_n sg_n(z_n) \quad (7.51)$$

$$\dot{\hat{D}} = \eta(|z_n| - \delta_n)f_n \quad (7.52)$$

where c_n, γ and η are three positive design parameters, Γ is a positive definite matrix, $\hat{\vartheta}$, $\hat{a}$ and $\hat{D}$ are estimates of $\vartheta = 1/\theta$, a and D. We define Lyapunov function as

$$V = \sum_{i=1}^{n} \frac{1}{n-i+2}(|z_i| - \delta_i)^{n-i+2}f_i + \frac{1}{2}\tilde{a}^T \Gamma^{-1}\tilde{a} + \frac{\theta}{2\gamma}\tilde{\vartheta}^2 + \frac{1}{2\eta}\tilde{D}^2 \quad (7.53)$$

Then the derivative of V is given by

$$\dot{V} = \dot{V}_i + (|z_n| - \delta_n)^2 f_n sg_n(z_n)\dot{z}_n + \tilde{a}^T \Gamma^{-1}\dot{\tilde{a}} + \frac{\theta}{\gamma}\tilde{\vartheta}\dot{\tilde{\vartheta}} + \frac{1}{\eta}\tilde{D}\dot{\tilde{D}}$$

$$\leq -\sum_{i=1}^{n} c_i(|z_i| - \delta_i)^{2(n-i+1)}f_i + \tilde{a}^T \Gamma^{-1}\big(\Gamma Y(|z_n| - \delta_n)f_n sg_n(z_n) - \dot{\hat{a}}\big)$$

$$-\frac{\theta}{\gamma}\tilde{\vartheta}\big(\gamma\bar{w}(|z_n| - \delta_n)f_n sg_n(z_n) + \dot{\hat{\vartheta}}\big) + \frac{1}{\eta}\tilde{D}\big(\eta(|z_n| - \delta_n)f_n - \dot{\hat{D}}\big)$$

$$= -\sum_{i=1}^{n} c_i(|z_i| - \delta_i)^{2(n-i+1)}f_i \quad (7.54)$$

where (7.8),(7.39),(7.42) and (7.48) to (7.52) have been used.

Theorem 7.2. *Consider the uncertain nonlinear system (7.1) satisfying Assumptions 1-2. With the application of controller (7.48) and the parameter update laws (7.50) -(7.52), the following statements hold:*

- *The resulting closed loop system is globally stable.*
- *The tracking error converges to δ_1 asymptotically, i.e.,*

$$\lim_{t\to\infty} |x(t) - y_r(t)| = \delta_1, \quad |z_1| \geq \delta_1 \quad (7.55)$$

- *The transient tracking error performance is given by*

$$\| |x(t) - y_r(t)| - \delta_1 \|_2 \leq c_1^{\frac{-1}{2n}}\Big(\frac{1}{2}\tilde{a}(0)^T\Gamma^{-1}\tilde{a}(0) + \frac{\theta}{2\gamma}\tilde{\vartheta}(0)^2 + \frac{1}{2\eta}\tilde{D}(0)^2\Big)^{\frac{1}{2n}} \quad (7.56)$$

with $z_i(0) = \delta_i, i = 1, \ldots, n,$

Proof: Based (7.54), we established that V is non increasing. Hence, $|z_i| - \delta_i$ ($i = 1, \ldots, n$), $\hat{\vartheta}, \hat{a}, \hat{D}$ are bounded. By applying the LaSalle-Yoshizawa theorem in Appendix B to (7.54), it further follows that $|z_i| - \delta_i \to 0, i = 1, \ldots, n$ as $t \to \infty$, which implies that $\lim_{t\to\infty} |x(t) - y_r(t)| = \delta_1$.

From (7.54) we establish that V is non increasing. Then we have

$$\| z_1 - \delta_1 \|_2^{2n} = \int_0^\infty |z_1(\tau) - \delta_1|^{2n} d\tau$$

$$\leq \frac{1}{c_1}(V(0) - V(\infty)) \leq \frac{1}{c_1}V(0) \qquad (7.57)$$

Thus, by setting $z_i(0) = \delta_i, i = 1, \ldots, n$, we obtain

$$V(0) = \frac{1}{2}\tilde{a}(0)^T \Gamma^{-1}\tilde{a}(0) + \frac{\theta}{2\gamma}\tilde{\vartheta}(0)^2 + \frac{1}{2\eta}\tilde{D}(0)^2, \qquad (7.58)$$

a decreasing function of γ, η and Γ, independent of c_1. This means that the bound resulting from (7.57) and (7.58) is

$$\| |x(t) - y_r(t)| - \delta_1 \|_2 \leq c_1^{\frac{-1}{2n}} \left(\frac{1}{2}\tilde{a}(0)^T \Gamma^{-1}\tilde{a}(0) + \frac{\theta}{2\gamma}\tilde{\vartheta}(0)^2 + \frac{1}{2\eta}\tilde{D}(0)^2\right)^{\frac{1}{2n}}$$

$$(7.59)$$

$$\triangle\triangle\triangle$$

Remark 7.4. From Theorem 7.2 the following conclusions can also be obtained:

- The transient performance depends on the initial estimate errors $\tilde{\vartheta}(0)$, $\tilde{a}(0)$, $\tilde{D}(0)$ and the explicit design parameters.
- The bound for $\| x(t) - y_r(t) \|_2$ is an explicit function of design parameters and thus computable. We can decrease the effects of the initial error estimates on the transient performance by increasing the adaptation gains c_1, γ, η and Γ.

Remark 7.5. To further improve system performance such as the tracking error, especially in the case without using sign functions, it is worthy to take the system hysteresis into account in the controller design, instead of only considering its effect like bounded disturbances.

7.2.3   An Illustrative Example

In this section, we illustrate the state feedback methodologies on the same example system in [43] which is described as:

$$\dot{x} = a\frac{1 - e^{-x(t)}}{1 + e^{-x(t)}} + bu(t), \quad u(t) = B(\omega(t)) \qquad (7.60)$$

where $u(t)$ represents the output of the the backlash described by (7.2). The actual parameter values are $b = 1$ and $a = 1$. Without control, i.e. $u(t) = 0$, (7.60) is unstable, because $\dot{x} = \frac{1 - e^{-x(t)}}{1 + e^{-x(t)}} > 0$ for $x > 0$, and $\dot{x} < 0$ for $x < 0$. The objective is to control the system state x to follow a desired trajectory $y_r(t) = 12.5\sin(2.3t)$ as in [43].

Two adaptive backstepping schemes are used.

In the simulation of Scheme I, the robust adaptive control law (7.17)-(7.21) was used, taking $c_1 = 10, \gamma = \Gamma = \eta = 0.4$. The initial values are chosen to $\hat{\vartheta}(0) = 0.8/3, \hat{a}(0) = 1.5, \hat{D}(0) = 2, x(0) = 1.05$ and $w(0) = 0$ which are the same as in [43]. The simulation results presented in Figure 7.1 and Figure 7.2 are system tracking error and input. The effectiveness of adaptive Scheme I is demonstrated by the fact that the tracking error is reduced to zero after a few periods of the reference input as shown in Figure 7.1.

In the simulation of Scheme II by using the robust adaptive control law (7.48)-(7.52), we choose $c_1, \gamma, \eta, \Gamma$ and the initial values to be same as above and $\delta_1 = 0.1$. The simulation results presented in Figure 7.3 and Figure 7.4 are system tracking error and input. The effectiveness of adaptive Scheme II is also

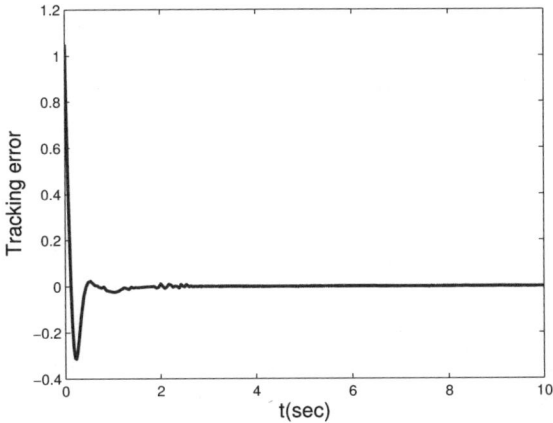

Fig. 7.1. Tracking error-Control Scheme I

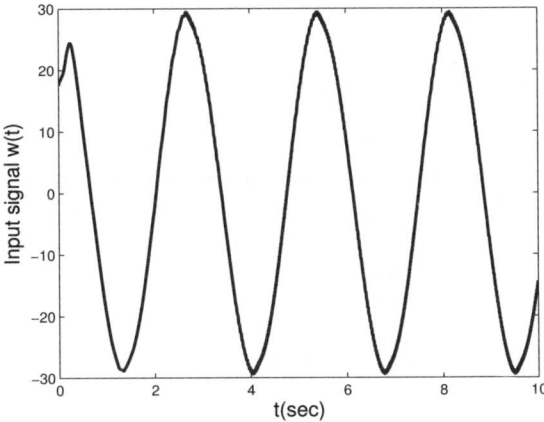

Fig. 7.2. Control signal $w(t)$-Control Scheme I

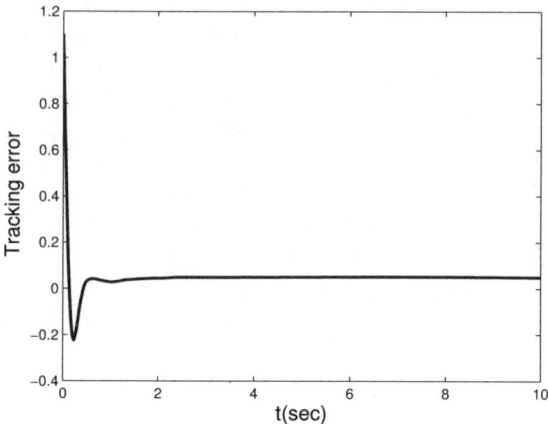

Fig. 7.3. Tracking error-Control Scheme II

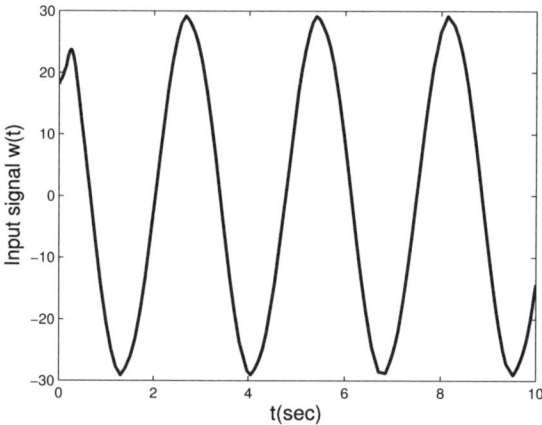

Fig. 7.4. Control signal $w(t)$-Control Scheme II

demonstrated by the fact that the tracking error is reduced to $\delta_1 = 0.1$ after a few periods of the reference input as shown in Figure 7.3.

As a conclusion, all the results verify our theoretical findings and show the effectiveness of the control schemes.

7.3 Output Feedback Control

In this section, a new scheme is proposed to address an output feedback control problem: control of a class of uncertain linear systems preceded by unknown backlash nonlinearity, where the backlash is modelled by a differential equation.

The controller designed by using backstepping technique consists of a new robust control law and a new estimator to estimate the unknown parameters as in [121]. For the implementation of the controller, no knowledge is assumed on the bounds of unknown system parameters and the effect contributed by the backlash. It is shown that all the signals are bounded. A bound for the truncated L_2 norm of the tracking error is obtained as a function of design parameters. Simulation studies also verify the effectiveness of the proposed scheme.

7.3.1   Plant Model

The class of single-input single-output linear systems is given by

$$y(s) = \frac{B(s)}{A(s)}u(s) = \frac{\bar{b}_m s^m + \ldots + \bar{b}_1 s + \bar{b}_0}{s^n + a_{n-1}s^{n-1} + \ldots + a_1 s + a_0}u(s) \qquad (7.61)$$

$$\frac{du}{dt} = \alpha \left| \frac{dw}{dt} \right| (cw - u) + B_1 \frac{dw}{dt} \qquad (7.62)$$

where the coefficients a_i and $\bar{b}_i$ are constant but unknown, u denotes a backlash nonlinearity, w is the design controller. (7.62) is treated as in previous section.

The control objective is for the system output y to asymptotically track a reference signal $y_r(t)$. Regarding the system and the reference signal, the following assumptions are made:

Assumption 1. The plant is minimum phase, i.e., the polynomial $B(s) = \bar{b}_m s^m + \ldots + \bar{b}_1 s + \bar{b}_0$ is Hurwitz.

Assumption 2. The relative degree ($\rho = n - m$) and an upper bound for the plant order (n) are known.

Assumption 3. The reference signal $y_r(t)$ and its first ρ derivatives are known and bounded, and, in addition, $y_r^{(\rho)}(t)$ is piecewise continuous.

Assumption 4. The sign of high-frequency gain (sign(b_m)), where $b_m = \bar{b}_m c$, is known.

7.3.2   State Estimation Filters

We start by representing the plant (7.61) as in the observer canonical form

$$\dot{x} = Ax - ya + \begin{bmatrix} 0_{(\rho-1)\times 1} \\ b \end{bmatrix} w + D(t) \qquad (7.63)$$

$$y = e_1^T x$$

where

$$A = \begin{bmatrix} 0 & & \\ \vdots & I_{n-1} & \\ 0 \ldots 0 & \end{bmatrix}, \quad b = \begin{bmatrix} \bar{b}_m c \\ \vdots \\ \bar{b}_0 c \end{bmatrix} = \begin{bmatrix} b_m \\ \vdots \\ b_0 \end{bmatrix}, \quad D(t) = \begin{bmatrix} 0_{(\rho-1)\times 1} \\ \bar{b}_m d(t) \\ \vdots \\ \bar{b}_0 d(t) \end{bmatrix}$$

$$a = [a_{n-1}, \ \ldots \ a_0]^T$$

In order to proceed, we rewrite (7.63) as

$$\dot{x} = Ax + F(y,w)^T\theta + D(t) \tag{7.64}$$
$$y = e_1^T x$$

where

$$F(y,w)^T = \left[\begin{bmatrix} 0_{(\rho-1)\times(m+1)} \\ I_{m+1} \end{bmatrix} w, \ -I_n y \right], \tag{7.65}$$

and the $p = n + m + 1-$dimensional parameter vector θ is defined by

$$\theta = \begin{bmatrix} b \\ a \end{bmatrix}. \tag{7.66}$$

For state estimation, we use the following filters

$$\dot{\lambda} = A_0\lambda + e_n w \tag{7.67}$$
$$\dot{\eta} = A_0\eta + e_n y \tag{7.68}$$
$$\Omega^T = [v_m, \ldots, v_1, v_0, \Xi] \tag{7.69}$$
$$v_j = A_0^j \lambda, \quad j = 0, \ldots, m \tag{7.70}$$
$$\Xi = -[A_0^{n-1}\eta, \ldots, A_0\eta, \eta] \tag{7.71}$$
$$\xi = -A_0^n \eta \tag{7.72}$$

where the vector $k = [k_1, \ldots, k_n]^T$ is chosen so that the matrix $A_0 = A - ke_1^T$ is Hurwitz. Hence there exists a P such that $PA_0 + A_0^T P = -2I$, $P = P^T > 0$. With these designed filters our state estimate is

$$\hat{x} = \xi + \Omega^T\theta \tag{7.73}$$

and the state estimation error $\epsilon = x - \hat{x}$ satisfies

$$\dot{\epsilon} = A_0\epsilon + D(t) \tag{7.74}$$

Let $V_\epsilon = \epsilon^T P\epsilon$. It can be shown that

$$\dot{V_\epsilon} = \epsilon^T(PA_0 + A_0^T P)\epsilon + 2\epsilon^T PD(t)$$
$$\leq -\epsilon^T\epsilon + \| PD(t) \|^2 \tag{7.75}$$

Then system (7.64) can be expressed as

$$\dot{y} = b_m v_{m,2} + \xi_2 + \bar{\delta}^T\theta + \epsilon_2 \tag{7.76}$$
$$\dot{v}_{m,i} = v_{m,i+1} - k_i v_{m,1}, \quad i = 2, \ldots, \rho - 1 \tag{7.77}$$
$$\dot{v}_{m,\rho} = v_{m,\rho+1} - k_\rho v_{m,1} + w \tag{7.78}$$

where

$$\delta = [v_{m,2}, v_{m-1,2}, \ldots, v_{0,2}, \Xi_{(2)} - ye_1^T]^T \tag{7.79}$$

$$\bar{\delta} = [0, v_{m-1,2}, \ldots, v_{0,2}, \Xi_{(2)} - ye_1^T]^T \tag{7.80}$$

and $v_{i,2}, \epsilon_2, \xi_2$ denote the second entries of v_i, ϵ, ξ respectively. All of its states are available for feedback.

7.3.3 Design of Adaptive Controllers

Design Procedure

As usual in backstepping approach, the following change of coordinates is made.

$$z_1 = y - y_r \tag{7.81}$$

$$z_i = v_{m,i} - \hat{\vartheta} y_r^{(i-1)} - \alpha_{i-1}, \ i = 2, 3, \ldots, \rho \tag{7.82}$$

where $\hat{\vartheta}$ is an estimate of $\vartheta = 1/b_m$ and α_{i-1} is the virtual control at the *ith step* and will be determined in later discussion.

To illustrate the backstepping procedures, only the first and the last steps of the design, i.e. *steps 1 and n* below, are elaborated in details.

• *Step 1:* We start with the equation for the tracking error z_1 obtained from (7.64) and (7.81) that

$$\dot{z}_1 = b_m v_{m,2} + \xi_2 + \bar{\delta}^T \theta + \epsilon_2 - \dot{y}_r \tag{7.83}$$

By substituting (7.82) for $i = 2$ into (7.83) and using $\tilde{\vartheta} = \frac{1}{b_m} - \hat{\vartheta}$, we get

$$\dot{z}_1 = b_m \alpha_1 + \xi_2 + \bar{\delta}^T \theta + \epsilon_2 - b_m \tilde{\vartheta} \dot{y}_r + b_m z_2 \tag{7.84}$$

We design the virtual control law α_1 as

$$\alpha_1 = \hat{\vartheta} \bar{\alpha}_1 \tag{7.85}$$

$$\bar{\alpha}_1 = -c_1 z_1 - d_1 z_1 - \xi_2 - \bar{\delta}^T \hat{\theta} \tag{7.86}$$

where c_1 and d_1 are positive design parameters, $\hat{\theta}$ is the estimate of θ. From (7.84) and (7.85) we have

$$\dot{z}_1 = -c_1 z_1 - d_1 z_1 + \epsilon_2 + \bar{\delta}^T \tilde{\theta} - b_m (\dot{y}_r + \bar{\alpha}_1) \tilde{\vartheta} + b_m z_2$$
$$= -(c_1 + d_1) z_1 + \epsilon_2 + (\delta - \hat{\vartheta}(\dot{y}_r + \bar{\alpha}_1) e_1)^T \tilde{\theta} - b_m (\dot{y}_r + \bar{\alpha}_1) \tilde{\vartheta} + \hat{b}_m z_2 \tag{7.87}$$

where $\tilde{\theta} = \theta - \hat{\theta}$, we have

$$b_m \alpha_1 = b_m \hat{\vartheta} \bar{\alpha}_1 = \bar{\alpha}_1 - b_m \tilde{\vartheta} \bar{\alpha}_1 \tag{7.88}$$

$$\bar{\delta}^T \tilde{\theta} + b_m z_2 = \bar{\delta}^T \tilde{\theta} + \tilde{b}_m z_2 + \hat{b}_m z_2$$
$$= \bar{\delta}^T \tilde{\theta} + (v_{m,2} - \hat{\vartheta} \dot{y}_r - \alpha_1) e_1^T \tilde{\theta} + \hat{b}_m z_2$$
$$= (\delta - \hat{\vartheta}(\dot{y}_r + \bar{\alpha}_1) e_1)^T \tilde{\theta} + \hat{b}_m z_2 \tag{7.89}$$

We consider the Lyapunov function

$$V_1 = \frac{1}{2}z_1^2 + \frac{1}{2}\tilde{\theta}^T \Gamma^{-1}\tilde{\theta} + \frac{|b_m|}{2\gamma}\tilde{\vartheta}^2 + \frac{1}{2d_1}V_\epsilon \qquad (7.90)$$

where Γ is a positive definite design matrix and γ is a positive design parameter. We examine the derivative of V_1

$$\dot{V}_1 \le z_1\dot{z}_1 - \tilde{\theta}^T \Gamma^{-1}\dot{\hat{\theta}} - \frac{|b_m|}{\gamma}\tilde{\vartheta}\dot{\hat{\vartheta}} - \frac{1}{2d_1}\epsilon^T\epsilon + \frac{1}{2d_1} \parallel PD(t) \parallel^2$$

$$\le -c_1z_1^2 + \hat{b}_m z_1 z_2 - d_1 z_1^2 + z_1\epsilon_2 - |b_m|\tilde{\vartheta}\frac{1}{\gamma}[\gamma\text{sign}(b_m)(\dot{y}_r + \bar{\alpha}_1)z_1 + \dot{\hat{\vartheta}}]$$

$$+\tilde{\theta}^T\Gamma^{-1}[\Gamma(\delta - \hat{\vartheta}(\dot{y}_r + \bar{\alpha}_1)e_1)z_1 - \dot{\hat{\theta}}] - \frac{1}{2d_1}\epsilon^T\epsilon + \frac{1}{2d_1} \parallel PD(t) \parallel^2$$

$$(7.91)$$

Now we choose

$$\dot{\hat{\vartheta}} = -\gamma\text{sign}(b_m)(\dot{y}_r + \bar{\alpha}_1)z_1 - \gamma l_\vartheta(\hat{\vartheta} - \vartheta_0) \qquad (7.92)$$

$$\tau_1 = (\delta - \hat{\vartheta}(\dot{y}_r + \bar{\alpha}_1)e_1)z_1 \qquad (7.93)$$

where l_ϑ, ϑ_0 are positive design constants.

From the choice, the following useful property can be obtained:

$$l_\vartheta\tilde{\vartheta}(\hat{\vartheta} - \vartheta_0) = -l_\vartheta(\hat{\vartheta} - \vartheta)(\frac{1}{2}(\hat{\vartheta} - \vartheta) + \frac{1}{2}(\hat{\vartheta} + \vartheta) - \vartheta_0)$$

$$\le -\frac{1}{2}l_\vartheta\tilde{\vartheta}^2 + \frac{1}{2}l_\vartheta(\vartheta - \vartheta_0)^2 \qquad (7.94)$$

Then the following derivation for the derivative of V_1 can be carried out by using (7.92)-(7.94)

$$\dot{V}_1 \le -c_1 z_1^2 + \hat{b}_m z_1 z_2 - \frac{|b_m|}{2}l_\vartheta\tilde{\vartheta}^2 - \frac{1}{4d_1}\epsilon^T\epsilon + \frac{|b_m|}{2}l_\vartheta(\vartheta - \vartheta_0)^2$$

$$+\tilde{\theta}^T(\tau_1 - \Gamma^{-1}\dot{\hat{\theta}}) + \frac{1}{2d_1} \parallel PD(t) \parallel^2 \qquad (7.95)$$

Remark 7.6. Note that a new term $\gamma l_\vartheta(\hat{\vartheta} - \vartheta_0)$ is introduced in the parameter update law (7.92) compared with the traditional estimator using backstepping. This term is used to mitigate the backlash effect for system stability as shown in later discussion.

• *Step i (i = 2, ..., ρ):* Choose virtual control laws

$$\alpha_2 = -\hat{b}_m z_1 - \left[c_2 + d_2\left(\frac{\partial\alpha_1}{\partial y}\right)^2\right]z_2 + \beta_2 + \frac{\partial\alpha_1}{\partial\hat{\theta}}\Gamma\tau_2 + \frac{\partial\alpha_1}{\partial\hat{\theta}}\Gamma l_\theta(\hat{\theta} - \theta_0) \quad (7.96)$$

$$\alpha_i = -z_{i-1} - \left[c_i + d_i\left(\frac{\partial\alpha_{i-1}}{\partial y}\right)^2\right]z_i + \beta_i + \frac{\partial\alpha_{i-1}}{\partial\hat{\theta}}\Gamma\tau_i + \frac{\partial\alpha_{i-1}}{\partial\hat{\theta}}\Gamma l_\theta(\hat{\theta} - \theta_0)$$

$$-\left(\sum_{k=2}^{i-1} z_k\frac{\partial\alpha_{k-1}}{\partial\hat{\theta}}\right)\Gamma\frac{\partial\alpha_{i-1}}{\partial y}\delta \qquad (7.97)$$

where $c_i, i = 3, ..., \rho$ are positive design parameters, and

$$\tau_i = \tau_{i-1} - \frac{\partial \alpha_{i-1}}{\partial y} \delta z_i \tag{7.98}$$

$$\beta_i = \frac{\partial \alpha_{i-1}}{\partial y}(\xi_2 + \delta^T \hat{\theta}) + \frac{\partial \alpha_{i-1}}{\partial \eta}(A_0 \eta + e_n y) + k_i v_{m,1} + \sum_{j=1}^{i-1} \frac{\partial \alpha_{i-1}}{\partial y_r^{(j-1)}} y_r^{(j)}$$

$$+ (y_r^{(i-1)} + \frac{\partial \alpha_{i-1}}{\partial \hat{\vartheta}})\dot{\hat{\vartheta}} + \sum_{j=1}^{m+i-1} \frac{\partial \alpha_{i-1}}{\partial \lambda_j}(-k_j \lambda_1 + \lambda_{j+1}) \tag{7.99}$$

Then the adaptive controller and parameter update law are finally given by

$$w = \alpha_\rho - v_{m,\rho+1} + \hat{\vartheta} y_r^{(\rho)} \tag{7.100}$$

$$\dot{\hat{\theta}} = \Gamma \tau_\rho + \Gamma l_\theta(\hat{\theta} - \theta_0) \tag{7.101}$$

where l_θ and θ_0 are positive design constants.

Remark 7.7. Again note that $\frac{\partial \alpha_{i-1}}{\partial \hat{\theta}} \Gamma l_\theta(\hat{\theta} - \theta_0)$ and $\Gamma l_\theta(\hat{\theta} - \theta_0)$ are added in the virtual control (7.97) and in the parameter update law $\dot{\hat{\theta}}$ (7.101), respectively. These terms are employed to ensure system stability and its performance in the presence of backlash effects as shown below.

Stability Analysis

We define the final Lyapunov function V_ρ as

$$V_\rho = \sum_{i=1}^{\rho} \frac{1}{2} z_i^2 + \frac{1}{2} \tilde{\theta}^T \Gamma^{-1} \tilde{\theta} + \frac{|b_m|}{2\gamma} \tilde{\vartheta}^2 + \sum_{i=1}^{\rho} \frac{1}{2d_i} V_\epsilon \tag{7.102}$$

Note that

$$\Gamma \tau_{i-1} - \dot{\hat{\theta}} = \Gamma \tau_{i-1} - \Gamma \tau_i + \Gamma \tau_i - \dot{\hat{\theta}}$$

$$= \Gamma \frac{\partial \alpha_{i-1}}{\partial y} \delta z_i + (\Gamma \tau_i - \dot{\hat{\theta}}) \tag{7.103}$$

$$l_\theta \tilde{\theta}^T(\hat{\theta} - \theta_0) \leq -\frac{1}{2} l_\theta \parallel \tilde{\theta} \parallel^2 + \frac{1}{2} l_\theta \parallel \theta - \theta_0 \parallel^2 \tag{7.104}$$

From (7.97) - (7.101), the derivative of the last Lyapunov function satisfies

$$\dot{V}_\rho = \sum_{i=1}^{\rho} z_i \dot{z}_i - \tilde{\theta}^T \Gamma^{-1} \dot{\hat{\theta}} - \frac{|b_m|}{\gamma} \tilde{\vartheta} \dot{\hat{\vartheta}} + \sum_{i=1}^{\rho} \frac{1}{2d_i} \dot{V}_\epsilon$$

$$\leq -\sum_{i=1}^{\rho} c_i z_i^2 - \tilde{\theta}^T \Gamma^{-1}(\dot{\hat{\theta}} - \Gamma \tau_\rho) - \sum_{i=1}^{\rho} \frac{1}{4d_i} \epsilon^T \epsilon - \frac{|b_m|}{2} l_\vartheta \tilde{\vartheta}^2 + \frac{|b_m|}{2} l_\vartheta(\vartheta - \vartheta_0)^2$$

$$+ (\sum_{k=2}^{\rho} z_k \frac{\partial \alpha_{k-1}}{\partial \hat{\theta}})[\Gamma \tau_\rho + \Gamma l_\theta(\hat{\theta} - \theta_0) - \dot{\hat{\theta}}] + \sum_{i=1}^{\rho} \frac{1}{2d_i} \parallel PD(t) \parallel^2$$

$$\leq -\sum_{i=1}^{\rho} c_i z_i^2 - \frac{1}{2} l_\theta \parallel \tilde{\theta} \parallel^2 + \frac{1}{2} l_\theta \parallel \theta - \theta_0 \parallel^2 + \sum_{i=1}^{\rho} \frac{1}{2d_i} \parallel PD(t) \parallel^2$$

$$- \frac{|b_m|}{2} l_\vartheta \tilde{\vartheta}^2 + \frac{|b_m|}{2} l_\vartheta (\vartheta - \vartheta_0)^2 - \sum_{i=1}^{\rho} \frac{1}{4d_i} \epsilon^T \epsilon$$

$$\leq -\sum_{i=1}^{\rho} c_i z_i^2 - \frac{|b_m|}{2} l_\vartheta \tilde{\vartheta}^2 - \frac{1}{2} l_\theta \parallel \tilde{\theta} \parallel^2 - \sum_{i=1}^{\rho} \frac{1}{4d_i} \epsilon^T \epsilon + M^* \qquad (7.105)$$

where

$$M^* = M + \sum_{i=1}^{\rho} \frac{1}{2d_i} \parallel P \parallel^2 D_{max}^2 \qquad (7.106)$$

$$M = \frac{|b_m|}{2} l_\vartheta (\vartheta - \vartheta_0)^2 + \frac{1}{2} l_\theta \parallel \theta - \theta_0 \parallel^2 \qquad (7.107)$$

In (7.106), D_{max} denotes the bound of $D(t)$ which may not be available. Notice that

$$-\sum_{i=1}^{\rho} c_i z_i^2 - \frac{|b_m|}{2} l_\vartheta \tilde{\vartheta}^2 - \frac{1}{2} l_\theta \parallel \tilde{\theta} \parallel^2 - \sum_{i=1}^{\rho} \frac{1}{4d_i} \epsilon^T \epsilon \leq -f_- \bar{V}_\rho \qquad (7.108)$$

and

$$V_\rho = \sum_{i=1}^{\rho} \frac{1}{2} z_i^2 + \frac{1}{2} \tilde{\theta}^T \Gamma^{-1} \tilde{\theta} + \frac{|b_m|}{2\gamma} \tilde{\vartheta}^2 + \sum_{i=1}^{\rho} \frac{1}{2d_i} V_\epsilon \leq f_+ \bar{V}_\rho \qquad (7.109)$$

where

$$\bar{V}_\rho = \sum_{i=1}^{\rho} z_i^2 + \tilde{\theta}^T \tilde{\theta} + \tilde{\vartheta}^2 + \sum_{i=1}^{\rho} \epsilon^T \epsilon \qquad (7.110)$$

$$f_- = min\{c_i, \frac{|b_m|}{2} l_\vartheta, \frac{1}{2} l_\theta, \frac{1}{4d_i}\} \qquad (7.111)$$

$$f_+ = max\{\frac{1}{2}, \frac{1}{2}\lambda_{max}(\Gamma), \frac{|b_m|}{2\gamma}, \frac{1}{2d_i}\lambda_{max}(P)\} \qquad (7.112)$$

where $\lambda_{max}(P)$ and $\lambda_{max}(\Gamma)$ are the maximum eigenvalues of P and Γ, respectively. Therefore, from (7.105) we obtain

$$\dot{V}_\rho \leq -f^* V_\rho + M^* \qquad (7.113)$$

By direct integrations of the differential inequality (7.113), we have

$$V_\rho \leq V_\rho(0) e^{-f^* t} + \frac{M^*}{f^*} (1 - e^{-f^* t}) \leq V_\rho(0) + \frac{M^*}{f^*} \qquad (7.114)$$

where $f^* = f^-/f^+$. This shows that V_ρ is uniformly bounded. Thus $z_i, \hat{\vartheta}, \hat{\theta}$ and ϵ are bounded. Since z_1 and y_r are bounded, y is also bounded. Then from (7.67) and (7.68) we can show that λ, η and x are bounded as in Chapter 2. Therefore boundedness of all signals in the system is ensured as formally stated in the following Theorem.

Theorem 7.1. *Consider the closed-loop adaptive system consisting of the plant (7.61) under Assumptions 1-4, the controller (7.100), the estimator (7.92), (7.101), and the filters (7.67) and (7.68). All the signals in the system are globally uniformly bounded.*

We now derive a bound for the vector $z(t)$ where $z(t) = [z_1, z_2, \ldots, z_\rho]^T$. Firstly, the following definitions are made.

$$c_0 = min_{1 \leq i \leq \rho} c_i, \quad d_0 = \sum_{i=1}^{\rho} \frac{1}{2d_i} \tag{7.115}$$

$$\| z \|_{[0,T]} = \sqrt{\frac{1}{T} \int_0^T z(t)^2 dt} \tag{7.116}$$

Then from (7.105), we have

$$\dot{V}_\rho \leq -c_0 \| z \|^2 + M^* \tag{7.117}$$

Integrating both sides, we obtain

$$\| z \|_{[0,T]} \leq \frac{1}{c_0} [\frac{|V_\rho(0) - V_\rho(T)|}{T} + M + \frac{1}{T} d_0 \| P \|^2 \int_0^T D(t)^2 dt] \tag{7.118}$$

On the other hand, from (7.113), we have

$$\frac{|V_\rho(0) - V_\rho(T)|}{T}$$

$$\leq \frac{1 - e^{-f^* T}}{T} \frac{M}{f^*} + V_\rho(0)) + \frac{1}{T} d_0 \| P \|^2 \int_0^T e^{-f^*(T-t)} D(t)^2 dt$$

$$\leq M + f^* V_\rho(0) + \frac{\| P \|^2}{T} d_0 \int_0^T D(t)^2 dt, \quad \forall T \geq 0, \tag{7.119}$$

where we have used the fact that $e^{-f^*(T-t)} \leq 1$ and $\frac{1-e^{-f^* T}}{T} \leq f^*$.
By setting $z_i(0) = 0$, the initial value of the Lyapunov function is

$$V_\rho(0) = \frac{1}{2} \| \tilde{\theta}(0) \|_{\Gamma^{-1}}^2 + \frac{|b_m|}{2\gamma} |\tilde{\vartheta}(0)|^2 + d_0 |\epsilon(0)|_P^2 \tag{7.120}$$

Using (7.111) and (7.112), the fact that $f^*/c_0 \leq 2$, the a bound resulting from (7.118) - (7.120) is given by

$$\| z \|_{[0,T]} \leq \| \tilde{\theta}(0) \|_{\Gamma^{-1}}^2 + \frac{|b_m|}{\gamma} |\tilde{\vartheta}(0)|^2 + \frac{2}{T c_0} d_0 \| P \|^2 \int_0^T D(t)^2 dt$$

$$+ 2d_0 |\epsilon(0)|_P^2 + \frac{1}{c_0} (|b_m| l_\vartheta (\vartheta - \vartheta_0)^2 + l_\theta \| \theta - \theta_0 \|^2) \tag{7.121}$$

Remark 7.8. Regarding the above bound, the following conclusions can be drawn:

- The transient performance in the sense of truncated norm given in (7.121) depends on the initial estimate errors $\tilde{\theta}(0), \tilde{\vartheta}(0)$ and $\epsilon(0)$. The closer the initial estimates to the true values, the better the transient performance.
- This bound can also be systematically reduced by increasing Γ, γ, c_0 and decreasing $d_0, l_\vartheta, l_\theta$.

7.3.4   Extension to Nonlinear Systems

We consider the following class of nonlinear systems

$$\dot{x} = Ax + \psi(y) + \Phi^T(y)\theta + \bar{b}u \tag{7.122}$$

$$\frac{du}{dt} = \alpha \left| \frac{dw}{dt} \right| (cw - u) + B_1 \frac{dw}{dt} \tag{7.123}$$

$$y = e_1^T x \tag{7.124}$$

where

$$A = \begin{bmatrix} 0 & I_{n-1} \\ 0 & 0 \end{bmatrix}, \bar{b} = \begin{bmatrix} 0_{n-m-1} \\ \bar{b}_m \\ \vdots \\ \bar{b}_0 \end{bmatrix} \tag{7.125}$$

$$\psi(y) = \begin{bmatrix} \psi_1(y) \\ \vdots \\ \psi_n(y) \end{bmatrix}, \Phi^T(y) = \begin{bmatrix} \phi_1^T(y) \\ \vdots \\ \phi_n^T(y) \end{bmatrix}, e_1 = \begin{bmatrix} 1 \\ 0 \\ \vdots \\ 0 \end{bmatrix} \tag{7.126}$$

where $x \in R^n$, $u \in R^1$ and $y \in R^1$ are the state, input and output of the system, respectively, $\theta \in R^r$ and $\bar{b} \in R^n$ are unknown constant vectors, $\phi_i(y) \in R^r$ and $\psi_i(y) \in R$ are known smooth functions, u denotes a backlash nonlinearity, w is the input to the backlash element, and backlash (7.123) is treated as in previous section.

The control objective is to design an output feedback control law for $w(t)$ to ensure that all closed-loop signals are bounded and the plant output $y(t)$ tracks a given reference signal $y_r(t)$ under the following assumptions:

Assumption 1: The sign of b_m is known and the relative degree $\rho = n - m$ is fixed and known. The polynomial $B(s) = b_m s^m + \ldots + b_1 s + b_0$ is stable.

Assumption 2. The reference signal y_r and its ρth order derivatives are continuous, known and bounded.

Then we rewritten system (7.122) as follows

$$\dot{x} = Ax + \psi(y) + \Phi^T(y)\theta + bw + D(t) \tag{7.127}$$

where

$$b = \begin{bmatrix} 0_{n-m-1} \\ \bar{b}_m c \\ \vdots \\ \bar{b}_0 c \end{bmatrix} = \begin{bmatrix} 0_{n-m-1} \\ b_m \\ \vdots \\ b_0 \end{bmatrix}, \ D(t) = \begin{bmatrix} 0_{(\rho-1)\times 1} \\ \bar{b}_m d(t) \\ \vdots \\ \bar{b}_0 d(t) \end{bmatrix} \tag{7.128}$$

We employ the filters as follows

$$\hat{x}(t) = \xi_0 + \xi^T \theta + \sum_{i=0}^{m} b_i v_i \tag{7.129}$$

$$\dot{\lambda} = A_0 \lambda + e_n w, \ i = 0, 1, \ldots, m \tag{7.130}$$

$$\dot{\xi}_0 = A_0 \xi_0 + ky + \psi(y) \tag{7.131}$$

$$\dot{\xi}^T = A_0 \xi^T + \Phi^T(y) \tag{7.132}$$

$$v_i = A_0^i \lambda, \ i = 0, \ldots, m \tag{7.133}$$

where $k = [k_1, \ldots, k_n]^T$ such that all eigenvalues of $A_0 = A - ke_1^T$ are at some desired stable locations. With the designed filters, the state estimation error $\epsilon = x - \hat{x}$ satisfies

$$\dot{\epsilon} = A_0 \epsilon + D(t) \tag{7.134}$$

Then system can be expressed as

$$\dot{y} = b_m v_{m,2} + \psi_1(y) + \xi_2 + \delta^T \Theta + \epsilon_2 \tag{7.135}$$

$$\dot{v}_{m,i} = v_{m,i+1} - k_i v_{m,1}, \quad i = 2, \ldots, \rho - 1 \tag{7.136}$$

$$\dot{v}_{m,\rho} = v_{m,\rho+1} - k_\rho v_{m,1} + w \tag{7.137}$$

where

$$\Theta = [b_m, \ldots, b_0, \theta^T]^T \tag{7.138}$$

$$\delta = [v_{m,2}, v_{m-1,2}, \ldots, v_{0,2}, \xi_2^T + \phi_1^T]^T \tag{7.139}$$

$$\bar{\delta} = [0, v_{m-1,2}, \ldots, v_{0,2}, \xi_2^T + \phi_1^T]^T \tag{7.140}$$

and $v_{i,2}, \epsilon_2, \xi_2$ denote the second entries of v_i, ϵ, ξ respectively. All of its states are available for feedback.

The controller design is achieved by following the design procedures in Chapter 2 and are summarized in Table 7.1, $\hat{\Theta}$ where $c_i, i = 1, \ldots, \rho$ are positive design parameters, $\hat{\Theta}$ is the estimate of Θ, $\hat{\vartheta}$ is an estimate of $\vartheta = 1/b_m$, l_ϑ, ϑ_0 and l_Θ are positive design constants, and Θ_0 is a positive constant vectors. The result is stated in the following theorem.

Theorem 7.2. *Consider the closed-loop nonlinear system consisting of the plant (7.122) under Assumptions 1-2, the controller (7.143), the estimator (7.148), (7.149), and the filters (7.130), (7.131) and (7.132). All the signals in the system are globally uniformly bounded.*

Table 7.1. Adaptive Backstepping Controller

Change of coordinates:	

$$z_1 = y - y_r \tag{7.141}$$

$$z_i = v_{m,i} - \hat{\vartheta} y_r^{(i-1)} - \alpha_{i-1}, \ i = 2, 3, \ldots, \rho \tag{7.142}$$

Adaptive Control Laws:

$$w = \alpha_\rho - v_{m,\rho+1} + \hat{\vartheta} y_r^{(\rho)} \tag{7.143}$$

$$\alpha_1 = \hat{\vartheta} \bar{\alpha}_1 \tag{7.144}$$

$$\bar{\alpha}_1 = -c_1 z_1 - d_1 z_1 - \psi_1(y) - \xi_2 - \bar{\delta}^T \hat{\Theta} \tag{7.145}$$

$$\alpha_2 = -\hat{b}_m z_1 - \left[c_2 + d_2 \left(\frac{\partial \alpha_1}{\partial y} \right)^2 \right] z_2 + \beta_2 + \frac{\partial \alpha_1}{\partial \hat{\Theta}} \Gamma \tau_2$$

$$+ \frac{\partial \alpha_1}{\partial \hat{\Theta}} \Gamma l_\Theta (\hat{\Theta} - \Theta_0) \tag{7.146}$$

$$\alpha_i = -z_{i-1} - \left[c_i + d_i \left(\frac{\partial \alpha_{i-1}}{\partial y} \right)^2 \right] z_i + \beta_i + \frac{\partial \alpha_{i-1}}{\partial \hat{\Theta}} \Gamma \tau_i$$

$$+ \frac{\partial \alpha_{i-1}}{\partial \hat{\Theta}} \Gamma l_\Theta (\hat{\Theta} - \Theta_0) - \left(\sum_{k=2}^{i-1} z_k \frac{\partial \alpha_{k-1}}{\partial \hat{\Theta}} \right) \Gamma \frac{\partial \alpha_{i-1}}{\partial y} \delta \tag{7.147}$$

Parameter Update Laws:

$$\dot{\hat{\vartheta}} = -\gamma \text{sign}(b_m)(\dot{y}_r + \bar{\alpha}_1) z_1 - \gamma l_\vartheta (\hat{\vartheta} - \vartheta_0) \tag{7.148}$$

$$\dot{\hat{\Theta}} = \Gamma \tau_\rho + \Gamma l_\Theta (\hat{\Theta} - \Theta_0) \tag{7.149}$$

Tuning functions:

$$\tau_1 = (\delta - \hat{\vartheta}(\dot{y}_r + \bar{\alpha}_1) e_1) z_1 \tag{7.150}$$

$$\tau_i = \tau_{i-1} - \frac{\partial \alpha_{i-1}}{\partial y} \delta z_i \tag{7.151}$$

$$\beta_i = \frac{\partial \alpha_{i-1}}{\partial y} (\xi_2 + \delta^T \hat{\Theta}) + \frac{\partial \alpha_{i-1}}{\partial \eta} (A_0 \eta + e_n y) + k_i v_{m,1} + \sum_{j=1}^{i-1} \frac{\partial \alpha_{i-1}}{\partial y_r^{(j-1)}} y_r^{(j)}$$

$$+ \left(y_r^{(i-1)} + \frac{\partial \alpha_{i-1}}{\partial \hat{\vartheta}} \right) \dot{\hat{\vartheta}} + \sum_{j=1}^{m+i-1} \frac{\partial \alpha_{i-1}}{\partial \lambda_j} (-k_j \lambda_1 + \lambda_{j+1}) \tag{7.152}$$

7.3.5 Simulation Studies

Design Example 1: Output Feedback Control

We illustrate the output feedback method on a simple linear systems

$$y(s) = \frac{\bar{b}}{s^2 + as} u(s) \tag{7.153}$$

$$u(t) = B(w(t)) \tag{7.154}$$

where $\bar{b} = 1, a = 1.2$, $B(w)$ is the backlash described by (7.2) with parameters $\alpha = 1, c = 3.1635, B_1 = 0.345$. These parameters are not needed to be known in the controller design. The objective is to control the system output y to follow a desired trajectory $y_r(t) = 2\sin(2t)$. The filters from (7.67) and (7.68) are implemented as

$$\dot{\eta} = A_0\eta + e_2 y \tag{7.155}$$

$$\dot{\lambda} = A_0\lambda + e_2 w \tag{7.156}$$

$$\Xi = -[A_0\eta, \eta], \quad \xi = -A_0^2\eta \quad v = \lambda \tag{7.157}$$

$$A_0 = \begin{bmatrix} -k_1 & 1 \\ -k_2 & 0 \end{bmatrix} \tag{7.158}$$

The adaptive control laws α_1 (7.85), $w(t)$ (7.100), and parameter update laws $\hat{\vartheta}$ (7.92), $\hat{\theta}$ (7.101) were used, where $\hat{\vartheta}$ and $\hat{\theta}$ are estimates of $\vartheta = 1/\bar{b}c$ and $\theta = [\bar{b}c, a]^T$, respectively. The design parameters are chosen as $c_1 = 15, c_2 = 5, d_1 = d_2 = 0.1, \gamma = 2, \Gamma = I_2, l_\vartheta = 0.1, l_\theta = 0.1, k_1 = 6, k_2 = 8$. The initials are set $y(0) = 1, \hat{\vartheta}(0) = 0.2, \hat{\theta}(0) = [2, 0.5]^T$. The simulation results presented in the Figure 7.5 shows the system output y and the desired trajectory signal. Figure 7.6 shows the control signal $w(t)$. These simulation results verify that our proposed scheme is effective to cope with backlash nonlinearity.

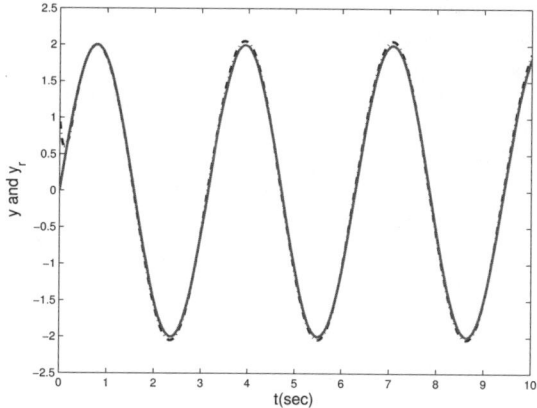

Fig. 7.5. Output y(dashed) and trajectory y_r(solid line)

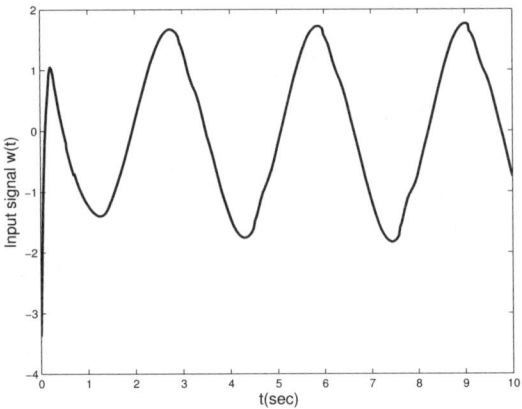

Fig. 7.6. Control signal $w(t)$

Design Example 2: Valve Control Mechanism

In this section, we illustrate our proposed scheme on valve control mechanism shown in Figure 7.7 as in Chapter 6, where the backlash is in the control inflow $u = B(v)$ and the output is the liquid level h. The transfer function $h(p)$ can be expressed as

$$h(p) = G(p)(u(p) - d(p)) = \frac{k}{p}(u(p) - d(p)) \qquad (7.159)$$

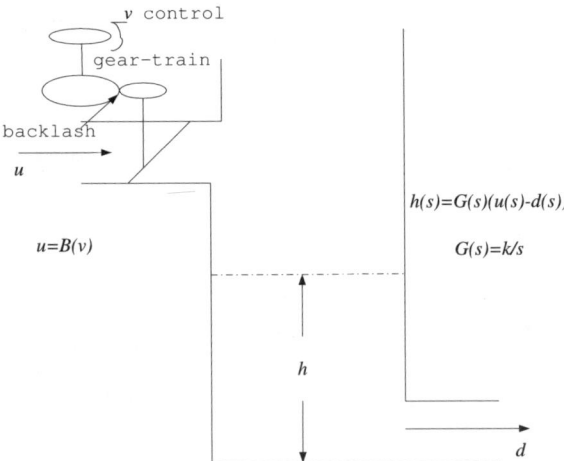

Fig. 7.7. Backlash in the valve control mechanism of a liquid tank

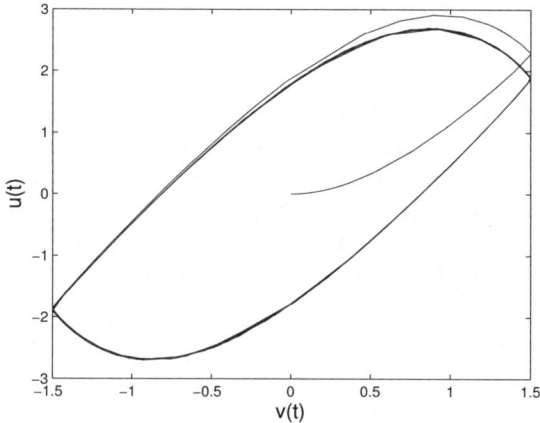

Fig. 7.8. Input backlash

where d is the uncontrolled outflow which is treated as a disturbance and $p = \frac{d}{dt}$. The backlash is expressed as follows

$$\frac{du}{dt} = \alpha \left| \frac{dv}{dt} \right| (cv - u) + B_1 \frac{dv}{dt} \tag{7.160}$$

where α, c and B_1 are constants. The true parameters are set as $k = 2, \alpha = 1, c = 3, B_1 = 0.2$. The objective is to control the liquid level to $10m$.

The adaptive control law is designed by Scheme II as follows

$$v = \hat{\vartheta} \bar{w} \tag{7.161}$$

$$\bar{w} = -c_1 (|z_1| - \delta_1) sg_1 - sg_1 \hat{D} \tag{7.162}$$

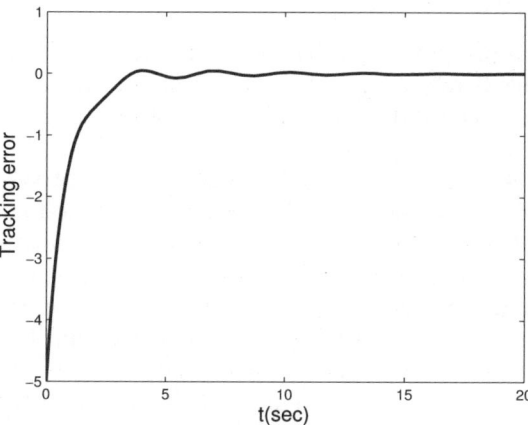

Fig. 7.9. Tracking error with controller designed by our proposed Scheme II

$$\dot{\hat{\vartheta}} = -\gamma \bar{w}(|z_1| - \delta_1)f_1 sg_1 \tag{7.163}$$

$$\dot{\hat{D}} = \eta(|z_1| - \delta_1)f_1 \tag{7.164}$$

where $z_1 = h - y_r, y_r = 2$, $\hat{D}$ is an estimate of D which is the bound of $kd_1(t) - kd(t)$. In the simulation, the design parameters are chosen as $c_1 = 2, \gamma = 1, \eta = 2, \delta_1 = 0.01$ and the initial value is chosen as $h(0) = 5, \hat{\vartheta}(0) = 0.15$. The disturbance is selected as $\sin(2t)$. Figure 7.8 is the response of input backlash. The simulation results presented in Figure 7.9 and Figure 7.10 show the tracking error of controlled liquid level and the input control obtained by the designed controller.

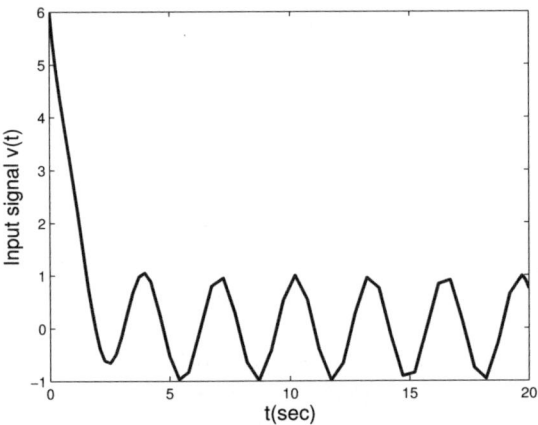

Fig. 7.10. Control signal with controller designed by our proposed Scheme II

7.4   Summary

In this chapter, we present two types of robust adaptive backstepping control algorithms: state feedback control of a class of nonlinear systems with unknown backlash and output feedback control of a class of linear systems with unknown backlash.

For state feedback control, two backstepping adaptive controller design schemes are developed. In the first scheme, a sign function is involved and this can ensure perfect tracking. To avoid possible chattering caused by the sign function, we propose an alternative smooth control law and the tracking error is still ensured to approach a prescribed bound in this case. The developed backstepping controls do not require the model parameters within known intervals and the knowledge on the bound of 'disturbance-like' term is not required. Besides showing global stability, we also give an explicit bound on the L_2 performance of the tracking error in terms of design parameters.

For output feedback control, an adaptive control scheme with certain modifications to the existing backstepping control design is proposed to achieve tracking. For the implementation of the controller, no a priori knowledge on the bounds of all the unknown parameters and the backlash effect is required. It is shown that adaptive control system is globally stable in the sense that all the signals are bounded. Also a truncated L_2 bound is derived for the tracking error as a function of the design parameters. Simulation results verify the effectiveness of the proposed schemes. We also extend the results to nonlinear system.

8 Inverse Control of Systems with Backlash Nonlinearity

We consider a class of uncertain dynamic nonlinear systems preceded by unknown backlash nonlinearity. The control design is achieved by introducing a smooth inverse function of the backlash and using it in the controller design with backstepping technique. For the design and implementation of the controller, no knowledge is assumed on the unknown system parameters. It is shown that the proposed controller not only can guarantee stability, but also transient performance.

8.1 Introduction

Backlash exists in a wide range of physical systems and devices, such as biology optics, electro-magnetism, mechanical actuators, electronic relay circuits and other areas. Such nonlinearity is usually poorly known and often limits system performance. Control of systems with backlash nonlinearity is an important area of control system research and typically challenging. For backlash nonlinearity, several adaptive control schemes have recently been proposed, see for examples [45, 49, 122]. In [28, 44, 55] an inverse nonlinearity was constructed. In the controller design, the term multiplying the control and the uncertain parameters of the system and nonsmooth nonlinearity must be within known bounded intervals. Backlash compensation using neural network and fuzzy logic has also been used in feedback control systems [72]. The system states and uncertain weights must be within known compact sets. With these developed schemes, the transient performance is usually not guaranteed. In [52], variable structure control was proposed to stabilize the nonlinear plants by using a quasistatic and dynamic description of the nonlinearity, where the parameters of nonlinearity are bounded by known constants. In [43] a dynamic backlash model is defined to pattern a backlash nonlinearity rather than constructing an inverse model to mitigate the effects of the backlash. However in [43], the term multiplying the control and the uncertain parameters of the system must be within known intervals and the 'disturbance-like' term must be bounded with known bound. Projection was used to handle the 'disturbance-like' term and unknown parameters. System stability was established and the tracking error was shown to converge

J. Zhou & C. Wen: Adapt. Backstepping Ctrl. of Uncertain Systems, LNCIS 372, pp. 125–138, 2008.
springerlink.com

to a residual. In [120], a state feedback backstepping design was developed to deal with the backlash nonlinearity, where the effect of backlash was treated as a bounded disturbance and an estimate was used to estimate its bound. The detailed characteristic of backlash was not considered in the controller design.

In this chapter, we develop a simple output feedback adaptive control scheme for a class of nonlinear systems, in the presence of unknown backlash actuator nonlinearity as in [123, 124]. To further improve system performance, we take the system backlash into account in the controller design, instead of only considering its effect like bounded disturbances as in [43, 120]. An efficient smooth adaptive inverse was developed to compensate the effect of the backlash in controller design with backstepping approach. Furthermore, the over-parametrization problem is also solved by using the concept of tuning functions. A state observer is proposed for output feedback control. To avoid possible chattering caused by the sign function, we propose a smooth control law and the tracking error is ensured to approach a prescribed bound. In our design, the term multiplying the control and the system parameters are not assumed to be within known intervals. Besides showing stability of the system, transient performance in terms of L_2 norm of the tracking error is derived to be an explicit function of design parameters and thus our scheme allows designers to obtain the closed loop behavior by tuning design parameters in an explicit way.

8.2   Problem Statement

8.2.1   System Model

We consider the following class of nonlinear systems

$$\dot{x} = Ax + \psi(y) + \sum_{i=1}^{r} \theta_i \phi_i(y) + bu \tag{8.1}$$

$$y = e_1^T x, \ u = B(v) \tag{8.2}$$

where

$$A = \begin{bmatrix} 0 & I_{n-1} \\ 0 & 0 \end{bmatrix}, \theta = \begin{bmatrix} \theta_1 \\ \vdots \\ \theta_r \end{bmatrix}, b = \begin{bmatrix} 0 \\ b_m \\ \vdots \\ b_0 \end{bmatrix} \tag{8.3}$$

$$\psi(y) = \begin{bmatrix} \psi_1(y) \\ \vdots \\ \psi_n(y) \end{bmatrix}, e_1 = \begin{bmatrix} 1 \\ 0 \\ \vdots \\ 0 \end{bmatrix} \tag{8.4}$$

where $x \in R^n$, $u \in R^1$ and $y \in R^1$ are the states, input and output of the system, respectively, $\theta \in R^r$ and $b \in R^n$ and are unknown constant vectors, $\phi_i(y) \in R^n$ and $\psi_i(y) \in R$ are known smooth functions. The actuator nonlinearity $u = B(v)$ is described as a backlash characteristic.

The control objective is to design an output feedback control law for $v(t)$ to ensure that all closed-loop signals are bounded and the plant output $y(t)$ tracks a given reference signal $y_r(t)$ under the following assumptions:

Assumption 1. The sign of b_m is known and the relative degree $\rho = n - m$ is fixed and known. The polynomial $B(s) = b_m s^m + \ldots + b_1 s + b_0$ is stable.

Assumption 2. The reference signal y_r and its $(n-1)$th order derivatives are known and bounded.

8.2.2 Backlash Characteristic

Traditionally, a backlash nonlinearity in Figure 8.1 can be described by

$$\dot{u}(t) = \begin{cases} m\dot{v}(t) & if \ \dot{v}(t) > 0 \ and \ u(t) = m(v(t) - B_r) \ or \\ & if \ \dot{v}(t) < 0 \ and \ u(t) = m(v(t) - B_l) \\ 0 & otherise \end{cases} \tag{8.5}$$

where $m \geq m_0$ is the slope of the lines, with m_0 being a small positive constant, and $B_r > B_l$ are constant parameters.

The essence of compensating backlash effect is to employ a backlash inverse. However the inverse of (8.5) is itself nonsmooth and may not be amenable to controller design.

In this chapter, we propose a new smooth inverse for the backlash as follows

$$v = BI(u) = \frac{1}{m}u + B_r\chi_r(\dot{u}) + B_l\chi_l(\dot{u}) \tag{8.6}$$

where

$$\chi_r(\dot{u}) = \frac{e^{(k\dot{u})}}{e^{(k\dot{u})} + e^{(-k\dot{u})}} \tag{8.7}$$

$$\chi_l(\dot{u}) = \frac{e^{(-k\dot{u})}}{e^{(k\dot{u})} + e^{(-k\dot{u})}} \tag{8.8}$$

where k is a positive constant. χ_r and χ_l have the following properties

$$\chi_r(\dot{u}) \to 1 \ as \ \dot{u} \to \infty; \quad \chi_r(\dot{u}) \to 0 \ as \ \dot{u} \to -\infty; \tag{8.9}$$

$$\chi_l(\dot{u}) \to 0 \ as \ \dot{u} \to \infty; \quad \chi_l(\dot{u}) \to 1 \ as \ \dot{u} \to -\infty. \tag{8.10}$$

Note that the larger the value k, the closer χ_r to 1 and 0 when $\dot{u} \to \infty$ and $\dot{u} \to -\infty$. Also the larger the k, the closer χ_l to 0 and 1 when $\dot{u} \to \infty$ and $\dot{u} \to -\infty$. Such an inverse is shown in Figure 8.2.

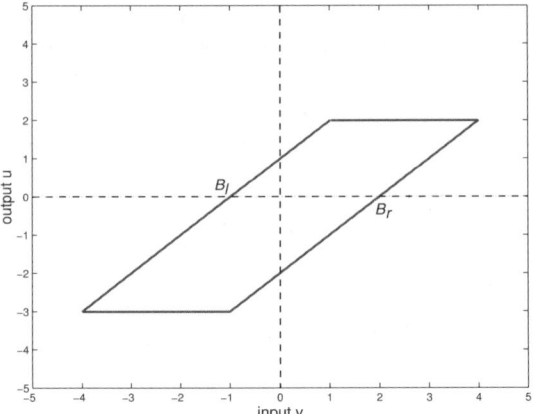

Fig. 8.1. Backlash

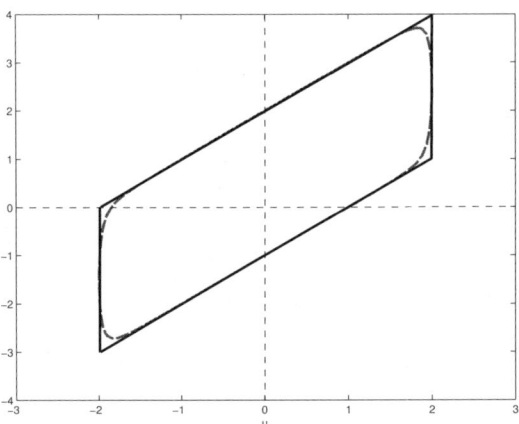

Fig. 8.2. Backlash Inverse (dashed: smooth inverse; solid: hard inverse)

Remark 8.1. Note that we will obtain an efficient adaptive backlash inverse in (8.15) when the backlash parameters are unknown. We will take the system backlash into account in the controller design, instead of only considering its effect like bounded disturbances in [43] and [120].

Remark 8.2. Note that the use of smooth functions $\chi_r(\dot{u})$ and $\chi_l(\dot{u})$ are continuous and differentiable. This is different from the inverse in [49], where the inverse indicator functions are nonsmooth. The latter case may cause chattering phenomenon in the recursive backstepping control.

To design an adaptive controller for the system, we re-parameterize the backlash as in [49] as follows

$$u(t) = \sigma_r(t)m\big(v(t) - B_r\big) + \sigma_l(t)m\big(v(t) - B_l\big) + \sigma_s(t)u_s \qquad (8.11)$$

where u_s is a generic constant corresponding to the value at any active inner segment characterized by $\frac{u_s}{m} + B_l \leq v(t) \leq \frac{u_s}{m} + B_r$.

$$\sigma_r(t) = \begin{cases} 1 \; if \; \dot{u}(t) > 0 \\ 0 \; otherwise \end{cases} \tag{8.12}$$

$$\sigma_l(t) = \begin{cases} 1 \; if \; \dot{u}(t) < 0 \\ 0 \; otherwise \end{cases} \tag{8.13}$$

$$\sigma_s(t) = \begin{cases} 1 \; if \; \dot{u}(t) = 0 \\ 0 \; otherwise \end{cases} \tag{8.14}$$

These functions satisfy that $\sigma_r(t) + \sigma_l(t) + \sigma_s(t) = 1$.

As parameters m, B_r, B_l are unknown and u is unavailable, the actual $v(t)$ is designed as

$$v(t) = \widehat{BI}(u_d) = \frac{1}{\hat{m}}[u_d + \widehat{mB_r}\chi_r(\dot{u}_d) + \widehat{mB_l}\chi_l(\dot{u}_d)] \tag{8.15}$$

where $\hat{m}$, $\widehat{mB_r}$ and $\widehat{mB_l}$ are estimates of m, mB_r and mB_l, $u_d(t)$ is the actual control input. Then corresponding control input $u_d(t)$ is given by

$$u_d(t) = \hat{m}v(t) - \widehat{mB_r}\chi_r(\dot{u}_d) - \widehat{mB_l}\chi_l(\dot{u}_d) \tag{8.16}$$

The resulting error between u and u_d is

$$u(t) - u_d(t) = \tilde{m}v - \widetilde{mB_r}\chi_r(\dot{u}_d) - \widetilde{mB_l}\chi_l(\dot{u}_d) + d_b(t) \tag{8.17}$$

where $d_b(t) = u_s\sigma_s - mv\sigma_s + mB_r(\chi_r(\dot{u}_d) - \sigma_r) + mB_l(\chi_l(\dot{u}_d) - \sigma_l)$.

Proposition. The un-parameterizable part $d_b(t)$ of the backlash inverse control error $u(t) - u_d(t)$ is bounded for any $t \geq 0$.

Proof: There are three cases to be examined:
Case 1: $\sigma_r(t) = 1, \sigma_l(t) = \sigma_s(t) = 0$.

$$d_b(t) = mB_r(\chi_r(\dot{u}_d) - 1) + mB_l\chi_l(\dot{u}_d) \tag{8.18}$$

$$Thus \; |d_b(t)| \leq m(B_r - B_l) \tag{8.19}$$

Case 2: $\sigma_l(t) = 1, \sigma_r(t) = \sigma_s(t) = 0$.

$$d_b(t) = mB_l(\chi_l(\dot{u}_d) - 1) + mB_r\chi_r(\dot{u}_d) \tag{8.20}$$

$$Thus \; |d_b(t)| \leq m(B_r - B_l) \tag{8.21}$$

Case 3: $\sigma_s(t) = 1, \sigma_r(t) = \sigma_l(t) = 0$.

$$|d_b(t)| \leq m(B_r - B_l + B_s) \tag{8.22}$$

where B_s depends on the motion of $v(t)$ and $u(t)$ on the inner segment when $\sigma_s(t) = 1$ and B_s is in the internal (B_l, B_r). From these expressions, it is clear that $d_b(t)$ is bounded.

8.3   State Estimation Filters

We employ the filters similar to those in [125] as follows

$$\hat{x}(t) = \xi_0 + \sum_{i=1}^{r} \theta_i \xi_i + \sum_{i=0}^{m} b_i \eta_i \tag{8.23}$$

$$\dot{\eta}_i = A_0 \eta_i + e_{n-i} u, \ i = 0, 1, \ldots, m \tag{8.24}$$

$$\dot{\xi}_0 = A_0 \xi_0 + ky + \psi(y) + \chi \tag{8.25}$$

$$\dot{\xi}_i = A_0 \xi_i + \phi_i(y), \ i = 1, \ldots, r \tag{8.26}$$

where $k = [k_1, \ldots, k_n]^T$ such that all eigenvalues of $A_0 = A - ke_1^T$ are at some desired stable locations and χ is a design signal specified later. It can be shown that the state estimation error $\epsilon = x(t) - \hat{x}(t)$ satisfies $\dot{\epsilon} = A_0 \epsilon - \chi$.

Note that the signal $u(t)$ is not available. Thus the signal η in (8.24) needs to be re-parameterized. Let p denote $\frac{d}{dt}$. With $\Delta(p) = det(pI - A_0)$, we express $\eta(t)$ as

$$\eta_i(t) = [\eta_{i1}(t), \eta_{i2}(t), ..., \eta_{in}(t)]^T$$
$$= [q_{i1}(p), q_{i2}(p), ..., q_{in}(p)]^T \frac{1}{\Delta(p)} u(t), i = 0, ..., m \tag{8.27}$$

for some known polynomials $q_{ij}(p), i = 0, ..., m, j = 1, ..., n$. Using (8.16) and (8.17), we have

$$u(t) = mv - mB_r \chi_r(\dot{u}_d) - mB_l \chi_l(\dot{u}_d) + d_b(t) = \beta^T \hat{\omega}(t) + d_b(t) \tag{8.28}$$

where

$$\beta = [m, mB_r, mB_l]^T \tag{8.29}$$

$$\hat{\omega}(t) = [v, -\chi_r(\dot{u}_d), -\chi_l(\dot{u}_d)]^T \tag{8.30}$$

With (8.27), we obtain

$$\eta_{ij}(t) = \beta^T \hat{\omega}_{ij}(t) + d_{ij}(t) \tag{8.31}$$

where

$$\hat{\omega}_{ij}(t) = \frac{q_{ij}(p)I_3}{\Delta(p)} \hat{\omega}(t), \ \ d_{ij}(t) = \frac{q_{ij}(p)}{\Delta(p)} d_b(t) \tag{8.32}$$

and I_3 is a 3×3 identity matrix. Based on (8.31), $\hat{\omega}_i$ is available for controller design in place of u. Denoting the second components of ξ_0, ξ_i as $\xi_{02}, \xi_{i2}, i = 1, \ldots, r$, we have

$$\hat{x}_2 = \xi_{02} + \sum_{i=1}^{r} \theta_i \xi_{i2} + \sum_{i=0}^{m} b_i \beta^T \hat{\omega}_{i2}(t) + \sum_{i=0}^{m} b_i d_{i2}(t) \tag{8.33}$$

$$\hat{\omega}_{i2}(t) = \frac{(p^{m+1} + k_1 p^m)I_3}{p^n + k_1 p^{n-1} + \ldots + k_{n-1}p + k_n} \hat{\omega}(t) \tag{8.34}$$

8.4   Design of Adaptive Controllers

As usual in backstepping approach, the following change of coordinates is made.

$$z_1 = y - y_r, \quad z_i = \hat{\beta}^T \hat{\omega}_{m2}^{(i-2)} - \hat{e} y_r^{(i-1)} - \alpha_{i-1}, \quad i = 2, 3, \ldots, \rho \quad (8.35)$$

where $\hat{\beta}$ and $\hat{e}$ are estimates of β and $e = 1/b_m$, respectively, α_{i-1} is the virtual control at the *ith step* and will be determined in later discussion.

As in [120], we define functions $sg_i(z_i)$ and $f_i(z_i)$ as follows

$$sg_i(z_i) = \begin{cases} \dfrac{z_i}{|z_i|} & |z_i| \geq \delta_i \\[2mm] \dfrac{z_i^{(2q+1)}}{(\delta_i^2 - z_i^2)^{\rho-i+2} + |z_i|^{(2q+1)}} & |z_i| < \delta_i \end{cases} \quad (8.36)$$

$$f_i(z_i) = \begin{cases} 1 & |z_i| \geq \delta_i \\ 0 & |z_i| < \delta_i \end{cases} \quad (8.37)$$

where $\delta_i (i = 1, \ldots, \rho)$ is a positive design parameter and $q = round\{(\rho-i+2)/2\}$, where $round\{x\}$ is the nearest integer to x. Clearly $2q + 1 \geq \rho - i + 2$. It can be shown that $sg_i(z_i)$ is $(\rho - i + 1)th$ order differentiable.

Remark 8.3. The standard adaptive backstepping design approach [1] results in difficulties in designing robust adaptive controllers with the capability of estimating the bound of disturbances for the case that relative degree $\rho > 1$. In this chapter, function sg_i is used to overcome the difficulties by counteracting the disturbance in each step and also avoid using sign function which is non-differentiable and may cause the chattering.

Note that, the first and the last steps of the design are quite different from the approach in [120], due to output feedback and the use of estimated inverse backlash parameters in control design, and are elaborated in details. The results of other steps, i.e. step i, $i = 2, \ldots, \rho-1$ are only presented without elaboration.

• *Step 1:* We start with the equation for the tracking error z_1 given (8.1) and (8.33) to obtain

$$\dot{z}_1 = \xi_{02} + \theta^T(\xi_{(2)} + \Phi_1(y)) + b_m\hat{\beta}^T\hat{\omega}_{m2}(t) - b_m\hat{e}\dot{y}_r + b_m z_2 + b_m\alpha_1$$
$$+ \sum_{i=0}^{m-1} b_i\hat{\beta}^T\hat{\omega}_{i2}(t) + d(t) + \psi_1(y) + \epsilon_2 \quad (8.38)$$

where $\xi_{(2)} = [\xi_{12}, \ldots, \xi_{r2}]^T$, $d(t) = \sum_{i=0}^{m} b_i d_{i2}(t)$, and $\Phi_1(y) = [\phi_{11}(y), \ldots, \phi_{r1}(y)]^T$. From proposition, there exists a positive constant D such that $|d(t)| \leq D$.

Remark 8.4. The unknown bound D of $d(t)$ will be estimated online and thus it is not assumed to be known in contrast to [41], [49] and [126]. With our proposed scheme, only one estimator will be used to estimate its bound in the backstepping design to overcome the over-parametrization problem. This is also different from [125], where a number of estimators are used for the same variable D.

Now we select the virtual control law α_1 as

$$\alpha_1 = \hat{e}\bar{\alpha}_1 \tag{8.39}$$

$$\bar{\alpha}_1 = -(c_1 + \frac{\hat{b}_m^2}{4})(|z_1| - \delta_1)^n sg_1 - \xi_{02} - \hat{\Theta}^T \varphi(t) - \hat{D}sg_1$$
$$-(\delta_2 + 1)\sqrt{\hat{b}_m^2 + \delta_0} \cdot sg_1 \tag{8.40}$$

where c_1 is a positive constant, δ_0 is a small positive real number, $\hat{b}_m, \hat{D}$ and $\hat{\Theta}$ are estimates of b_m, D and $\Theta^T = [\theta^T, b_0\beta^T, ..., b_{m-1}\beta^T]$, and $\varphi(t) = [\xi_{(2)} + \Phi_1(y), \hat{\omega}_{02}(t), ..., \hat{\omega}_{(m-1)2}(t)]^T$. Then the choice of (8.39) and (8.40) results in the following system

$$\dot{z}_1 = -(c_1 + \frac{\hat{b}_m^2}{4})(|z_1| - \delta_1)^p sg_1(z_1) + \tilde{\Theta}^T \varphi(t) + b_m z_2 - b_m(\bar{\alpha}_1 + \dot{y}_r)\tilde{e}$$
$$-b_m\tilde{\beta}^T \hat{\omega}_{m2}(t) + d(t) - \hat{D}sg_1 + \epsilon_2 - (\delta_2 + 1)\sqrt{\hat{b}_m^2 + \delta_0} \cdot sg_1 \tag{8.41}$$

We define a positive definite function V_1 as

$$V_1 = \frac{1}{\rho + 1}(|z_1| - \delta_1)^{\rho+1} f_1 + \frac{1}{2}|b_m|\tilde{\beta}^T \Gamma_\beta^{-1} \tilde{\beta} + \frac{1}{2}\tilde{\Theta}^T \Gamma_\Theta^{-1} \tilde{\Theta}$$
$$+ \frac{|b_m|}{2\gamma_e}\tilde{e}^2 + \frac{1}{2\gamma_d}\tilde{D}^2 + \frac{1}{2l_1}\epsilon^T P\epsilon \tag{8.42}$$

where $\tilde{\Theta} = \Theta - \hat{\Theta}, \tilde{\beta} = \beta - \hat{\beta}, \tilde{e} = e - \hat{e}, \tilde{D} = D - \hat{D}, \Gamma_\Theta, \Gamma_\beta$ are positive definite matrices, γ_e, γ_d are positive constants, and $P = P^T > 0$ satisfies the equation $PA_0 + A_0^T P = -2I$. Let $\beta_i = \bar{e}_i^T \theta, \ i = 1, ..., 3$, where $\bar{e}_i \in R^3$ is an identity vector. We select the adaptive update laws as

$$\dot{\hat{\beta}}_i = \bar{e}_i^T \tau_\beta, i = 2, 3, \quad \dot{\hat{\beta}}_i = Proj(\bar{e}_i^T \tau_\beta), i = 1 \tag{8.43}$$
$$\tau_\beta = -\text{sign}(b_m)\Gamma_\beta\hat{\omega}_{m2}(t)(|z_1| - \delta_1)^p f_1 sg_1 \tag{8.44}$$
$$\dot{\hat{e}} = -\text{sign}(b_m)\gamma_e(\bar{\alpha}_1 + \dot{y}_r)(|z_1| - \delta_1)^p f_1 sg_1 \tag{8.45}$$

where $Proj(.)$ is a smooth projection operation to ensure the estimate $\hat{m}(t) \geq m_0$. Such an operation can be found in [1].

Then from (8.41) to (8.45) and using $\dot{\epsilon} = A_0\epsilon - \chi$ and the property that $-\tilde{\beta}^T \Gamma_\beta^{-1} Proj(\tau_\beta) \leq -\tilde{\beta}^T \Gamma_\beta^{-1} \tau_\beta$, we obtain the time derivative of V_1 as

$$\dot{V}_1 = (|z_1| - \delta_1)^p f_1 sg_1 \dot{z}_1 - |b_m|\tilde{\beta}^T \Gamma_\beta^{-1} \dot{\hat{\beta}} - \tilde{\Theta}^T \Gamma_\Theta^{-1} \dot{\hat{\Theta}}$$
$$- \frac{|b_m|}{\gamma_e}\tilde{e}\dot{\hat{e}} - \frac{1}{\gamma_d}\tilde{D}\dot{\hat{D}} + \frac{1}{l_1}\epsilon^T P\dot{\epsilon}$$

$$\leq -(c_1 + \frac{\hat{b}_m^2}{4})(|z_1| - \delta_1)^{2\rho} f_1 + \tilde{D}\big[(|z_1| - \delta_1)^p f_1 - \frac{1}{\gamma_d}\dot{\hat{D}}\big] - \frac{1}{l_1}\epsilon^T \epsilon$$

$$+\tilde{\Theta}^T \big[\varphi(|z_1| - \delta_1)^p f_1 sg_1 - \Gamma_\Theta^{-1}\dot{\hat{\Theta}}\big] + \epsilon^T \big[e_2(|z_1| - \delta_1)^p f_1 sg_1 - \frac{1}{l_1}P\chi\big]$$

$$-|b_m|\tilde{\beta}^T\left[\text{sign}(b_m)\hat{\omega}_{m2}(|z_1| - \delta_1)^\rho f_1 sg_1 + \Gamma_\beta^{-1}\dot{\hat{\beta}}\right]$$

$$-|b_m|\tilde{e}\left[\text{sign}(b_m)(\bar{\alpha}_1 + \dot{y}_r)(|z_1| - \delta_1)^\rho f_1 sg_1 + \frac{1}{\gamma_e}\dot{\hat{e}}\right]$$

$$+(|z_1| - \delta_1)^\rho f_1 sg_1\left[b_m z_2 - (\delta_2 + 1)\sqrt{\hat{b}_m^2 + \delta_0} sg_1\right]$$

$$\leq -(c_1 + \frac{\hat{b}_m^2}{4})(|z_1| - \delta_1)^{2\rho} f_1 + \tilde{\Theta}^T(\tau_{\Theta 1} - \Gamma_\Theta^{-1}\dot{\hat{\Theta}}) + \tilde{D}(\tau_{D1} - \frac{1}{\gamma_d}\dot{\hat{D}})$$

$$+(|z_1| - \delta_1)^\rho f_1 sg_1\left(b_m z_2 - (\delta_2 + 1)\sqrt{\hat{b}_m^2 + \delta_0} sg_1\right)$$

$$-\frac{1}{l_1}\epsilon^T\epsilon + \epsilon^T(\tau_{\chi 1} - \frac{1}{l_1}P\chi) \tag{8.46}$$

$$\tau_{\Theta 1} = \phi(|z_1| - \delta_1)^\rho f_1 sg_1 \tag{8.47}$$

$$\tau_{\chi 1} = e_2(|z_1| - \delta_1)^\rho f_1 sg_1 \tag{8.48}$$

$$\tau_{D1} = (|z_1| - \delta_1)^\rho f_1 \tag{8.49}$$

where $e_2 = [0, 1, 0, \ldots, 0]^T \in R^n$.

• *Step i*, $i = 2, \ldots, \rho$: As detailed in [120], we choose

$$\alpha_i = -(c_i + 1)(|z_i| - \delta_i)^{\rho-i+1} sg_i - g_i - (\delta_{i+1} + 1)sg_i + \frac{\partial\alpha_{i-1}}{\partial y}\hat{\Theta}^T\varphi$$

$$+\frac{\partial\alpha_{i-1}}{\partial y}\hat{\vartheta}^T\hat{\omega}_{m2}(t) + \sqrt{\|\frac{\partial\alpha_{i-1}}{\partial y}\|^2 + \delta_0} \cdot \hat{D}sg_i + \frac{\partial\alpha_{i-1}}{\partial\hat{\Theta}}\Gamma_\Theta\tau_{\Theta i}$$

$$+\frac{\partial\alpha_{i-1}}{\partial\xi_0}l_1 P^{-1}\tau_{\chi i} + \frac{\partial\alpha_{i-1}}{\partial\hat{\vartheta}}\Gamma_\vartheta\tau_{\vartheta i} + \sum_{k=2}^{i-1}(|z_k| - \delta_k)^{\rho-k+1}f_k sg_k$$

$$\left[-\frac{\partial\alpha_{k-1}}{\partial\hat{\Theta}}\frac{\partial\alpha_{i-1}}{\partial y}\varphi - \frac{\partial\alpha_{k-1}}{\partial\xi_0}\frac{\partial\alpha_{i-1}}{\partial y}l_1 P^{-1}e_2 - \frac{\partial\alpha_{k-1}}{\partial\hat{D}}\frac{\partial\alpha_{i-1}}{\partial y}sg_i\right]$$

$$-\sum_{k=3}^{i-1}(|z_k| - \delta_k)^{\rho-k+1}f_k sg_k\frac{\partial\alpha_{k-1}}{\partial\hat{\vartheta}}\frac{\partial\alpha_{i-1}}{\partial y}\hat{\omega}_{m2} + \frac{\partial\alpha_{i-1}}{\partial\hat{D}}\frac{1}{\gamma_d}\tau_{Di} \tag{8.50}$$

$$\dot{\hat{b}}_m = \gamma_b(|z_1| - \delta_1)^\rho f_1 sg_1 z_2 \tag{8.51}$$

$$\tau_{Di} = \tau_{Di-1} - \sqrt{\|\frac{\partial\alpha_{i-1}}{\partial y}\|^2 + \delta_0} \cdot (|z_i| - \delta_i)^{\rho-i+1}f_i \tag{8.52}$$

$$\tau_{\Theta i} = \tau_{\Theta i-1} - \frac{\partial\alpha_{i-1}}{\partial y}\varphi(|z_i| - \delta_i)^{\rho-i+1}f_i sg_i \tag{8.53}$$

$$\tau_{\chi i} = \tau_{\chi i-1} - \frac{\partial\alpha_{i-1}}{\partial y}(|z_i| - \delta_i)^{\rho-i+1}f_i sg_i e_2 \tag{8.54}$$

$$\tau_{\vartheta i} = \tau_{\vartheta i-1} - \frac{\partial\alpha_{i-1}}{\partial y}\hat{\omega}_{m2}(|z_i| - \delta_i)^{\rho-i+1}f_i sg_i \tag{8.55}$$

$$V_i = \sum_{k=1}^{i} \frac{1}{\rho - k + 2} (|z_k| - \delta_k)^{\rho - k + 2} f_k + \frac{1}{2} |b_m| \tilde{\beta}^T \Gamma_\beta^{-1} \tilde{\beta}$$
$$+ \frac{1}{2} \tilde{\Theta}^T \Gamma_\Theta \tilde{\Theta} + \frac{|b_m|}{2\gamma_e} \tilde{e}^2 + \frac{1}{2} \tilde{\vartheta}^T \Gamma_\vartheta^{-1} \tilde{\vartheta} + \frac{1}{2\gamma_b} \tilde{b}_m^2 + \frac{1}{2l_1} \epsilon^T P \epsilon \qquad (8.56)$$

where $\hat{\vartheta}$ is an estimate of $\vartheta = b_m \beta$, $\tilde{\vartheta} = \vartheta - \hat{\vartheta}$, $\tilde{b}_m = b_m - \hat{b}_m$, g_i contains all known terms, c_i is a positive constant, δ_i is a small positive constant, γ_b is a positive constant, Γ_ϑ is a positive definite matrix.

Step ρ: Using (8.16) and (8.34), we have

$$\hat{\beta}^T \hat{\omega}_{m2}^{(\rho-1)} = \hat{\beta}^T \frac{(p^n + k_1 p^{n-1}) I_3}{p^n + k_1 p^{n-1} + \ldots + k_{n-1} p + k_n} \hat{\omega}(t)$$
$$= u_d(t) + \omega_0 \qquad (8.57)$$

where ω_0 is given by

$$\omega_0 = -\frac{(k_2 p^{n-2} + \ldots + k_{n-1} p + k_n) I_3}{p^n + k_1 p^{n-1} + \ldots + k_{n-1} p + k_n} \hat{\omega}(t) \qquad (8.58)$$

With this equation, the derivative of $z_n = -\hat{\theta}^T \hat{\omega}_{m2}^{(\rho-2)} - \hat{e} y_r^{(\rho-1)} - \alpha_{\rho-1}$ is

$$\dot{z}_\rho = u_d + g_\rho - \frac{\partial \alpha_{\rho-1}}{\partial y} \Theta^T \varphi - \frac{\partial \alpha_{\rho-1}}{\partial y} \vartheta^T \hat{\omega}_{m2}(t) - \frac{\partial \alpha_{\rho-1}}{\partial \hat{\Theta}} \dot{\hat{\Theta}} - \frac{\partial \alpha_{\rho-1}}{\partial \hat{D}} \dot{\hat{D}}$$
$$- \frac{\partial \alpha_{\rho-1}}{\partial \hat{\vartheta}} \dot{\hat{\vartheta}} - \frac{\partial \alpha_{\rho-1}}{\partial \xi_0} \chi - \frac{\partial \alpha_{\rho-1}}{\partial y} d(t) - \frac{\partial \alpha_{\rho-1}}{\partial y} \epsilon_2 \qquad (8.59)$$

where β_ρ contains all known terms. Define a positive definite Lyapunov function V_ρ as

$$V_\rho = V_{\rho-1} + \frac{1}{2} (|z_\rho| - \delta_\rho)^2 f_\rho + \frac{1}{2\gamma_d} \tilde{D}_\rho^2 \qquad (8.60)$$

We choose the update laws for $\hat{\vartheta}$, $\hat{\Theta}$, $\hat{D}$

$$\dot{\hat{\Theta}} = \Gamma_\Theta \tau_{\Theta\rho} \qquad (8.61)$$

$$\dot{\hat{\vartheta}} = \Gamma_\vartheta \tau_{\vartheta\rho} \qquad (8.62)$$

$$\dot{\hat{D}} = \gamma_d \tau_{D\rho} \qquad (8.63)$$

and the design signal χ as

$$\chi = l_1 P^{-1} \tau_{\chi\rho} \qquad (8.64)$$

Finally the control law is given by

$$u_d = \alpha_\rho \qquad (8.65)$$

$$v(t) = \widehat{BI}(u_d) = \frac{1}{\hat{m}} [u_d + \widehat{mB_r} \chi_r(\dot{u}_d) + \widehat{mB_l} \chi_l(\dot{u}_d)] \qquad (8.66)$$

With this choice and similar steps in step 1 for $\dot{V}_1$, the derivative of V_n becomes

$$
\dot{V}_\rho \leq -\sum_{i=1}^{\rho} c_i(|z_i| - \delta_i)^{2(\rho-i+1)} f_i + \tilde{\vartheta}^T(\tau_{\vartheta\rho} - \Gamma_\vartheta^{-1}\dot{\hat{\vartheta}}) + \frac{1}{\gamma_d}\tilde{D}(\gamma_d\tau_{D\rho} - \dot{\hat{D}})
$$

$$
+\epsilon^T(\tau_{\chi\rho} - \frac{1}{l_1}P\chi) + \sum_{k=2}^{\rho}(|z_k| - \delta_k)^{\rho-k+1} f_k\Big[\frac{\partial\alpha_{k-1}}{\partial\hat{\vartheta}}(\Gamma_\vartheta\tau_{\vartheta\rho} - \dot{\hat{\vartheta}})
$$

$$
+\frac{\partial\alpha_{k-1}}{\partial\hat{D}}(\gamma_d\tau_{D\rho} - \dot{\hat{D}}) + \frac{\partial\alpha_{k-1}}{\partial\xi_0}(l_1P^{-1}\tau_{\chi\rho} - \chi)\Big] - \frac{1}{l_1}\epsilon^T\epsilon
$$

$$
+\tilde{\Theta}^T(\tau_{\Theta\rho} - \Gamma_\Theta^{-1}\dot{\hat{\Theta}}) + \sum_{k=3}^{\rho}(|z_k| - \delta_k)^{\rho-k+1} f_k\frac{\partial\alpha_{k-1}}{\partial\hat{\Theta}}(\Gamma_\Theta\tau_{\Theta\rho} - \dot{\hat{\Theta}})
$$

$$
= -\sum_{i=1}^{\rho} c_i(|z_i| - \delta_i)^{2(\rho-i+1)} f_i - \frac{1}{l_1}\epsilon^T\epsilon \tag{8.67}
$$

From (8.67), we get the following Lemma.

Lemma 8.1. *The adaptive controller designed above ensures that $z_1,\ldots,z_\rho$, $\hat{\Theta}$, $\hat{e}$, $\hat{b}_m$, $\hat{\vartheta}$, $\hat{D}$ and ϵ are all bounded.*

With Lemma 8.1, all the signals in the closed-loop can be shown to be bounded and a bound can be established for the tracking error, as stated in the following theorem.

Theorem 8.1. *Consider the system consisting of the parameter estimators given by (8.43), (8.45), (8.51) and (8.61)-(8.63), adaptive controllers designed using (8.65)-(8.66) with virtual control laws (8.39) and (8.50), and plant (8.1) with a backlash nonlinearity (8.5). The system is stable in the sense that all signals in the closed loop are bounded. Furthermore*

- *The tracking error converges $[\delta_1, -\delta_1]$ asymptotically, i.e.,*

$$
\lim_{t\to\infty} |y(t) - y_r(t)| = \delta_1 \tag{8.68}
$$

- *The transient tracking error performance is given by*

$$
\| |y(t) - y_r(t)| - \delta_1 \|_2
$$

$$
\leq \frac{1}{c_1^{1/2\rho}}\Big(\frac{1}{2}\tilde{\Theta}(0)^T\Gamma_\Theta^{-1}\tilde{\Theta}(0) + \frac{|b_m|}{2\Gamma_\beta}\tilde{\beta}(0)^2 + \frac{1}{2\Gamma_\vartheta}\tilde{\vartheta}(0)^2
$$

$$
+\frac{|b_m|}{2\gamma_e}\tilde{e}(0)^2 + \frac{1}{2\gamma_d}\tilde{D}(0)^2 + \frac{1}{2\gamma_b}\tilde{b}_m(0)^2 + \frac{1}{2l_1}\epsilon(0)^2\Big)^{1/2\rho} \tag{8.69}
$$

with $z_i(0) = 0, i = 1,\ldots,\rho$,

Proof: From Lemma 8.1, we have that $z_1,\ldots,z_\rho,\hat{\beta},\hat{\Theta},\hat{e},\hat{b}_m,\hat{\vartheta},\hat{D},\epsilon$ are bounded. Following similar approaches to those in [125], we can obtain the boundedness of $\alpha_i, i = 1, \ldots, \rho$, χ and u_d, and so are $v = \widehat{BI}(u_d)$ and $u = DI(v)$. It follows that $\hat{\omega} \in L^\infty$. From (8.24), we have that η is bounded. Then $\hat{x}$ is bounded from (8.23) and finally $x(t) = \hat{x}(t) + \epsilon(t)$ is bounded from (8.23-8.24). Thus all signals in the closed-loop are bounded. The tracking error performance can be obtained from (8.67) following similar approaches to those in [120].

Remark 8.5. From Theorem 8.1 the following conclusions can be obtained:

- The transient performance depends on the initial estimate errors $\tilde{e}(0)$, $\tilde{\beta}(0)$, $\tilde{\Theta}(0)$, $\tilde{\vartheta}(0)$, $\tilde{D}(0)$, $\tilde{b}_m(0)$ and the explicit design parameters. The closer the initial estimates $\hat{e}(0)$, $\hat{\beta}(0)$, $\hat{\Theta}(0)$, $\hat{\vartheta}(0)$, $\hat{b}_m(0)$ and $\hat{D}(0)$ to the true values $e, \beta, \Theta, \vartheta, b_m$ and D, the better the transient performance.
- The bound for $\| y(t) - y_r(t) \|_2$ is an explicit function of design parameters and thus computable. We can decrease the effects of the initial error estimates on the transient performance by increasing the adaptation gains $c_1, \gamma_d, \gamma_e, \gamma_b$ and $\Gamma_\beta, \Gamma_\Theta, \Gamma_\vartheta$.
- The value of δ_1 can be chosen as small as possible according to the desired accuracy, since the output tracking error will converge to $[-\delta_1, \delta_1]$. Notes that δ_1 may influence the control input through the effects of sg_1, f_1 and their derivatives in the backstepping design.

8.5 Simulation Study

In this section, we illustrate the above methodology on the following nonlinear system

$$\dot{x}_1 = x_2, \quad \dot{x}_2 = u + a\frac{1 - e^{-x_1(t)}}{1 + e^{-x_1(t)}}$$

$$y = x_1, \quad u = B(v) \tag{8.70}$$

where u represents the output of the backlash nonlinearity as in (8.5), parameter a and backlash parameters m, B_r, B_l are unknown, but $m \geq 0.1$. The actual parameter values are chosen as $a = 1, m = 1, B_r = 0.5, B_l = -0.8$ for simulation. The objective is to control the system output y to follow a desired trajectory $y_r(t) = 10\sin(2.5t)$. Firstly we choose the backlash inverse $v(t) = \widehat{BI}(u_d)$ and the filters

$$\dot{\xi}_0 = A_0\xi_0 + ky + \chi, \quad \dot{\xi}_1 = A_0\xi_1 + Y_1 \tag{8.71}$$

$$\dot{\eta} = A_0\eta + e_2 u, \quad \hat{\omega}_2 = \frac{p + k_1}{p^2 + k_1 p + k_2}I_3[\hat{\omega}] \tag{8.72}$$

$$\text{where } Y_1 = [0, \frac{1 - e^{-x_1(t)}}{1 + e^{-x_1(t)}}]^T \tag{8.73}$$

$$k = [k_1, k_2]^T = [1, 3]^T \tag{8.74}$$

$$A_0 = \begin{bmatrix} -k_1 & 1 \\ -k_2 & 0 \end{bmatrix} = \begin{bmatrix} -1 & 1 \\ -3 & 0 \end{bmatrix} \tag{8.75}$$

Then we apply our designed control to the plant. In the simulations, we choose $c = 2, e_0 = 1, \delta_1 = 0.01, \gamma = 1, c_1 = c_2 = c, \Gamma_a = \gamma_d = \gamma, \Gamma_\theta = \gamma I_3$ and the initial parameters $\hat{a}(0) = 1.2, \hat{D}(0) = 0.4, \hat{\theta}(0) = [1, 0.4, -0.6]^T$. The initial state is chosen as $y(0) = 0.6$. The parameters and the initial states are the same as in [120]. For comparison, the scheme in [120] and our proposed scheme are both applied to the system. In [120], state feedback control is used and the effect of backlash is considered as a disturbance. The newly developed scheme studies output feedback control and a smooth backlash inverse is used to compensate the effect of backlash. The simulation results presented in the Figure 8.3 and Figure 8.4 are the tracking error $y - y_r$ and the controller input $v(t)$. Clearly,

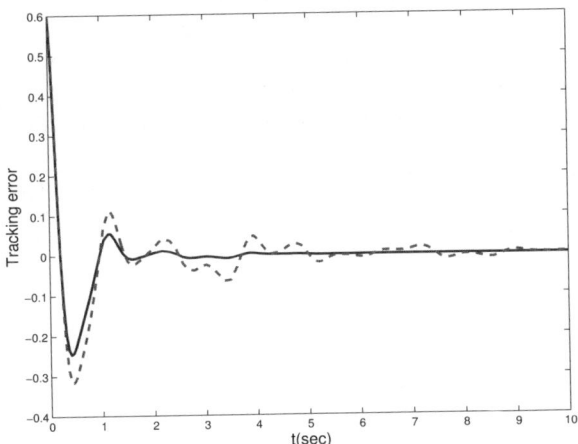

Fig. 8.3. Tracking error (solid line: proposed scheme; dash line: scheme in [120])

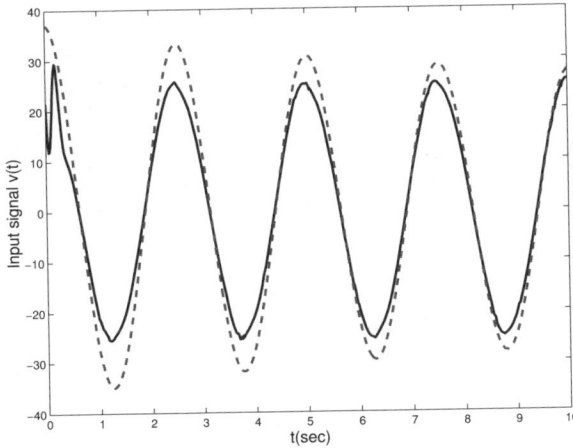

Fig. 8.4. Control signal $v(t)$ (solid line: proposed scheme; dash line: scheme in [120])

the simulation results verify our theoretical findings and show the effectiveness of our control scheme. The newly developed scheme gives better performance compared with [120].

8.6 Conclusion

This chapter presents an output feedback backstepping adaptive controller design scheme for a class of uncertain nonlinear SISO system preceded by uncertain backlash actuator nonlinearity. We propose a new smooth adaptive inverse to compensate the effect of the unknown backlash. Such an inverse can avoid possible chattering phenomenon which may be caused by nonsmooth inverse. The inverse function is employed in the backstepping controller design. The over-parametrization problem is solved by using the concept of tuning functions. For the design and implementation of the controller, no knowledge is assumed on the unknown system parameters. Besides showing stability, we also give an explicit bound on the L_2 performance of the tracking error in terms of design parameters. Simulation results illustrate the effectiveness of our proposed scheme.

9 Stabilization of Interconnected Systems with Backlash Nonlinearity

Due to the difficulty of handling both hysteresis and interactions between subsystems, there are limited results available on decentralized stabilization of unknown interconnected systems with hysteresis, even though the problem is practical and important. In this chapter, we provide solutions to this challenging problem by proposing two new schemes to design decentralized output feedback adaptive controllers using backstepping approach. For each subsystem, a general transfer function with arbitrary relative degree is considered. The interactions between subsystems are allowed to satisfy a nonlinear bound with certain structural conditions. In the first scheme, no knowledge is assumed on the bounds of unknown system parameters. In case that the uncertain parameters are inside known compact sets, we propose an alternative scheme where a projection operation is employed in the adaptive laws. In both schemes, the effect of the hysteresis and the effects due to interactions are taken into consideration in devising local control laws. It is shown that the designed local adaptive controllers can ensure all the signals in the closed loop system bounded. A root mean square type of bound is obtained for the system states as a function of design parameters. This implies that the transient system performance can be adjusted by choosing suitable design parameters. With Scheme II, the proposed control laws allow arbitrarily strong interactions provided their upper bounds are available. In the absence of hysteresis, perfect stabilization is ensured and the L_2 norm of the system states is also shown to be bounded by a function of design parameters when the second scheme is applied.

9.1 Introduction

In the control of a large scale system, one usually faces poor knowledge on the plant parameters and interactions between subsystems. Thus adaptive control technique in this case is an appropriate strategy to be employed. If some subsystems are distributed distantly, it is difficult for a centralized controller to gather feedback signals from these subsystems. Also the design and implementation of the centralized controller are complicated. Therefore decentralized controllers, designed independently for local subsystems and using local available signals for feedback, are proposed to overcome such problems. Such decentralized

J. Zhou & C. Wen: Adapt. Backstepping Ctrl. of Uncertain Systems, LNCIS 372, pp. 139–164, 2008.
springerlink.com © Springer-Verlag Berlin Heidelberg 2008

controllers, however, should be robust against the ignored interactions. In the context of decentralized adaptive control, only a limited number of results have been obtained, see for examples [20, 22, 24, 25, 32, 33, 96, 97, 98, 99, 105]. The scheme presented in [105] is the first result using backstepping technique to relax the requirement on the relative degree of subsystems. But the result is only applicable to interactions satisfying a first-order type of bound and transient performance is not established.

Hysteresis can be represented by both dynamic input-output and static constitutive relationships. It exists in a wide range of physical systems and materials, such as electro-magnetism [127], piezoelectric actuators [112], brakes [49], electronic circuits [128], motors [129], smart materials [119], and so on [130]. When a plant is preceded by the hysteresis nonlinearity, the system usually exhibits undesirable inaccuracies or oscillations and even instability due to the combined effects of the non-differentiable and non-memoryless character of the hysteresis and the plant. Hysteresis nonlinearity is one of the key factors limiting both static and dynamic performance of feedback control systems. The development of control techniques to mitigate the effects of hysteresis is typically challenging and has recently attracted significant attention, [43, 44, 45, 55, 56, 120, 131, 132]. In [131], a model derivation for smart materials using physical principles leads to a hysteresis operator at the input end of a linear system. Adaptive recursive identification and inverse control are addressed. In [44], [55] and [56] an inverse hysteresis nonlinearity was constructed. An adaptive hysteresis inverse cascaded with the plant was employed to cancel the effects of hysteresis. In [43], a dynamic hysteresis model is used to pattern a backlash-like hysteresis rather than constructing an inverse model to mitigate the bounded effects of the hysteresis. In the paper, an adaptive state feedback control scheme is developed for a class of nonlinear systems. In the design, the term multiplying the control and the uncertain parameters of the system must be within known compact sets and a bound for the effect from hysteresis must also be available, in order to implement the projection operation in the estimator. If the hysteresis effect is not bounded by the given bound, system stability cannot be ensured. In [120], a state feedback control for a special structure of nonlinear systems with backlash-like hysteresis is developed using backstepping methodology. System stability was established and the tracking error was shown to converge to a residual.

Due to difficulties in considering the effects of interconnections, extension of single loop results to multi-loop interconnecting systems is challenging, which is why the number of available results is still limited, especially for the case when the relative degree of each subsystem is greater than two. In the presence of hysteresis in unknown interconnected systems, there is one result [133] available for decentralized stabilization so far. In this chapter, we develop two output feedback decentralized backstepping adaptive stabilizers for a class of interconnected systems with arbitrary subsystem relative degrees and with the input of each subsystem preceded by unknown backlash-like hysteresis modelled by a differential equation as in [43], [112] and [130]. The interactions between subsystems are allowed to satisfy a nonlinear bound. The effects of both hysteresis and interactions are taken into consideration in the development of local control laws. For each subsystem, we consider a general transfer function. In Scheme I, the term multiplying the control and the system parameters are not assumed to be within known intervals. Compared with conventional backstepping approaches, two new terms are added in the parameter updating laws in order to ensure boundedness of estimates. In Scheme II, we assume uncertain parameters are inside some known bounded intervals, which is apriori information available. Thus we use projection operation in the adaptive laws. It is established that the designed local controllers with both schemes can ensure all the signals in the closed loop system bounded. Besides stability, a root mean square type of bound is also obtained for system states as a function of design parameters. This implies that the transient system performance can be adjusted by choosing suitable design parameters. With Scheme II, arbitrarily strong interactions can be accommodated provided their upper bounds are available. In the absence of hysteresis, perfect stabilization is ensured and the L_2 norm of the system states is also shown to be bounded by a function of design parameters when Scheme II is used.

9.2   Problem Formulation

A system consisting of N interconnected subsystems of order n_i modelled below is considered.

$$\dot{x}_{oi} = A_{oi}x_{oi} + b_{oi}u_i + \sum_{j=1}^{N} \bar{f}_{ij}(t, y_j) \tag{9.1}$$

$$y_i = c_{oi}^T x_{oi}, \; for \; i = 1, \ldots, N \tag{9.2}$$

where $x_{oi} \in R^{n_i}$, $u_i \in R^1$ and $y_i \in R^1$ are the states, input and output of the ith subsystem, respectively, $\bar{f}_{ij}(t, y_j) \in R^{n_i}$ denotes the nonlinear interactions from the jth subsystem to the ith subsystem for $j \neq i$, or a nonlinear un-modelled part of the ith subsystem for $j = i$. The matrices and vectors in (9.1) and (9.2) have appropriate dimensions, and their elements are constant but unknown.

Usually each loop has a backlash-like hysteresis nonlinearity and u_i is the output of such hysteresis described by

$$u_i(t) = BH_i(w_i(t)) \tag{9.3}$$

where $w_i(t)$ is the input of the hysteresis, $BH_i(\cdot)$ is the backlash hysteresis operator.

In this chapter, we consider a hysteresis described by a continuous-time dynamic model

$$\frac{du_i}{dt} = \alpha_i' \left|\frac{dw_i}{dt}\right|(c_i'w_i - u_i) + h_i \frac{dw_i}{dt} \tag{9.4}$$

where α_i', c_i' and h_i are constants, $c_i' > 0$ is the slope of the lines satisfying $c_i' > h_i$. This equation can be solved explicitly

$$u_i(t) = c_i'w_i(t) + \bar{d}_i(t) \tag{9.5}$$

$$\bar{\bar{d}}_i(t) = [u_i(0) - c'_i w_i(0)]e^{-\alpha'_i (w_i - w_i(0))sgn\dot{w}_i}$$

$$+e^{-\alpha'_i w_i sgn\dot{w}_i} \int_{w_i(0)}^{w_i} [h_i - c'_i]e^{\alpha'_i \xi(sgn\dot{w}_i)}d\xi \tag{9.6}$$

The solution indicates that dynamic equation (9.4) can be used to model a class of backlash-like hysteresis as shown in Figure 9.1, where $\alpha'_i = 1, c'_i = 3.1635, h_i = 0.345$, the input signal $w_i(t) = 6.5\sin(2.3t)$ and the initial condition $u_i(0) = 0$. For $\bar{\bar{d}}_i(t)$, it is bounded clearly from Figure 9.1 and the bound is unknown.

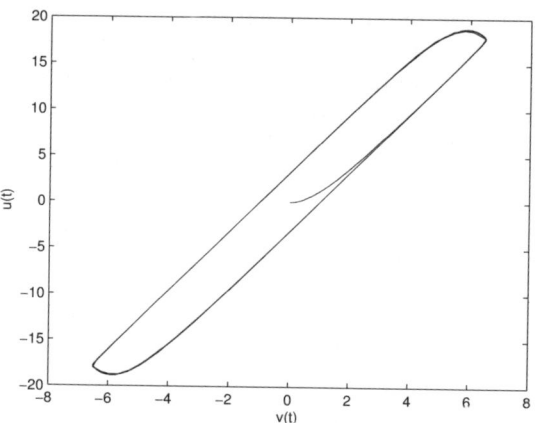

Fig. 9.1. Hysteresis curves

Remark 9.1. A number of different methods of modelling hysteresis are available in literature [119], [130] and [134]. The hysteresis model of this chapter focuses on the fact that the output can only change its characteristics when the input changes direction. This model uses a phenomenological approach, postulating an integral operator or differential equation to model the relation. The works in [135, 136, 137] show that such a model is useful in applied electro-magnetics because the functions and parameters can be fine-tuned to match experimental results in a given situation. This hysteresis nonlinearity is the key factor limiting both static and dynamic performance of feedback control systems.

Now substituting (9.5) to (9.1) gives

$$\dot{x}_{oi} = A_{oi}x_{oi} + \bar{b}_{oi}w_i + \sum_{j=1}^{N} \bar{f}_{ij}(t, y_j) + \bar{d}_i(t) \tag{9.7}$$

$$y_j = c_{oi}^T x_{oi} \tag{9.8}$$

where $\bar{b}_{oi} = b_{oi}c'_i$ and $\bar{d}_i(t) = b_{oi}\bar{\bar{d}}_i(t)$. For each local system, we make the following assumptions.

Assumption 1. n_i is known;

Assumption 2. The triple $(A_{oi}, \bar{b}_{oi}, c_{oi})$ are completely controllable and observable;

Assumption 3. In the transfer function

$$G_i(s) = c_{oi}^T(sI - A_{oi})^{-1}\bar{b}_{oi} = \frac{N_i(s)}{D_i(s)}$$

$$= \frac{b_i^{m_i} s^{m_i} + \ldots + b_i^1 s + b_i^0}{s^{n_i} + a_i^{n_i-1} s^{n_i-1} + \ldots + a_i^1 s + a_i^0} \tag{9.9}$$

$N_i(s)$ is a Hurwitz polynomial. The sign of $b_i^{m_i}$ and the relative degree $\rho_i (= n_i - m_i)$ of $G_i(s)$ are known;

Assumption 4. The nonlinear interaction terms satisfy

$$\| \bar{f}_{ij}(t, y_j) \| \le \bar{\gamma}_{ij} |y_j \psi_j(y_j)| \tag{9.10}$$

where $\| \cdot \|$ denotes the Euclidean norm, $\bar{\gamma}_{ij}$ are constants denoting the strength of the interaction, and $\psi_j(y_j), j = 1, 2, \ldots, N$ are known nonlinear functions and differentiable at least ρ_i times.

Remark 9.2. Assumption 4 means that the effects of the nonlinear interactions to a local subsystem from other subsystems or its unmodelled part is bounded by a function of the output of this subsystem. With this condition, it is possible for the designed local controllers to stabilize the interconnected systems with strong interactions. In fact, this assumption is much more relaxed version of the linear bounding conditions used in [31, 97, 138, 139].

The control objective is to design totally decentralized adaptive controllers for system (9.1) and (9.4) satisfying Assumptions 1-4 so that the closed-loop system is stable and the system performance in certain sense is adjustable by design parameters.

9.3   Local State Estimation Filters

In this section, a filter using only local input and output will be designed to estimate the states of each unknown local system in the presence of both interaction and hysteresis. To achieve this, each local system model given in (9.1) is transformed to a more suitable form. From Assumption 2, there exists a nonsingular matrix T_i, such that under transformation $x_{oi} = T_i x_i$, (9.7) and (9.8) can be transformed to

$$\dot{x}_i = A_i x_i + a_i y_i + \begin{bmatrix} 0 \\ b_i \end{bmatrix} w_i + f_i + d_i \tag{9.11}$$

$$y_i = (e_{n_i}^1)^T x_i, \quad for \ i = 1, \ldots, N \tag{9.12}$$

where

$$A_i = \begin{bmatrix} 0 & & \\ \vdots & I_{n_i-1} & \\ 0 & \cdots & 0 \end{bmatrix}, \quad a_i = \begin{bmatrix} -a_i^{n_i-1} \\ \vdots \\ -a_i^0 \end{bmatrix}, \quad b_i = \begin{bmatrix} b_i^{m_i} \\ \vdots \\ b_i^0 \end{bmatrix} \tag{9.13}$$

$$f_i = \sum_{j=1}^{N} T_i^{-1} \bar{f}_{ij}, \quad d_i = T_i^{-1} \bar{d}_i(t) \tag{9.14}$$

and e_j^k denotes the kth coordinate vector in $\Re^j$. Similar transformations can be found in [1] and [84]. For state estimation, by following the standard procedures as in [105], we can obtain

$$\dot{v}_i^j = A_i^0 v_i^j + e_{n_i}^{(n_i-j)} w_i, \quad j = 0, \ldots, m_i \tag{9.15}$$

$$\dot{\eta}_i = A_i^0 \eta_i + e_{n_i}^{n_i} y_i \tag{9.16}$$

$$\Omega_i^T = [v_i^{m_i}, \ldots, v_i^1, v_i^0, \Xi_i] \tag{9.17}$$

$$\Xi_i = -[(A_i^0)^{n_i-1} \eta_i, \ldots, A_i^0 \eta_i, \eta_i] \tag{9.18}$$

$$\xi_i^{n_i} = -(A_i^0)^{n_i} \eta_i \tag{9.19}$$

where the vector $k_i = [k_i^1, \ldots, k_i^{n_i}]^T$ is chosen so that the matrix $A_i^0 = A_i - k_i(e_{n_i}^1)^T$ is Hurwitz. Hence there exists a P_i such that $P_i A_i^0 + (A_i^0)^T P_i = -2I$, $P_i = P_i^T > 0$. With these designed filters our state estimate is

$$\hat{x}_i = \xi_i^{n_i} + \Omega_i^T \theta_i \tag{9.20}$$

$$\theta_i^T = [b_i^T, a_i^T] \tag{9.21}$$

and the state estimation error $\epsilon_i = x_i - \hat{x}_i$ satisfies

$$\dot{\epsilon}_i = A_i^0 \epsilon_i + f_i + d_i \tag{9.22}$$

Let $V_{\epsilon_i} = \epsilon_i^T P_i \epsilon_i$. It can be shown that

$$\dot{V}_{\epsilon_i} = \epsilon_i^T [P_i A_i^0 + (A_i^0)^T P_i] \epsilon_i + 2\epsilon_i^T P_i (f_i + d_i)$$
$$\leq -\epsilon_i^T \epsilon_i + 2 \parallel P_i d_i \parallel^2 + 2 \parallel P_i f_i \parallel^2 \tag{9.23}$$

Then system (9.11) can be expressed as

$$\dot{y}_i = b_i^{m_i} v_i^{m_i,2} + \xi_i^{n_i,2} + \bar{\delta}_i^T \theta_i + \epsilon_i^2 + f_i^1 + d_i^1 \tag{9.24}$$

$$\dot{v}_i^{m_i,q} = v_i^{m_i,q+1} - k_i^q v_i^{m_i,1}, \quad q = 2, \ldots, \rho_i - 1 \tag{9.25}$$

$$\dot{v}_i^{m_i,\rho_i} = v_i^{m_i,\rho_i+1} - k_i^{\rho_i} v_i^{m_i,1} + w_i \tag{9.26}$$

where

$$\delta_i = [v_i^{m_i,2}, v_i^{m_i-1,2}, \ldots, v_i^{0,2}, \Xi_i^{(2)} - y_i(e_{n_i}^1)^T]^T \tag{9.27}$$

$$\bar{\delta}_i = [0, v_i^{m_i-1,2}, \ldots, v_i^{0,2}, \Xi_i^{(2)} - y_i(e_{n_i}^1)^T]^T \tag{9.28}$$

and $v_i^{m_i,2}, \epsilon_i^2, \xi_i^{n_i,2}, \Xi_i^2$ denote the second entries of $v_i^{m_i}, \epsilon_i, \xi_i^{n_i}, \Xi_i$ respectively, f_i^1 and d_i^1 are the first elements of vectors f_i and d_i. All states of the local filters in (9.15) and (9.16) are available for feedback.

9.4   Design of Adaptive Controllers

In this section, we develop two adaptive backstepping design schemes. The system parameters b_{m_i}, θ_i are uncertain parameters. In Scheme I, there is no apriori information required from these parameters and thus they can be allowed totally uncertain. To ensure the boundedness of parameter estimates, two new terms are added in the adaptive law compared with conventional backstepping approaches. In Scheme II, we assume uncertain parameters are inside known compact sets, which is apriori information available. A projection operation, which is to replace the role of the newly added two terms in Scheme I, is used in the adaptive laws in this case. To illustrate the backstepping procedures, only the first scheme is elaborated in details.

9.4.1   Control Scheme I

As usual in backstepping approach, the following change of coordinates is made.

$$z_i^1 = y_i \tag{9.29}$$
$$z_i^q = v_i^{m_i,q} - \alpha_i^{q-1}, \quad q = 2, 3, \ldots, \rho_i \tag{9.30}$$

where α_i^{q-1} is the virtual control at the qth step of the ith loop and will be determined in later discussion. To illustrate the controller design procedures, we now give a brief description on the first step.

• *Step 1:* We start with the equations for the stabilization error z_i^1 obtained from (9.24), (9.29) and (9.30) to get

$$\dot{z}_i^1 = b_i^{m_i} \alpha_i^1 + \xi_i^{n_i,2} + \bar{\delta}_i^T \theta_i + \epsilon_i^2 + f_i^1 + d_i^1 + b_i^{m_i} z_i^2 \tag{9.31}$$

The virtual control law α_i^1 is designed as

$$\alpha_i^1 = \hat{p}_i \bar{\alpha}_i^1 \tag{9.32}$$
$$\bar{\alpha}_i^1 = -\frac{3}{2} c_i^1 z_i^1 - l_i^1 z_i^1 - l_i^* z_i^1 \left(\psi_i(z_i^1) \right)^2 - \xi_i^{n_i,2} - \bar{\delta}_i^T \hat{\theta}_i \tag{9.33}$$

where c_i^1, l_i^1 and l_i^* are positive design parameters, $\hat{\theta}_i$ is the estimate of θ_i, $\hat{p}_i$ is the estimate of $p_i = 1/b_i^{m_i}$.

Remark 9.3. The term $l_i^* z_i^1 \left(\psi_i(z_i^1) \right)^2$ in (9.33) is designed to compensate the effects of interactions from other subsystems or the un-modelled part of its own subsystem. Note that the scheme in [105] does not have such a term and thus the result of [105] is not applicable to the systems considered here.

From (9.31) and (9.32) we have

$$\dot{z}_i^1 = -\frac{3}{2} c_i^1 z_i^1 - l_i^1 z_i^1 - l_i^* z_i^1 \left(\psi_i(z_i^1) \right)^2 + \epsilon_i^2 + \bar{\delta}_i^T \tilde{\theta}_i - b_i^{m_i} \bar{\alpha}_i^1 \tilde{p}_i + b_i^{m_i} z_i^2 + f_i^1 + d_i^1$$
$$= -\frac{3}{2} c_i^1 z_i^1 - l_i^1 z_i^1 - l_i^* z_i^1 \left(\psi_i(z_i^1) \right)^2 + \epsilon_i^2 + (\delta_i - \hat{p}_i \bar{\alpha}_i^1 e_{n_i+m_i+1}^1)^T \tilde{\theta}_i$$
$$- b_i^{m_i} \bar{\alpha}_i^1 \tilde{p}_i + \hat{b}_i^{m_i} z_i^2 + f_i^1 + d_i^1 \tag{9.34}$$

where $\tilde{\theta}_i = \theta_i - \hat{\theta}_i$. Using $\tilde{p}_i = p_i - \hat{p}_i$,

$$b_i^{m_i}\alpha_i^1 = b_i^{m_i}\hat{p}_i\bar{\alpha}_i^1 = \bar{\alpha}_i^1 - b_i^{m_i}\tilde{p}_i\bar{\alpha}_i^1 \tag{9.35}$$

$$\begin{aligned}
\bar{\delta}_i^T\tilde{\theta}_i + b_i^{m_i}z_i^2 &= \bar{\delta}_i^T\tilde{\theta}_i + \tilde{b}_i^{m_i}z_i^2 + \hat{b}_i^{m_i}z_i^2 \\
&= \bar{\delta}_i^T\tilde{\theta}_i + (v_i^{m_i,2} - \alpha_i^1)(e_{n_i+m_i+1}^1)^T\tilde{\theta}_i + \hat{b}_i^{m_i}z_i^2 \\
&= (\delta_i - \hat{p}_i\bar{\alpha}_i^1 e_{n_i+m_i+1}^1)^T\tilde{\theta}_i + \hat{b}_i^{m_i}z_i^2
\end{aligned} \tag{9.36}$$

We consider the Lyapunov function

$$V_i^1 = \frac{1}{2}(z_i^1)^2 + \frac{1}{2}\tilde{\theta}_i^T\Gamma_i^{-1}\tilde{\theta}_i + \frac{|b_i^{m_i}|}{2\gamma_i'}(\tilde{p}_i)^2 + \frac{1}{2\bar{l}_i^1}V_{\epsilon_i} \tag{9.37}$$

where Γ_i is a positive definite design matrix and γ_i' is a positive design parameter. We now examine the derivative of V_i^1

$$\begin{aligned}
\dot{V}_i^1 &= z_i^1\dot{z}_i^1 - \tilde{\theta}_i^T\Gamma_i^{-1}\dot{\hat{\theta}}_i - \frac{|b_i^{m_i}|}{\gamma_i'}\tilde{p}_i\dot{\hat{p}}_i + \frac{1}{2\bar{l}_i^1}\dot{V}_{\epsilon_i} \\
&\leq -\frac{3}{2}c_i^1(z_i^1)^2 + \hat{b}_i^{m_i}z_i^1z_i^2 - |b_i^{m_i}|\tilde{p}_i\frac{1}{\gamma_i'}[\gamma_i'sgn(b_i^{m_i})\bar{\alpha}_i^1z_i^1 + \dot{\hat{p}}_i] \\
&\quad -l_i^*(z_i^1)^2(\psi_i(z_i^1))^2 + \tilde{\theta}_i^T\Gamma_i^{-1}[\Gamma_i(\delta_i - \hat{p}_i\bar{\alpha}_i^1 e_{n_i+m_i+1}^1)z_i^1 - \dot{\hat{\theta}}_i] - \frac{1}{2\bar{l}_i^1}\epsilon_i^T\epsilon_i \\
&\quad +\frac{1}{\bar{l}_i^1}(\parallel P_id_i\parallel^2 + \parallel P_if_i\parallel^2) - l_i^1(z_i^1)^2 + (f_i^1 + d_i^1 + \epsilon_i^2)z_i^1
\end{aligned} \tag{9.38}$$

Now we choose

$$\dot{\hat{p}}_i = -\gamma_i'sgn(b_i^{m_i})\bar{\alpha}_i^1z_i^1 - \gamma_i'l_i^p(\hat{p}_i - p_i^0) \tag{9.39}$$

$$\tau_i^1 = (\delta_i - \hat{p}_i\bar{\alpha}_i^1 e_{n_i+m_i+1}^1)z_i^1 \tag{9.40}$$

where l_i^p and p_i^0 are two positive design constants.

From the choice, the following useful property can be obtained:

$$\begin{aligned}
l_i^p\tilde{p}_i(\hat{p}_i - p_i^0) &= -l_i^p(\hat{p}_i - p_i)[\frac{1}{2}(\hat{p}_i - p_i) + \frac{1}{2}(\hat{p}_i + p_i) - p_i^0] \\
&= -\frac{1}{2}l_i^p(\tilde{p}_i)^2 - \frac{1}{2}l_i^p(\hat{p}_i)^2 + \frac{1}{2}l_i^p(p_i)^2 + l_i^p\hat{p}_ip_i^0 - l_i^p p_i p_i^0 \\
&= -\frac{1}{2}l_i^p(\tilde{p}_i)^2 - \frac{1}{2}l_i^p(\hat{p}_i)^2 + l_i^p\hat{p}_ip_i^0 - \frac{1}{2}l_i^p(p_i^0)^2 + \frac{1}{2}l_i^p(p_i^0)^2 \\
&\quad -l_i^p p_i p_i^0 + \frac{1}{2}l_i^p(p_i)^2 \\
&= -\frac{1}{2}l_i^p(\tilde{p}_i)^2 + \frac{1}{2}l_i^p(p_i - p_i^0)^2 - \frac{1}{2}l_i^p(\hat{p}_i - p_i^0)^2 \\
&\leq -\frac{1}{2}l_i^p(\tilde{p}_i)^2 + \frac{1}{2}l_i^p(p_i - p_i^0)^2
\end{aligned} \tag{9.41}$$

Let $l_i^1 = 3\bar{l}_i^1$. Note that

$$-\bar{l}_i^1(z_i^1)^2 + f_i^1 z_i^1 \leq \frac{1}{4\bar{l}_i^1} \parallel f_i^1 \parallel^2 \tag{9.42}$$

$$-\bar{l}_i^1(z_i^1)^2 + d_i^1 z_i^1 \leq \frac{1}{4\bar{l}_i^1} \parallel d_i^1 \parallel^2 \tag{9.43}$$

$$-\bar{l}_i^1(z_i^1)^2 + \epsilon_i^2 z_i^1 - \frac{1}{4\bar{l}_i^1}\epsilon_i^T \epsilon_i \leq -\bar{l}_i^1(z_i^1)^2 + \epsilon_i^2 z_i^1 - \frac{1}{4\bar{l}_i^1}(\epsilon_i^2)^2$$

$$= -\bar{l}_i^1(z_i^1 - \frac{1}{2\bar{l}_i^1}\epsilon_i^2)^2 \leq 0 \tag{9.44}$$

Then the following derivation for the derivative of V_i^1 can be carried out by using (9.39)-(9.44)

$$\dot{V}_i^1 \leq -\frac{3}{2}c_i^1(z_i^1)^2 + \hat{b}_i^{m_i} z_i^1 z_i^2 - \frac{|b_i^{m_i}|}{2}l_i^p(\tilde{p}_i)^2 - \frac{1}{4\bar{l}_i^1}\epsilon_i^T \epsilon_i + \frac{|b_i^{m_i}|}{2}l_i^p(p_i - p_i^0)^2$$

$$-l_i^*\left(z_i^1\psi_i(z_i^1)\right)^2 + \tilde{\theta}_i^T(\tau_i^1 - \Gamma_i^{-1}\dot{\hat{\theta}}_i) + \frac{1}{\bar{l}_i^1} \parallel P_i d_i \parallel^2 + \frac{1}{4\bar{l}_i^1} \parallel d_i^1 \parallel^2$$

$$+\frac{1}{\bar{l}_i^1} \parallel P_i f_i \parallel^2 + \frac{1}{4\bar{l}_i^1} \parallel f_i^1 \parallel^2 \tag{9.45}$$

• *Step q ($q = 2, \ldots, \rho_i$, $i = 1, \ldots, N$)*: Choose virtual control laws

$$\alpha_i^2 = -\hat{b}_i^{m_i} z_i^1 - \left[c_i^2 + l_i^2\left(\frac{\partial\alpha_i^1}{\partial y_i}\right)^2\right]z_i^2 + \bar{B}_i^2 + \frac{\partial\alpha_i^1}{\partial\hat{\theta}_i}\Gamma_i\tau_i^2 + \frac{\partial\alpha_i^1}{\partial\hat{\theta}_i}\Gamma_i l_i^\theta(\hat{\theta}_i - \theta_i^0) \tag{9.46}$$

$$\alpha_i^q = -z_i^{q-1} - \left[c_i^q + l_i^q\left(\frac{\partial\alpha_i^{q-1}}{\partial y_i}\right)^2\right]z_i^q + \bar{B}_i^q + \frac{\partial\alpha_i^{q-1}}{\partial\hat{\theta}_i}\Gamma_i\tau_i^q + \frac{\partial\alpha_i^{q-1}}{\partial\hat{\theta}_i}\Gamma_i l_i^\theta(\hat{\theta}_i - \theta_i^0)$$

$$-\left(\sum_{k=2}^{q-1} z_i^k \frac{\partial\alpha_i^{k-1}}{\partial\hat{\theta}_i}\right)\Gamma_i\frac{\partial\alpha_i^{q-1}}{\partial y_i}\delta_i \tag{9.47}$$

$$\tau_i^q = \tau_i^{q-1} - \frac{\partial\alpha_i^{q-1}}{\partial y_i}\delta_i z_i^q \tag{9.48}$$

where $c_i^q, l_i^q, q = 3, \ldots, \rho_i$ are positive design parameters, and $\bar{B}_i^q, q = 2, \ldots, \rho_i$ denotes some known terms and its detailed structure can be found in [1].

Then the adaptive controller and parameter update laws are finally given by

$$w_i = \alpha_i^{\rho_i} - v_i^{m_i, \rho_i+1} \tag{9.49}$$

$$\dot{\hat{\theta}}_i = \Gamma_i\tau_i^{\rho_i} + \Gamma_i l_i^\theta(\hat{\theta}_i - \theta_i^0) \tag{9.50}$$

where l_i^θ and θ_i^0 are positive design constants. Note that if ψ_i is ρ_i-th order differentiable, then $\alpha_i^{\rho_i}$ will be differentiable. So w_i is differentiable. Thus u_i is well defined and continuous from (9.4).

The designed adaptive controllers are summarized in Table 9.1.

Table 9.1. Adaptive Backstepping Control Scheme I

Adaptive Control Laws:

$$\alpha_i^1 = \hat{p}_i \bar{\alpha}_i^1 \tag{9.51}$$

$$\bar{\alpha}_i^1 = -\frac{3}{2} c_i^1 z_i^1 - l_i^1 z_i^1 - l_i^* z_i^1 \big(\psi_i(z_i^1) \big)^2 - \xi_i^{n_i,2} - \bar{\delta}_i^T \hat{\theta}_i \tag{9.52}$$

$$\alpha_i^2 = -\hat{b}_i^{m_i} z_i^1 - \big[c_i^2 + l_i^2 \big(\frac{\partial \alpha_i^1}{\partial y_i} \big)^2 \big] z_i^2 + \bar{B}_i^2 + \frac{\partial \alpha_i^1}{\partial \hat{\theta}_i} \Gamma_i \tau_i^2 + \frac{\partial \alpha_i^1}{\partial \hat{\theta}_i} \Gamma_i l_i^\theta (\hat{\theta}_i - \theta_i^0) \tag{9.53}$$

$$\alpha_i^q = -z_i^{q-1} - \big[c_i^q + l_i^q \big(\frac{\partial \alpha_i^{q-1}}{\partial y_i} \big)^2 \big] z_i^q + \bar{B}_i^q + \frac{\partial \alpha_i^{q-1}}{\partial \hat{\theta}_i} \Gamma_i \tau_i^q$$

$$+ \frac{\partial \alpha_i^{q-1}}{\partial \hat{\theta}_i} \Gamma_i l_i^\theta (\hat{\theta}_i - \theta_i^0) - \big(\sum_{k=2}^{q-1} z_i^k \frac{\partial \alpha_i^{k-1}}{\partial \hat{\theta}_i} \big) \Gamma_i \frac{\partial \alpha_i^{q-1}}{\partial y_i} \delta_i$$

$$q = 2, \ldots, \rho_i, \; i = 1, \ldots, N \tag{9.54}$$

$$w_i = \alpha_i^{\rho_i} - v_i^{m_i, \rho_i + 1} \tag{9.55}$$

Parameter Update Laws:

$$\dot{\hat{p}}_i = -\gamma_i' sgn(b_i^{m_i}) \bar{\alpha}_i^1 z_i^1 - \gamma_i' l_i^p (\hat{p}_i - p_i^0) \tag{9.56}$$

$$\dot{\hat{\theta}}_i = \Gamma_i \tau_i^{\rho_i} + \Gamma_i l_i^\theta (\hat{\theta}_i - \theta_i^0) \tag{9.57}$$

$$with \;\; \tau_i^q = \tau_i^{q-1} - \frac{\partial \alpha_i^{q-1}}{\partial y_i} \delta_i z_i^q, \; \tau_i^1 = (\delta_i - \hat{p}_i \bar{\alpha}_i^1 e_{n_i + m_i + 1}^1) z_i^1 \tag{9.58}$$

Remark 9.4. From the analysis above, terms $\gamma_i' l_i^p (\hat{p}_i - p_i^0)$ and $\Gamma_i l_i^\theta (\hat{\theta}_i - \theta_i^0)$ in the adaptive controllers are used to handle the effects of hysteresis in order to ensure the boundedness of the parameter estimates. If projection operation is used as in Scheme II, such terms are not needed.

Remark 9.5. When going through the details of the design procedures, we note that in the equations concerning $\dot{z}_i^q, q = 1, 2, \ldots, \rho_i$, just functions f_i^1 from the interactions and d_i^1 due to the hysteresis effect appear, and they are always together with ϵ_i^2. This is because only $\dot{y}_i$ from the plant model (9.11) was used in the calculation of $\dot{\alpha}_i^q$ for steps $q = 2, \ldots, \rho_i$.

Remark 9.6. From our analysis, it can be noted that the design method can also be applied to system with perturbations satisfying similar boundedness properties to (9.10).

9.4.2   Control Scheme II

In this section, we assume uncertain parameters p_i and θ_i are inside known compact sets, which is the apriori information available as follows.

Assumption 5. Parameters p_i and θ_i and are inside known compact sets Ω_{p_i} and Ω_{θ_i} in which all the transfer functions with structure of (9.9) satisfy Assumption 2.

Thus we can use a smooth projection operation in the adaptive laws to ensure the estimates belonging to the compact sets for all the time. Such an operation can be found in [1]. As shown in Appendix C, the projection operation can ensure that the estimated parameter $\hat{p}_i(t) \in \Omega_{p_i}$ for all t, if $\hat{p}_i(0) \in \Omega_{p_i}$ and the estimated parameter vector $\hat{\theta}_i(t) \in \Omega_{\theta_i}$ for all t, if $\hat{\theta}_i(0) \in \Omega_{\theta_i}$. Thus, the boundedness of $\hat{\theta}_i$ and $\hat{p}_i$ are guaranteed for all t. Therefore, in this case, we do not need terms $\gamma_i' l_i^p (\hat{p}_i - p_i^0)$ and $\Gamma_i l_i^\theta (\hat{\theta}_i - \theta_i^0)$ in the controller design as in Scheme I.

As the controller design is similar to Scheme I, we only present the resulting control laws as summarized in Table 9.2.

Table 9.2. Adaptive Backstepping Control Scheme II

Change of Coordinates:
$$z_i^1 = y_i \qquad (9.59)$$ $$z_i^q = v_i^{m_i,q} - \alpha_i^{q-1}, \ q = 2, 3, \ldots, \rho_i \qquad (9.60)$$
Adaptive Control Laws:
$$\alpha_i^1 = \hat{p}_i \bar{\alpha}_i^1 \qquad (9.61)$$ $$\bar{\alpha}_i^1 = -\frac{3}{2} c_i^1 z_i^1 - l_i^1 z_i^1 - l_i^* z_i^1 \big(\psi_i(z_i^1)\big)^2 - \xi_i^{n_i,2} - \bar{\delta}_i^T \hat{\theta}_i \qquad (9.62)$$ $$\alpha_i^2 = -\hat{b}_i^{m_i} z_i^1 - \big[c_i^2 + l_i^2 \big(\frac{\partial \alpha_i^1}{\partial y_i}\big)^2\big] z_i^2 + \bar{B}_i^2 + \frac{\partial \alpha_i^1}{\partial \hat{\theta}_i} \Gamma_i \tau_i^2 \qquad (9.63)$$ $$\alpha_i^q = -z_i^{q-1} - \big[c_i^q + l_i^q \big(\frac{\partial \alpha_i^{q-1}}{\partial y_i}\big)^2\big] z_i^q + \bar{B}_i^q + \frac{\partial \alpha_i^{q-1}}{\partial \hat{\theta}_i} \Gamma_i \tau_i^q$$ $$-\big(\sum_{k=2}^{q-1} z_i^k \frac{\partial \alpha_i^{k-1}}{\partial \hat{\theta}_i}\big) \Gamma_i \frac{\partial \alpha_i^{q-1}}{\partial y_i} \delta_i, \ q = 2, \ldots, \rho_i, i = 1, \ldots, N \quad (9.64)$$ $$w_i = \alpha_i^{\rho_i} - v_i^{m_i,\rho_i+1} \qquad (9.65)$$
Parameter Update Laws:
$$\dot{\hat{p}}_i = Proj\big\{ - \gamma_i' sgn(b_i^{m_i}) \bar{\alpha}_i^1 z_i^1 \big\} \qquad (9.66)$$ $$\dot{\hat{\theta}}_i = Proj\big\{\Gamma_i \tau_i^{\rho_i}\big\} \qquad (9.67)$$ $$with \ \tau_i^q = \tau_i^{q-1} - \frac{\partial \alpha_i^{q-1}}{\partial y_i} \delta_i z_i^q, \ \tau_i^1 = (\delta_i - \hat{p}_i \bar{\alpha}_i^1 e_{n_i+m_i+1}^1) z_i^1 \qquad (9.68)$$

9.5   Stability Analysis

In this section, the stability of the overall closed-loop system consisting of the interconnected plants and decentralized controllers will be established.

9.5.1   Control Scheme I

Firstly, Define $z_i(t) = [z_i^1, z_i^2, \ldots, z_i^{\rho_i}]^T$. A mathematical model for each local closed-loop control system is derived from (9.34) and the rest of the design steps $2, \ldots, \rho_i$.

$$\dot{z}_i = A_{z_i} z_i + W_{\epsilon i}(\epsilon_i^2 + f_i^1 + d_i^1) + W_{\theta i}^T \tilde{\theta}_i - b_i^{m_i} \bar{\alpha}_i^1 \tilde{p}_i e_{\rho_i}^1 - l_i^* z_i^1 \big(\psi_i(z_i^1)\big)^2 e_{\rho_i}^1 \tag{9.69}$$

where A_{z_i} is a matrix having the structure as in the following.

$$A_{z_i} = \begin{bmatrix} -\frac{3}{2}c_i^1 - l_i^1 & \hat{b}_i^{m_i} & 0 & \ldots & 0 \\ -\hat{b}_i^{m_i} & -c_i^2 - l_i^2\big(\frac{\partial \alpha_i^1}{\partial y_i}\big)^2 & 1 + \sigma_i^{2,3} & \ldots & \sigma_i^{2,\rho_i} \\ 0 & -1 - \sigma_i^{2,3} & -c_i^3 - l_i^3\big(\frac{\partial \alpha_i^2}{\partial y_i}\big)^2 & \ldots & \sigma_i^{3,\rho_i} \\ \vdots & \vdots & \vdots & \vdots & \vdots \\ 0 & -\sigma_i^{2,\rho_i} & -\sigma_i^{3,\rho_i} & \ldots & -c_i^{\rho_i} - l_i^{\rho_i}\big(\frac{\partial \alpha_i^{\rho_i-1}}{\partial y_i}\big)^2 \end{bmatrix} \tag{9.70}$$

$$W_{\epsilon i} = \begin{bmatrix} 1 \\ -\frac{\partial \alpha_i^1}{\partial y_i} \\ \vdots \\ -\frac{\partial \alpha_i^{\rho_i-1}}{\partial y_i} \end{bmatrix}, \quad W_{\theta i}^T = W_{\epsilon i} \delta_i^T - \hat{p}_i \bar{\alpha}_i^1 e_{\rho_i}^1 {e_{\rho_i}^1}^T \tag{9.71}$$

where the terms $\sigma_i^{k,q}$ are due to the terms $\frac{\partial \alpha_i^{k-1}}{\partial \theta_i} \Gamma_i(\tau_i^q - \tau_i^{q-1})$ in the z_i^q equation.

To show the system stability, the variables of the filters in (9.16) and the zero dynamics of subsystems should be included in the Lyapunov function. Under a similar transformation as in [105], the variables ζ_i associated with the zero dynamics of the ith subsystem can be shown to satisfy

$$\dot{\zeta}_i = A_i^{b_i} \zeta_i + \bar{b}_i z_i^1 + \bar{f}_i \tag{9.72}$$

where the eigenvalues of the $m_i \times m_i$ matrix $A_i^{b_i}$ are the zeros of the Hurwitz polynomial $N_i(s)$, $\bar{b}_i \in R^{m_i}$ and $\bar{f}_i \in R^{m_i}$ denoting the effects of the transformed interactions.

Now we define a Lyapunov function of the overall decentralized adaptive control system as

$$V = \sum_{i=1}^{N} V_i \tag{9.73}$$

where

$$V_i = \sum_{q=1}^{\rho_i} \left(\frac{1}{2}(z_i^q)^2 + \frac{1}{2\bar{l}_i^q} \epsilon_i^T P_i \epsilon_i \right) + \frac{1}{2}\tilde{\theta}_i^T \Gamma_i^{-1} \tilde{\theta}_i + \frac{|b_i^{m_i}|}{2\gamma_i'} \tilde{p}_i^2$$

$$+ \frac{1}{2l_i^{\eta_i}} \eta_i^T P_i \eta_i + \frac{1}{2l_i^{\zeta_i}} \zeta_i^T P_i^{b_i} \zeta_i \qquad (9.74)$$

where $P_i^{b_i}$ satisfies $P_i^{b_i}(A_i^{b_i}) + (A_i^{b_i})^T P_i^{b_i} = -2I$, $l_i^{\eta_i}$ and $l_i^{\zeta_i}$ are constants satisfying

$$l_i^{\eta_i} \geq \frac{2 \parallel P_i e_{n_i}^{n_i} \parallel^2}{c_i^1}$$

$$l_i^{\zeta_i} \geq \frac{2 \parallel P_i^{b_i} \bar{b}_i \parallel^2}{c_i^1}$$

Note that

$$\Gamma_i \tau_i^{q-1} - \dot{\hat{\theta}}_i = \Gamma_i \tau_i^{q-1} - \Gamma_i \tau_i^q + \Gamma_i \tau_i^q - \dot{\hat{\theta}}_i$$

$$= \Gamma_i \frac{\partial \alpha_i^{q-1}}{\partial y_i} \delta z_i^q + (\Gamma_i \tau_i^q - \dot{\hat{\theta}}_i) \qquad (9.75)$$

$$l_i^\theta \tilde{\theta}_i^T (\hat{\theta}_i - \theta_i^0) = -l_i^\theta (\hat{\theta}_i - \theta_i)^T (\frac{1}{2}(\hat{\theta}_i - \theta_i) + \frac{1}{2}(\hat{\theta}_i + \theta_i) - \theta_i^0)$$

$$= -\frac{1}{2} l_i^\theta \parallel \tilde{\theta}_i \parallel^2 -\frac{1}{2} l_i^\theta \parallel \hat{\theta}_i \parallel^2 + \frac{1}{2} l_i^\theta \parallel \theta_i \parallel^2 + l_i^\theta \hat{\theta}_i^T \theta_i^0 - l_i^\theta \theta_i^T \theta_i^0$$

$$= -\frac{1}{2} l_i^\theta \parallel \tilde{\theta}_i \parallel^2 -\frac{1}{2} l_i^\theta \parallel \hat{\theta}_i \parallel^2 + l_i^\theta \hat{\theta}_i^T \theta_i^0 - \frac{1}{2} l_i^\theta \parallel \theta_i^0 \parallel^2$$

$$+ \frac{1}{2} l_i^\theta \parallel \theta_i^0 \parallel^2 - l_i^\theta \theta_i^T \theta_i^0 + \frac{1}{2} l_i^\theta \parallel \theta_i \parallel^2$$

$$= -\frac{1}{2} l_i^\theta \parallel \tilde{\theta}_i \parallel^2 + \frac{1}{2} l_i^\theta \parallel \theta_i - \theta_i^0 \parallel^2 - \frac{1}{2} l_i^\theta \parallel \hat{\theta}_i - \theta_i^0 \parallel^2$$

$$\leq -\frac{1}{2} l_i^\theta \parallel \tilde{\theta}_i \parallel^2 + \frac{1}{2} l_i^\theta \parallel \theta_i - \theta_i^0 \parallel^2 \qquad (9.76)$$

From (9.23), (9.45), (9.61-9.67), (9.72), (9.75) and (9.76), the derivative of V_i in (9.74) is given by

$$\dot{V}_i \leq -\sum_{q=1}^{\rho_i} c_i^q (z_i^q)^2 - \frac{1}{2} l_i^\theta \parallel \tilde{\theta}_i \parallel^2 + \frac{1}{2} l_i^\theta \parallel \theta_i - \theta_i^0 \parallel^2 - \frac{|b_i^{m_i}|}{2} l_i^p (\tilde{p}_i)^2 - \frac{1}{4\bar{l}_i^1} \epsilon_i^T \epsilon_i$$

$$+ \sum_{q=1}^{\rho_i} \frac{1}{\bar{l}_i^q} (\parallel P_i d_i \parallel^2 + \parallel P_i f_i \parallel^2) + \frac{|b_i^{m_i}|}{2} l_i^p (p_i - p_i^0)^2 - l_i^* (z_i^1)^2 (\psi_i(z_i^1))^2$$

$$+ \sum_{q=2}^{\rho_i} \left(-l_i^q (\frac{\partial \alpha_i^{q-1}}{\partial y_i})^2 (z_i^q)^2 + \frac{\partial \alpha_i^{q-1}}{\partial y_i} (f_i^1 + d_i^1 + \epsilon_i^2) z_i^q - \frac{1}{2\bar{l}_i^q} \epsilon_i^T \epsilon_i \right)$$

$$- \frac{1}{2} c_i^1 (z_i^1)^2 - \frac{1}{l_i^{\eta_i}} \parallel \eta_i \parallel^2 + \frac{1}{l_i^{\eta_i}} \eta_i^T P_i e_{n_i}^{n_i} y_i - \frac{1}{l_i^{\zeta_i}} \parallel \zeta_i \parallel^2 + \frac{1}{l_i^{\zeta_i}} \zeta_i^T P_i^{b_i} \bar{b}_i z_i^1$$

$$+ \frac{1}{l_i^{\zeta_i}} \zeta_i^T P_i^{b_i} \bar{f}_i + \frac{1}{4\bar{l}_i^1} (\parallel f_i^1 \parallel^2 + \parallel d_i^1 \parallel^2) \qquad (9.77)$$

Using the inequality $ab \leq (a^2 + b^2)/2$, we have

$$-\bar{l}_i^q \Big(\frac{\partial \alpha_i^{q-1}}{\partial y_i}\Big)^2 (z_i^q)^2 + \frac{\partial \alpha_i^{q-1}}{\partial y_i} f_i^1 z_i^q \leq \frac{1}{4\bar{l}_i^q} \parallel f_i^1 \parallel^2 \tag{9.78}$$

$$-\bar{l}_i^q \Big(\frac{\partial \alpha_i^{q-1}}{\partial y_i}\Big)^2 (z_i^q)^2 + \frac{\partial \alpha_i^{q-1}}{\partial y_i} d_i^1 z_i^q \leq \frac{1}{4\bar{l}_i^q} \parallel d_i^1 \parallel^2 \tag{9.79}$$

$$-\bar{l}_i^q \Big(\frac{\partial \alpha_i^{q-1}}{\partial y_i}\Big)^2 (z_i^q)^2 + \frac{\partial \alpha_i^{q-1}}{\partial y_i} \epsilon_i^2 z_i^q - \frac{1}{4\bar{l}_i^q} \epsilon_i^T \epsilon_i \leq 0 \tag{9.80}$$

$$-\frac{1}{2l_i^{\eta_i}} \parallel \eta_i \parallel^2 + \frac{1}{l_i^{\eta_i}} \eta_i^T P_i e_{n_i} z_i^1 - \frac{1}{4} c_i^1 (z_i^1)^2 \leq -\frac{\parallel \eta_i \parallel^2}{2(l_i^{\eta_i})^2}\Big(l_i^{\eta_i} - \frac{2 \parallel P_i e_{n_i} \parallel^2}{c_i^1}\Big)$$
$$\leq 0 \tag{9.81}$$

$$-\frac{1}{2l_i^{\zeta_i}} \parallel \zeta_i \parallel^2 + \frac{1}{l_i^{\zeta_i}} \zeta_i^T P_i^{b_i} \bar{b}_i z_i^1 - \frac{1}{4} c_i^1 (z_i^1)^2 \leq -\frac{\parallel \zeta_i \parallel^2}{2(l_i^{\zeta_i})^2}\Big(l_i^{\zeta_i} - \frac{2 \parallel P_i^{b_i} \bar{b}_i \parallel^2}{c_i^1}\Big)$$
$$\leq 0 \tag{9.82}$$

$$-\frac{1}{4l_i^{\zeta_i}} \parallel \zeta_i \parallel^2 + \frac{1}{l_i^{\zeta_i}} \parallel \zeta_i \parallel \parallel P_i^{b_i} \bar{f}_i \parallel \leq \frac{1}{l_i^{\zeta_i}} \parallel P_i^{b_i} \bar{f}_i \parallel^2 \tag{9.83}$$

Then, the derivative of the V_i satisfies

$$\dot{V}_i \leq -\sum_{q=1}^{\rho_i} c_i^q (z_i^q)^2 - \frac{1}{2} l_i^\theta \parallel \tilde{\theta}_i \parallel^2 - \frac{|b_i^{m_i}|}{2} l_i^p (\tilde{p}_i)^2 - \sum_{q=1}^{\rho_i} \frac{1}{4\bar{l}_i^q} \epsilon_i^T \epsilon_i - \frac{1}{2l_i^{\eta_i}} \parallel \eta_i \parallel^2$$

$$-\frac{1}{4l_i^{\zeta_i}} \parallel \zeta_i \parallel^2 - l_i^*(z_i^1)^2 \big(\psi_i(z_i^1)\big)^2 + \sum_{q=1}^{\rho_i} \frac{1}{\bar{l}_i^q}\big(\parallel P_i f_i \parallel^2 + \frac{1}{4} \parallel f_i \parallel^2\big)$$

$$+\frac{1}{l_i^{\zeta_i}} \parallel P_i^{b_i} \bar{f}_i \parallel^2 + M_i^* \tag{9.84}$$

where $D_{i,max}$ denotes the bound of $d_i(t)$, and

$$M_i^* = M_i + \sum_{q=1}^{\rho_i} \frac{1}{4\bar{l}_i^q}(4 \parallel P_i \parallel^2 + 1) D_{i,max}^2 \tag{9.85}$$

$$M_i = \frac{|b_i^{m_i}|}{2} l_i^p (p_i - p_i^0)^2 + \frac{1}{2} l_i^\theta \parallel \theta_i - \theta_i^0 \parallel^2 \tag{9.86}$$

Remark 9.7. Due to the presence of hysteresis, an extra term M_i^* appears in (9.84) compared to the analysis in [105]. The handling of M_i^* is elaborated after (9.91).

From Assumption 4, we can show that

$$\sum_{q=1}^{\rho_i} \frac{1}{\bar{l}_i^q}\big(\parallel P_i f_i \parallel^2 + \frac{1}{4} \parallel f_i \parallel^2\big) + \frac{1}{l_i^{\zeta_i}} \parallel P_i^{b_i} \bar{f}_i \parallel^2 \leq \sum_{j=1}^{N} \gamma_{ij} \big|z_j^1 \psi_j(z_j^1)\big|^2 \tag{9.87}$$

where $\gamma_{ij} = O(\bar{\gamma}_{ij}^2)$ indicating the coupling strength from the jth subsystem to the ith subsystem depending on $\bar{l}_i^q, l_i^{\zeta_i}, \parallel P_i \parallel, \parallel P_i^{b_i} \parallel$ and $\parallel T_j^{-1} \parallel, j = 1, 2, \ldots, N$. $O(\bar{\gamma}_{ij}^2)$ denotes that γ_{ij} and $O(\bar{\gamma}_{ij}^2)$ are in the same order mathematically. Clearly there exist a constant γ_{ij}^* such that for each constant γ_{ij} satisfying $\gamma_{ij} \leq \gamma_{ij}^*$,

$$l_i^* \geq \sum_{j=1}^N \gamma_{ji} \tag{9.88}$$

$$if \quad l_i^* \geq \sum_{j=1}^N \gamma_{ji}^* \tag{9.89}$$

Constant γ_{ij}^* stands for a upper bound of γ_{ij}. Now taking the summation of the first term in (9.84) into account and using (9.87) and (9.88), we get

$$\sum_{i=1}^N -[l_i^*(z_i^1)^2(\psi_i(z_i^1))^2 - \sum_{k=1}^{\rho_i} \frac{1}{l_i^k}(\parallel P_i f_i \parallel^2 + \frac{1}{4} \parallel f_i \parallel^2) - \frac{1}{l_i^{\zeta_i}} \parallel P_i^{b_i} \bar{f}_i \parallel^2]$$

$$\leq \sum_{i=1}^N -[l_i^* - \sum_{j=1}^N \gamma_{ji}]|z_i^1 \psi_i(z_i^1)|^2 \leq 0 \tag{9.90}$$

Then

$$\dot{V} \leq \sum_{i=1}^N \Big[-\sum_{q=1}^{\rho_i} c_i^q(z_i^q)^2 - \frac{1}{2} l_i^\theta \parallel \tilde{\theta}_i \parallel^2 - \frac{|b_i^{m_i}|}{2} l_i^p(\tilde{p}_i)^2 - \sum_{q=1}^{\rho_i} \frac{1}{4\bar{l}_i^q} \epsilon_i^T \epsilon_i$$

$$- \frac{1}{2l_i^{\eta_i}} \parallel \eta_i \parallel^2 - \frac{1}{4l_i^{\zeta_i}} \parallel \zeta_i \parallel^2 + M_i^* \Big] \tag{9.91}$$

Remark 9.8. The summation in (9.90) is one of the key steps in the stability analysis. Note that this results in the cancellation of the interaction effects from other subsystems. The approach in [105] cannot be applied here due to non-Lipschitz type nonlinear interactions.

Notice that

$$-\sum_{q=1}^{\rho_i} c_i^q(z_i^q)^2 - \frac{1}{2} l_i^\theta \parallel \tilde{\theta}_i \parallel^2 - \frac{|b_i^{m_i}|}{2} l_i^p(\tilde{p}_i)^2 - \sum_{q=1}^{\rho_i} \frac{1}{4\bar{l}_i^q} \epsilon_i^T \epsilon_i$$

$$- \frac{1}{2l_i^{\eta_i}} \parallel \eta_i \parallel^2 - \frac{1}{4l_i^{\zeta_i}} \parallel \zeta_i \parallel^2 \leq -f_i^- \bar{V}_i \tag{9.92}$$

and

$$V_i = \sum_{q=1}^{\rho_i} \frac{1}{2}(z_i^q)^2 + \frac{1}{2} \tilde{\theta}_i^T \Gamma_i^{-1} \tilde{\theta}_i + \frac{|b_i^{m_i}|}{2\gamma_i'}(\tilde{p}_i)^2 + \sum_{q=1}^{\rho_i} \frac{1}{2\bar{l}_i^q} \epsilon_i^T P_i \epsilon_i$$

$$+ \frac{1}{2l_i^{\eta_i}} \eta_i^T P_i \eta_i + \frac{1}{2l_i^{\zeta_i}} \zeta_i^T P_i^{b_i} \zeta_i \leq f_i^+ \bar{V}_i \tag{9.93}$$

where

$$\bar{V}_i = \sum_{q=1}^{\rho_i}(z_i^q)^2 + \tilde{\theta}_i^T\tilde{\theta}_i + (\tilde{p}_i)^2 + \sum_{q=1}^{\rho_i}\epsilon_i{}^T\epsilon_i + \eta_i^T\eta_i + \zeta_i^T\zeta_i \tag{9.94}$$

$$f_i^- = min\{c_i^q, \frac{1}{2}l_i^\theta, \frac{|b_i^{m_i}|}{2}l_i^p, \frac{1}{4\bar{l}_i^q}, \frac{1}{2l_i^{\eta_i}}, \frac{1}{4l_i^{\zeta_i}}\} \tag{9.95}$$

$$f_i^+ = max\{\frac{1}{2}, \frac{1}{2}\lambda_i^{q,max}(\Gamma_i), \frac{|b_i^{m_i}|}{2\gamma_i'}, \frac{1}{2min(\bar{l}_i^q, l_i^{\eta_i})}\lambda_i^{q,max}(P_i),$$

$$\frac{1}{2l_i^{\zeta_i}}\lambda_i^{q,max}(P_i^{b_i})\} \quad q = 1, \ldots, \rho_i \tag{9.96}$$

where $\lambda_i^{q,max}(P_i)$, $\lambda_i^{q,max}(P_i^{b_i})$ and $\lambda_i^{q,max}(\Gamma_i)$ are the maximum eigenvalues of $P_i, P_i^{b_i}$ and Γ_i, respectively. Therefore, from (9.91) we obtain

$$\dot{V} \leq -f^*V + M^* \tag{9.97}$$

where $f^* = \sum_{i=1}^{N} f_i^- / \sum_{i=1}^{N} f_i^+$, $M^* = \sum_{i=1}^{N} M_i^*$ is a bounded term. By direct integrations of the differential inequality (9.97), we have

$$V \leq V(0)e^{-f^*t} + \frac{M^*}{f^*}(1 - e^{-f^*t}) \leq V(0) + \frac{M^*}{f^*} \tag{9.98}$$

This shows that V is uniformly bounded. Thus $z_i^1, z_i^2, \ldots, z_i^{\rho_i}, \hat{p}_i, \hat{\theta}_i, \epsilon_i, \zeta_i, v_i^j, \eta_i$ and x_i are bounded. Therefore boundedness of all signals in the system is ensured as formally stated in the following theorem.

Theorem 9.1. *Consider the closed-loop adaptive system consisting of the plant (9.1) under Assumptions 1-4, the controller (9.55), the estimator (9.56), (9.57), and the filters (9.15) and (9.16). There exist a constant γ_{ij}^* such that for each constant γ_{ij} satisfying $\gamma_{ij} \leq \gamma_{ij}^*$, $i, j = 1, \ldots, N$, all the signals in the system are globally uniformly bounded.*

Remark 9.9. Parameter l_i^* can be chosen as any positive value and the condition that $\gamma_{ij} \leq \gamma_{ij}^*$ has the implication that the designed local controllers are able to stabilize any interconnected system with coupling strength satisfying (9.89). This implication is similar to the interpretations of the results in [22], [24],[32], [105], where sufficiently weak interactions are allowed. Thus the result is qualitative in nature, which shows the robustness of designed local controllers against interactions.

We now derive a bound for the vector $z_i(t)$ where $z_i(t) = [z_i^1, z_i^2, \ldots, z_i^{\rho_i}]^T$. Firstly, the following definitions are made.

$$c_i^0 = min_{1 \leq q \leq \rho_i} c_i^q, \quad d_0 = \sum_{i=1}^{N}\sum_{q=1}^{\rho_i}\frac{1}{4\bar{l}_i^q} \tag{9.99}$$

$$\| z_i \|_{[0,T]} = \sqrt{\frac{1}{T}\int_0^T \| z_i(t) \|^2 \, dt} \tag{9.100}$$

Note that definition (9.100) is similar to the root mean square value used in electric circuit.

Define $V_\rho = \sum_{i=1}^{N} \sum_{q=1}^{\rho_i} \left(\frac{1}{2}(z_i^q)^2 + \frac{1}{2l_i^q} \epsilon_i^T P_i \epsilon_i \right) + \frac{1}{2} \tilde{\theta}_i^T \Gamma_i^{-1} \tilde{\theta}_i + \frac{|b_i^{m_i}|}{2\gamma_i'}(\tilde{p}_i)^2$. Following similar analysis to (9.84), the derivative of V_ρ can be given as

$$\dot{V}_\rho \le -f^* V_\rho + M^* \le c_i^0 \parallel z_i \parallel^2 + M^* \qquad (9.101)$$

Integrating both sides, we obtain

$$\parallel z_i \parallel_{[0,T]} \le \frac{1}{c_i^0} [\frac{|V_\rho(0) - V_\rho(T)|}{T} + \sum_{i=1}^{N} M_i$$

$$+ d_0 \frac{1}{T} \sum_{i=1}^{N} \rho_i (4 \parallel P_i \parallel^2 + 1) \int_0^T (d_i(t))^2 dt] \qquad (9.102)$$

On the other hand, from (9.74), we have

$$\frac{|V_\rho(0) - V_\rho(T)|}{T}$$

$$\le \frac{1 - e^{-f^* T}}{T} (\frac{M}{f^*} + V_\rho(0)) + \frac{d_0}{T} \sum_{i=1}^{N} \rho_i (4 \parallel P_i \parallel^2 + 1) \int_0^T e^{-f^*(T-t)} (d_i(t))^2 dt$$

$$\le M + f^* V_\rho(0) + \frac{1}{T} d_0 \sum_{i=1}^{N} \rho_i (4 \parallel P_i \parallel^2 + 1) \int_0^T e^{-f^*(T-t)} (d_i(t))^2 dt$$

$$for \ all \ T \ge 0, \qquad (9.103)$$

where we have used the fact that $e^{-f^*(T-t)} \le 1$ and $\frac{1-e^{-f^* T}}{T} \le f^*$, and $M = \sum_{i=1}^{N} M_i$. A bound for $\parallel z_i \parallel_{[0,T]}$ is established

$$\parallel z_i \parallel_{[0,T]} \le 2V_\rho(0) + \frac{1}{c_i^0} \sum_{i=1}^{N} (|b_i^{m_i}| l_i^p (p_i - p_i^0)^2 + l_i^\theta \parallel \theta_i - \theta_i^0 \parallel^2)$$

$$+ \frac{1}{c_i^0} d_0 \sum_{i=1}^{N} 2\rho_i (4 \parallel P_i \parallel^2 + 1) D_{i,max}^2 \qquad (9.104)$$

using the fact that $f^*/c_i^0 \le 2$. The initial value of the Lyapunov function is

$$V_\rho(0) = \sum_{i=1}^{N} \frac{1}{2} [\parallel z_i(0) \parallel^2 + \parallel \tilde{\theta}_i(0) \parallel_{\Gamma_i^{-1}}^2 + \frac{|b_i^{m_i}|}{\gamma_i'} |\tilde{p}_i(0)|^2 + d_i^0 \parallel \epsilon_i(0) \parallel_{P_i}^2]$$

$$(9.105)$$

where $d_i^0 = \sum_{q=1}^{\rho_i} \frac{1}{l_i^q}$, $\parallel \tilde{\theta}_i(0) \parallel_{\Gamma_i^{-1}}^2 = \tilde{\theta}_i(0)^T \Gamma_i^{-1} \tilde{\theta}_i(0)$ and $\parallel \epsilon_i(0) \parallel_{P_i}^2 = \epsilon_i(0)^T P_i \epsilon_i(0)$.

Following similar ideas to [1] (page 455), where $z(0)$ is set to zero by appropriately initializing the reference trajectory for single loop case, we can set

$z_i^q, q = 2, \ldots, \rho_i$ to zero by suitably initializing our designed filters (9.15) and (9.16) as follows:

$$v_i^{m_i,q}(0) = \alpha_i^{q-1}\left(y_i(0), \hat{\theta}_i(0), \hat{p}_i(0), \eta_i(0), v_i^{m_i,q-1}(0), \bar{v}_i^{m_i-1,2}(0)\right),$$
$$q = 2, \ldots, \rho_i \tag{9.106}$$

Thus, by setting $z_i^q(0) = 0$, $q = 2, \ldots, \rho_i$, a bound for $\| z_i \|_{[0,T]}$ is established and stated in the following theorem.

Theorem 9.2. *Consider the initial values $z_i^q(0) = 0 (q = 2, \ldots, \rho_i, i = 1, \ldots, N)$, the bound $\| z_i \|_{[0,T]}$ satisfies*

$$\| z_i \|_{[0,T]} \leq \sum_{i=1}^{N} y_i(0) + \| \tilde{\theta}_i(0) \|_{\Gamma_i^{-1}}^2 + \frac{|b_i^{m_i}|}{\gamma_i'}|\tilde{p}_i(0)|^2 + d_i^0 \| \epsilon_i(0) \|_{P_i}^2$$

$$+ \frac{1}{c_i^0}\sum_{i=1}^{N}(|b_i^{m_i}|l_i^p(p_i - p_i^0)^2 + l_i^\theta \| \theta_i - \theta_i^0 \|^2)$$

$$+ \frac{1}{c_i^0}d_0\sum_{i=1}^{N}2\rho_i(4 \| P_i \|^2 + 1)D_{i,max}^2 \tag{9.107}$$

Proof: Using (9.95), (9.96) and (9.102) - (9.105), the fact that $f^*/c_i^0 \leq 2$, (9.107) can be obtained.

Remark 9.10. Regarding the above bound, the following conclusions can be drawn by noting that $\tilde{\theta}_i(0), \tilde{p}_i(0), \epsilon_i(0)$ and $y_i(0)$ are independent of c_i^0, Γ_i, γ_i', l_i^θ, l_i^p.

- The transient performance in the sense of truncated norm given in (9.107) depends on the initial estimation errors $\tilde{\theta}_i(0), \tilde{p}_i(0)$ and $\epsilon_i(0)$. The closer the initial estimates to the true values, the better the transient performance.
- This bound can also be systematically reduced down to a lower bound depending $y_i(0)$ by increasing Γ_i, γ_i', c_i^0 and decreasing l_i^p, l_i^θ.
- This bound is depending on the effect of hysteresis.

9.5.2   Control Scheme II

Now we define a Lyapunov function of the overall decentralized adaptive control system as

$$V = \sum_{i=1}^{N} V_i \tag{9.108}$$

where

$$V_i = \sum_{q=1}^{\rho_i}\left(\frac{1}{2}(z_i^q)^2 + \frac{1}{2l_i^q}V_{\epsilon_i}\right) + \frac{1}{2}\tilde{\theta}_i^T\Gamma_i^{-1}\tilde{\theta}_i + \frac{|b_i^{m_i}|}{2\gamma_i'}(\tilde{p}_i)^2$$

$$+ \frac{1}{2l_i^{\eta_i}}\eta_i^T P_i\eta_i + \frac{1}{2l_i^{\zeta_i}}\zeta_i^T P_i^{b_i}\zeta_i \tag{9.109}$$

Similar to the procedure of Scheme I, by using the properties that $-\tilde{\theta}^T \Gamma^{-1}$ $Proj(\tau) \leq -\tilde{\theta}^T \Gamma^{-1}\tau$, the derivative of the V_i satisfies

$$
\dot{V}_i \leq - \sum_{q=1}^{\rho_i} \left[c_i^q (z_i^q)^2 - \frac{1}{4\bar{l}_i^q} \epsilon_i^T \epsilon_i \right] - \frac{1}{2l_i^{\eta_i}} \eta_i^T \eta_i - \frac{1}{4l_i^{\zeta_i}} \zeta_i^T \zeta_i + M_i^*
$$

$$
+ \tilde{\theta}_i^T (\tau_i^{\rho_i} - \Gamma_i^{-1}\dot{\hat{\theta}}_i) - |b_i^{m_i}| \tilde{p}_i \frac{1}{\gamma_i} [\gamma_i' sgn(b_i^{m_i}) \bar{\alpha}_i^1 z_i^1 + \dot{p}_i]
$$

$$
- l_i^* (z_i^1)^2 (\psi_i(z_i^1))^2 + \sum_{q=1}^{\rho_i} \frac{1}{\bar{l}_i^q} (\parallel P_i f_i \parallel^2 + \frac{1}{4} \parallel f_i \parallel^2) + \frac{1}{l_i^{\zeta_i}} \parallel P_i^{b_i} \bar{f}_i \parallel^2
$$

$$
\leq - \sum_{q=1}^{\rho_i} \left[c_i^q (z_i^q)^2 - \frac{1}{4\bar{l}_i^q} \epsilon_i^T \epsilon_i \right] - \frac{1}{2l_i^{\eta_i}} \eta_i^T \eta_i - \frac{1}{4l_i^{\zeta_i}} \zeta_i^T \zeta_i + M_i^*
$$

$$
- l_i^* (z_i^1)^2 (\psi_i(z_i^1))^2 + \sum_{q=1}^{\rho_i} \frac{1}{\bar{l}_i^q} (\parallel P_i f_i \parallel^2 + \frac{1}{4} \parallel f_i \parallel^2) + \frac{1}{l_i^{\zeta_i}} \parallel P_i^{b_i} \bar{f}_i \parallel^2
$$

$$(9.110)$$

where

$$
M_i^* = \sum_{q=1}^{\rho_i} \frac{1}{4\bar{l}_i^q} (4 \parallel P_i \parallel^2 + 1) D_{i,max}^2 \tag{9.111}
$$

We choose

$$
l_i^* \geq \sum_{j=1}^{N} \gamma_{ji} \tag{9.112}
$$

$$
if \qquad l_i^* \geq \sum_{j=1}^{N} \gamma_{ji}^* \tag{9.113}
$$

where γ_{ij}^* stands for a upper bound of γ_{ij}, $\gamma_{ij} = O(\bar{\gamma}_{ij}^2)$ indicating the coupling strength from the jth subsystem to the ith subsystem, a constant γ_{ij}^* satisfying $\gamma_{ij} \leq \gamma_{ij}^*$. Similar to the proof in Scheme I, the derivative of the V satisfies

$$
\dot{V} \leq \sum_{i=1}^{N} \left[- \sum_{q=1}^{\rho_i} (c_i^q (z_i^q)^2 + \frac{1}{4\bar{l}_i^q} \epsilon_i^T \epsilon_i) - \frac{1}{2l_i^{\eta_i}} \eta_i^T \eta_i - \frac{1}{4l_i^{\zeta_i}} \zeta_i^T \zeta_i + M_i^* \right]
$$

$$(9.114)$$

This shows that $z_i^1, z_i^2, \ldots, z_i^{\rho_i}, \epsilon_i, \zeta_i, \lambda_i, \eta_i$ and x_i are bounded. With the projection operation, $\tilde{\theta}_i$ and $\tilde{p}_i$ are bounded. Therefore boundedness of all signals in the system is ensured as formally stated in the following theorem.

Theorem 9.3. *Consider the closed-loop adaptive system consisting of the plant (9.1) under Assumptions 1-5, the controller (9.65), the estimator (9.66), (9.67),* and the filters (9.15) and (9.16). There exist γ_{ij}^* such that for all $\gamma_{ij} \leq \gamma_{ij}^*$, $i, j = 1, \ldots, N$, all the signals in the system are uniformly bounded. A bound for $\| z_i \|_{[0,T]}$ is established as

$$\| z_i \|_{[0,T]} \leq \sum_{i=1}^{N} \left[y_i(0) + \| \tilde{\theta}_i(0) \|_{\Gamma_i^{-1}}^2 + \frac{|b_i^{m_i}|}{\gamma_i'} |\tilde{p}_i(0)|^2 + d_i^0 \| \epsilon_i(0) \|_{P_i}^2 \right.$$

$$\left. + \frac{1}{c_i^0} d_0 \sum_{i=1}^{N} 2\rho_i (4 \| P_i \|^2 + 1) D_{i,max}^2 \right] \tag{9.115}$$

by setting $z_i^q(0) = 0$, $q = 2, \ldots, \rho_i$, $i = 1, \ldots, N$.

Remark 9.11. The condition that $\gamma_{ij} \leq \gamma_{ij}^*$ now has the following two implications:

(1) If we know $\bar{\gamma}_{ij}$, then we can get an estimate of its bound γ_{ij}^* which depends on $\bar{l}_i^q, l_i^{\zeta_i}, \| P_i \|, \| P_i^{b_i} \|$ and the bound of $\| T_j^{-1} \|, j = 1, 2, \ldots, N$ and design l_i^* according to (9.112). This means that the coupling strength of the interconnection between subsystems can be allowed arbitrarily strong.
(2) If $\bar{\gamma}_{ij}$ is unknown, we have similar implication to Remark 9.9. In this case, l_i^* is chosen as any positive value and thus it is sufficient that only the local systems satisfy Assumption 2 instead of all the members in the given compact sets of Assumption 5.

If the system has no hysteresis, then $d_i(t) = 0$ and we have the following corollary.

Corollary 9.4. *Consider the closed-loop decentralized adaptive control system consisting of the plant (9.1) without input hysteresis under Assumptions 1-5 and the controller (9.65), the estimator (9.66) and (9.67), and the filters (9.15) and (9.16). All the states of the system asymptotically approach to zero and the bound* $\| z_i \|_2$ *is given by*

$$\| z_i \|_2 \leq \frac{1}{2\sqrt{c_i^0}} \left(\sum_{i=1}^{N} y_i(0) + \| \tilde{\theta}_i(0) \|_{\Gamma_i^{-1}}^2 + \frac{|b_i^{m_i}|}{\gamma_i'} |\tilde{p}_i(0)|^2 + d_i^0 |\epsilon_i(0)|_{P_i}^2 \right)^{1/2}$$

$$\tag{9.116}$$

by setting $z_i^q(0) = 0$, $q = 2, \ldots, \rho_i$, $i = 1, \ldots, N$.

Proof: In the absence of hysteresis the term $d_i(t) = 0$, so $M_i^* = 0$ in (9.114). We have

$$\dot{V} \leq \sum_{i=1}^{N} \left[-\sum_{q=1}^{\rho_i} c_i^q (z_i^q)^2 - \sum_{q=1}^{\rho_i} \frac{1}{4\bar{l}_i^q} \epsilon_i^T \epsilon_i - \frac{1}{2l_i^{\eta_i}} \eta_i^T \eta_i - \frac{1}{4l_i^{\zeta_i}} \zeta_i^T \zeta_i \right] \leq -c_i^0 \| z_i \|^2$$

$$\leq 0 \tag{9.117}$$

where $\| z_i \|_2^2 = \int_0^\infty \| z_i \|^2 d\tau$. This proves that the uniform stability and the uniform boundedness of $z_i^q, \hat{p}_i, \hat{\theta}_i, \epsilon_i, \zeta_i, \eta_i, v_i^j, x_i$ and u_i. It can be shown that both $\dot{V}$ and $\ddot{V}$ are bounded as well as $\dot{V}$ is integrable over $[0, \infty]$. Therefore, $\dot{V}$ tends to zero and thus the system states x_i converge to zero from (9.117). Also (9.116) can be obtained clearly.

Remark 9.12. In the absence of hysteresis, the L_2 norm of the system states is shown to be bounded by a function of design parameters. This implies that the transient system performance in terms of L_2 bounds can be adjusted by choosing suitable design parameters. This result further extends that presented in [105], where only first order interactions considered and no transient performance like (9.116) is available.

Remark 9.13. Following similar analysis for the L_2 bound and the approaches in [1], a bound on $\| z_i \|_\infty$ can also be established and this bound can be adjusted by choosing suitable design parameters.

9.6   An Illustrative Example

We consider the following interconnected system with three subsystems.

$$\dot{x}_1 = a_1 x_1 + b_1 u_1 + f_1, \; y_1 = x_1 \tag{9.118}$$

$$\dot{x}_2 = a_2 x_2 + b_2 u_2 + f_2, \; y_2 = x_2 \tag{9.119}$$

$$\dot{x}_3 = a_3 x_3 + b_3 u_3 + f_3, \; y_3 = x_3 \tag{9.120}$$

$$u_1 = BH_1(w_1), \; u_2 = BH_2(w_2), \; u_3 = BH_3(w_3) \tag{9.121}$$

where $a_1 = 1, b_1 = 1, a_2 = 0.5, b_2 = 1, a_3 = 2, b_3 = 1$, the nonlinear interaction terms $f_1 = y_2 + sin(y_2) + 0.2y_3, f_2 = 0.2y_1^2 + y_3, f_3 = y_1 + 0.5y_2^2$, $BH_1(w_1)$, $BH_2(w_2)$ and $BH_3(w_3)$ are the backlash hysteresis described by (9.4) with parameters $\alpha_1' = 1, c_1' = 2, h_1 = 0.2, \alpha_2' = 1, c_2' = 1, h_2 = 0.2, \alpha_3' = 1.2, c_3' = 1, h_3 = 0.3$. These parameters are not needed to be known in the controller

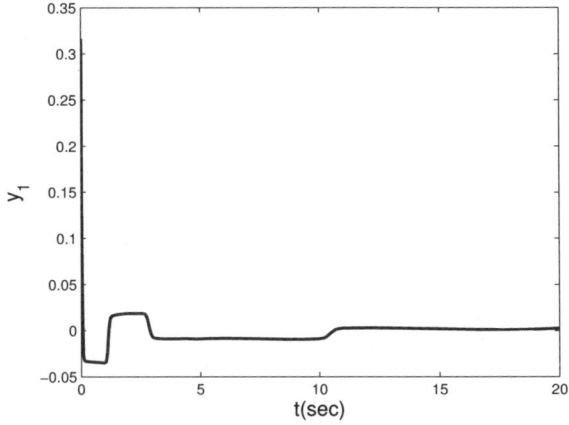

Fig. 9.2. Output y_1 with considering hysteresis using Scheme I

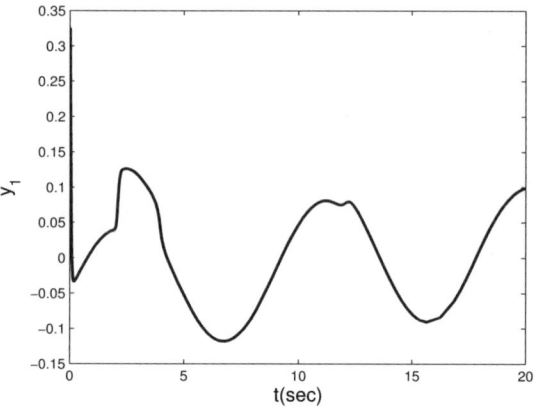

Fig. 9.3. Output y_1 without considering hysteresis

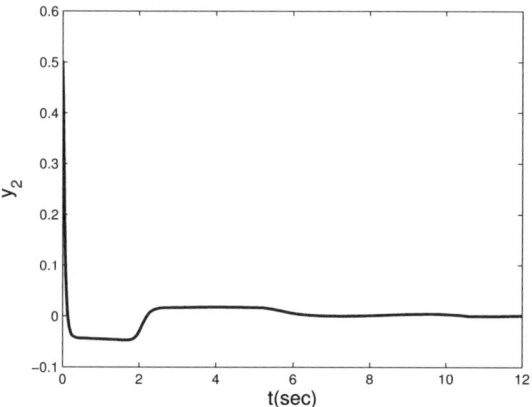

Fig. 9.4. Output y_2 with considering hysteresis using Scheme I

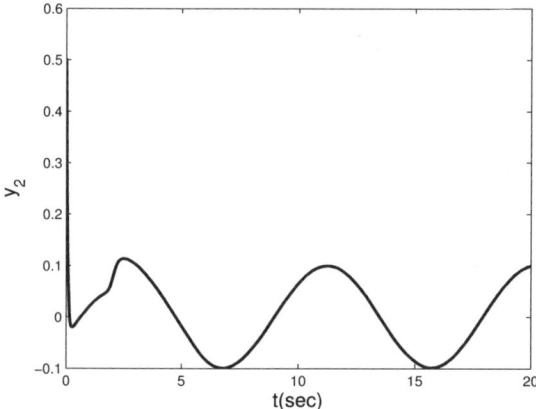

Fig. 9.5. Output y_2 without considering hysteresis

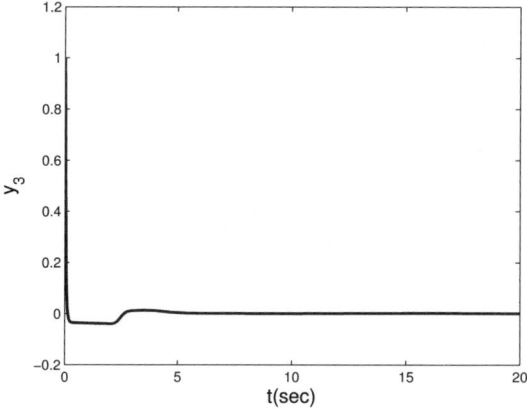

Fig. 9.6. Output y_3 with considering hysteresis using Scheme I

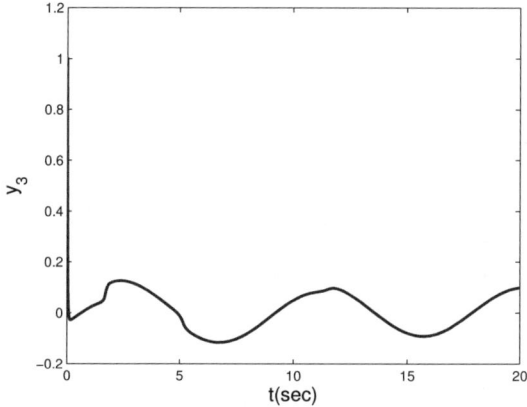

Fig. 9.7. Output y_3 without considering hysteresis

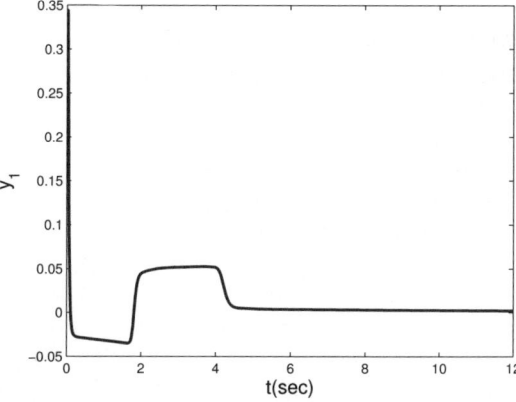

Fig. 9.8. Output y_1 using Scheme II in the presence of hysteresis

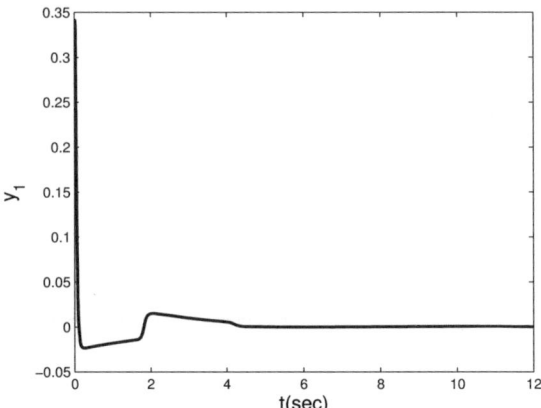

Fig. 9.9. Output y_1 using Scheme II in the absence of hysteresis

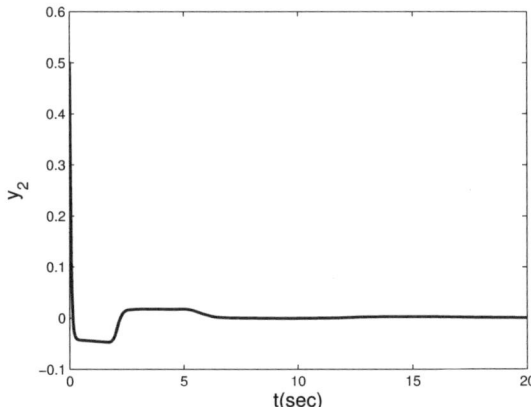

Fig. 9.10. Output y_2 using Scheme II

design. The objective is to stabilize system (9.118-9.120). The controllers with Scheme I and Scheme II are implemented, where $\hat{p}_i$ and $\hat{\theta}_i$ are estimates of $p_i = 1/b_i c_i'$ and $\theta_i = a_i, i = 1, 2$, respectively. The design parameters are chosen as $c_1^1 = c_2^1 = c_3^1 = 10, l_1^1 = l_1^2 = l_1^3 = 5, l_1^* = l_2^* = l_3^* = 5, \gamma_1 = 2, \gamma_2 = 2, \gamma_3 = 2, \Gamma_1 = \Gamma_2 = \Gamma_3 = 1$. The initials are set as $y_1(0) = 0.3, y_2(0) = 0.5, y_3(0) = 1.0$.

In order to illustrate the effects of hysteresis, we observe system performances by applying controllers designed without considering hysteresis and with our proposed Scheme I, respectively. The simulation results presented in Figures 9.2, 9.4, 9.6 and Figures 9.3, 9.5, 9.7 show the system outputs y_1, y_2 and y_3 with Scheme I and without considering hysteresis, respectively. Clearly, poor performance is observed if hysteresis is not taken into account in controller design. In fact, system stability is not even ensured theoretically in this case. When Scheme II is applied, we study the cases in the presence or absence of hysteresis. Figures 9.8-9.13 show

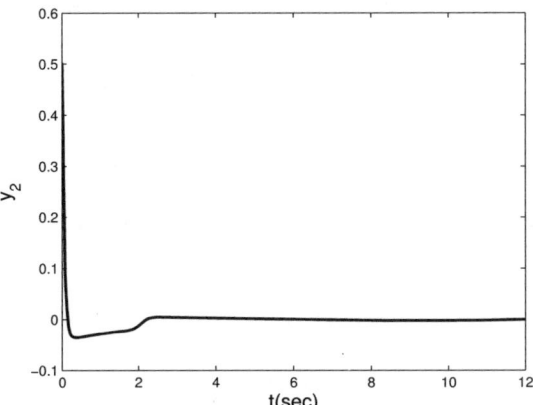

Fig. 9.11. Output y_2 using Scheme II

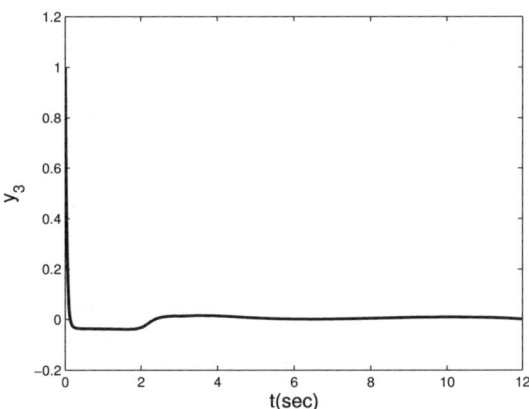

Fig. 9.12. Output y_3 using Scheme II

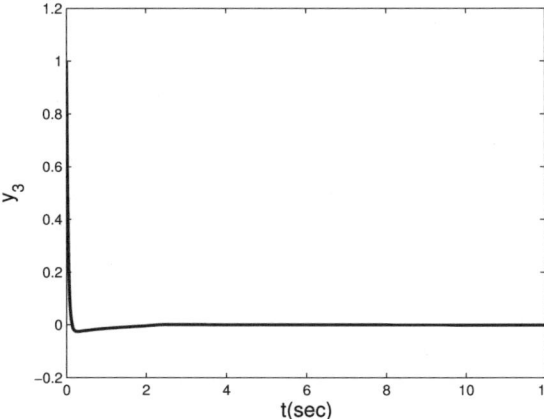

Fig. 9.13. Output y_3 using Scheme II

the system outputs, which show that $|y_i| \to 0$ in the absence of hysteresis. All the simulation results verify that our proposed two schemes are effective to cope with hysteresis nonlinearity and high order nonlinear interactions.

9.7 Conclusion

In this chapter, decentralized adaptive output feedback stabilization of a class of interconnected subsystems with the input of each loop preceded by unknown backlash-like hysteresis nonlinearity is considered. Each local adaptive controller is designed based on a general transfer function of the local subsystem with arbitrary relative degree by developing two adaptive control schemes. The effects of hysteresis and interactions are considered in the design. The nonlinear interactions between subsystems are allowed to satisfy higher-order nonlinear bounds. In Scheme I, the term multiplying the control and the system parameters are not assumed to be within known intervals. Two new terms are added in the parameter updating laws, compared to the standard backstepping approach. In Scheme II, uncertain parameters are assumed inside known compact sets. Thus we use projection operation in the adaptive laws. It is shown that the designed local adaptive controllers with both schemes stabilize the overall interconnected systems. We also derive an explicit bound on the root mean square performance of the system states in terms of design parameters. This implies that the transient system performance can be adjusted by choosing suitable design parameters. With Scheme II, in the absence of hysteresis, perfect stabilization is ensured and the L_2 norm of the system states is also shown to be bounded by a function of design parameters. The strengths can be allowed arbitrary strong if their upper bounds are available in this case. Simulation results illustrate the effectiveness of our schemes by comparing the cases with and without considering hysteresis in controller design, as well as examining the outputs in the presence and absence of hysteresis when Scheme II is employed.

10 Adaptive Control of Nonlinear Systems with Dead-Zone Nonlinearity

In this chapter, we present new adaptive schemes for uncertain nonlinear systems preceded by unknown dead-zone nonlinearity. Robust adaptive backstepping control algorithms are developed for state feedback tracking of a class of uncertain dynamic nonlinear systems preceded by unknown dead-zone nonlinearities, in the presence of bounded external disturbances. Output feedback tracking is also considered and this is achieved by introducing a new smooth inverse function of the dead-zone and using it in the controller design with backstepping technique. For the design and implementation of the controllers, no knowledge is assumed on the unknown system parameters and also the dead-zone. It is shown that the proposed controllers not only can guarantee stability, but also transient performance.

10.1 Introduction

Dead-zone, which can severely limit system performances, is one of the most important nonsmooth nonlinearities arisen in actuators, such as valves and DC servo motors and other devices. Therefore the effect of dead-zone should be taken into consideration in the design and analysis of control systems. In most practical motion systems, the dead-zone parameters are poorly known, and thus robust adaptive control techniques may be applied to design controllers. The study of adaptive control for systems with unknown dead-zone at the input was initiated in [50], where an adaptive scheme was proposed with full state measurement. An immediate method for the control of dead-zone is to construct an adaptive dead-zone inverse. This approach was used in [41, 42, 51, 52, 53, 73, 74], where the output of a dead-zone is measurable. In [41] a fuzzy pre-compensator was proposed in nonlinear industrial motion system. Such approaches promise to improve the tracking performance of motion system in presence of unknown dead-zones. An alternative approach based on sliding mode control is proposed in [52, 74]. In [51] and [57], an adaptive state feedback controller employs an adaptive inverse for a class of nonlinear systems. Dead-zone pre-compensation using neural network(NN) have also been used in feedback control systems [42],

J. Zhou & C. Wen: Adapt. Backstepping Ctrl. of Uncertain Systems, LNCIS 372, pp. 165–188, 2008.
springerlink.com   © Springer-Verlag Berlin Heidelberg 2008

where the uncertain NN weights must be within a known compact set. With this, the error resulted from using NN to approximate system functions will be bounded with known bounds. This assumption makes the control design and system analysis simpler. In [140], a robust adaptive control scheme is developed for a class of nonlinear systems without using the dead-zone inverse, where the dead-zone slopes in the positive and negative region are the same and the unknown system parameters are inside a known compact set. For all approaches mentioned above, strictly speaking, only local stability is ensured in the sense that the initial values of the parameter estimates, which can be considered as part of system state variables, must be chosen from the compact set. If the true parameters are outside the set, system stability cannot be ensured. Also the developed schemes cannot ensure the transient performance due to their design methods.

In this chapter, we consider state feedback control and output feedback control of a class of nonlinear systems.

10.2   State Feedback Backstepping Control

In this section, we consider a class of uncertain dynamic nonlinear systems preceded by unknown dead-zone nonlinearities, in the presence of bounded external disturbances. By using backstepping technique, robust adaptive control algorithms are developed as in [141]. Unlike some existing control schemes for systems with dead-zone, the developed backstepping controllers do not require the uncertain parameters within known intervals. Also no knowledge is assumed on the bound of the 'disturbance-like' term, a combination of the external disturbances and a term separated from the dead-zone model. It is shown that the proposed controllers not only can guarantee stability, but also transient performance.

10.2.1   Problem Statement

We consider a class of nonlinear systems with unknown dead-zone given as follows:

$$x^{(n)}(t) + \sum_{i=1}^{r} a_i Y_i\big(x(t), \dot{x}(t), \ldots, x^{(n-1)}(t)\big) = bu(v) + \bar{d}(t) \qquad (10.1)$$

where Y_i are known continuous linear or nonlinear functions, $\bar{d}(t)$ denotes bounded external disturbances, parameters a_i are unknown constants and control gain b is unknown constant, v is the control input, $u(v)$ denotes dead-zone nonlinearity described by

$$u = \begin{cases} m(v(t) - b_r) & v(t) \geq b_r \\ 0 & b_l < v(t) < b_r \\ m(v(t) - b_l) & v(t) \leq b_l \end{cases} \qquad (10.2)$$

where $b_r \geq 0, b_l \leq 0$ and $m > 0$ are constants, v is the input and u is the output. For plant (10.1) with dead-zone nonlinearity, the $u(t)$ can be expressed as

$$u(t) = mv(t) + d_1(v(t)) \tag{10.3}$$

where

$$d_1(v(t)) = \begin{cases} -mb_r & v(t) \geq b_r \\ -mv(t) & b_l < v(t) < b_r \\ -mb_l & v(t) \leq b_l \end{cases} \tag{10.4}$$

It is clear that $d_1(v(t))$ is bounded.

From the structure (10.3) of model (10.2), (10.1) becomes

$$x^{(n)}(t) + \sum_{i=1}^{r} a_i Y_i\big(x(t), \dot{x}(t), \ldots, x^{(n-1)}(t)\big) = \beta v(t) + d(t) \tag{10.5}$$

where $\beta = bm$ and $d(t) = bd_1(v(t) + \bar{d}(t)$. The effect of $d(t)$ is due to both external disturbances and $bd_1(v(t))$. We call $d(t)$ a 'disturbance-like' term for simplicity of presentation and use D to denote its bound.

Now equation (10.5) is rewritten in the following form

$$\dot{x}_1 = x_2$$

$$\vdots$$

$$\dot{x}_{n-1} = x_n$$

$$\dot{x}_n = -\sum_{i=1}^{r} a_i Y_i\big(x_1(t), x_2(t), \ldots, x_{(n-1)}(t)\big) + \beta v(t) + d(t)$$

$$= a^T Y + \beta v(t) + d(t) \tag{10.6}$$

where $x_1 = x, x_2 = \dot{x}, \ldots, x_n = x^{(n-1)}$, $a = [-a_1, -a_2, \ldots, -a_r]^T$ and $Y = [Y_1, Y_2, \ldots, Y_r]^T$.

For the development of control laws, the following assumptions are made.

Assumption 1. The uncertain parameters b and m are such that $\beta > 0$.

Assumption 2. The desired trajectory $y_r(t)$ and its $(n-1)$th order derivatives are known and bounded.

The control objectives are to design backstepping adaptive control laws such that

- The closed loop is globally stable in sense that all the signals in the loop are uniformly ultimately bounded;
- The tracking error $x(t) - y_r(t)$ is adjustable during the transient period by an explicit choice of design parameters and $\lim_{t \to \infty} x(t) - y_r(t) = 0$ or $\lim_{t \to \infty} |x(t) - y_r(t)| - \delta_1 = 0$ for an arbitrary specified bound δ_1.

10.2.2   Controller Design

We develop two adaptive backstepping designs as in Chapter 7. These schemes are now concisely summarized in the tables 10.1 and 10.2, where $c_i, i = 1, \ldots, n$, are positive design parameters, γ and η are two positive design parameters, Γ is a positive definite matrix, $\hat{e}$, $\hat{a}$ and $\hat{D}$ are estimates of $e = 1/\beta$, a and D, $\delta_i(i = 1, \ldots, n)$ are positive design parameters and $q = round\{(n - i + 2)/2\}$, $round\{x\}$ means the element of x to the nearest integer.

Table 10.1. Adaptive Backstepping Controller-Scheme I

Adaptive Control Laws:

$$\alpha_1 = -c_1 z_1 \tag{10.7}$$

$$\alpha_i = -c_i z_i - z_{i-1} + \dot{\alpha}_{i-1}(x_1, \ldots, x_{i-1}, y_r, \ldots, y_r^{(i-1)}) \tag{10.8}$$

$$\bar{v} = -c_n z_n - z_{n-1} - \hat{a}^T Y - sign(z_n)\hat{D} + y_r^{(n)} + \dot{\alpha}_{n-1}$$

$$v = \hat{e}\bar{v}, \quad i = 2, \ldots, n \tag{10.9}$$

Parameter Update Laws:

$$\dot{\hat{e}} = -\gamma \bar{v} z_n \tag{10.10}$$

$$\dot{\hat{a}} = \Gamma Y z_n \tag{10.11}$$

$$\dot{\hat{D}} = \eta |z_n| \tag{10.12}$$

Theorem 10.1. *Consider the uncertain nonlinear system (10.1) satisfying Assumptions 1-2. With the application of controller (10.9) and the parameter update laws (10.10) to (10.12), the following statements hold:*

- *The resulting closed loop system is globally stable.*
- *The asymptotic tracking is achieved, i.e.,*

$$\lim_{t \to \infty} [x(t) - y_r(t)] = 0 \tag{10.22}$$

- *The transient tracking error performance is given by*

$$\| x(t) - y_r(t) \|_2 \leq \frac{1}{\sqrt{c_1}} \left(\frac{1}{2} \tilde{a}(0)^T \Gamma^{-1} \tilde{a}(0) + \frac{\beta}{2\gamma} \tilde{e}(0)^2 + \frac{1}{2\eta} \tilde{D}(0)^2 \right)^{1/2} \tag{10.23}$$

Theorem 10.2. *Consider the uncertain nonlinear system (10.1) satisfying Assumptions 1-2. With the application of controller (10.18) and the parameter update laws (10.19) to (10.21), the following statements hold:*

Table 10.2. Adaptive Backstepping Controller-Scheme II

Functions:

$$sg_i(z_i) = \begin{cases} \dfrac{z_i}{|z_i|} & |z_i| \geq \delta_i \\[2mm] \dfrac{z_i^{(2q+1)}}{(\delta_i^2 - z_i^2)^{n-i+2} + |z_i|^{(2q+1)}} & |z_i| < \delta_i \end{cases}$$

$$(10.13)$$

$$f_i(z_i) = \begin{cases} 1 & |z_i| \geq \delta_i \\ 0 & |z_i| < \delta_i \end{cases} \quad i = 1, \ldots, n \qquad (10.14)$$

Adaptive Control Laws:

$$\alpha_1 = -(c_1 + \frac{1}{4})(|z_1| - \delta_1)^n sg_1(z_1) - (\delta_2 + 1)sg_1(z_1) \qquad (10.15)$$

$$\alpha_2 = -(c_2 + \frac{5}{4})(|z_2| - \delta_2)^{n-1} sg_2(z_2) + \dot\alpha_1 - (\delta_3 + 1)sg_2(z_2) \qquad (10.16)$$

$$\alpha_i = -(c_i + \frac{5}{4})(|z_i| - \delta_i)^{n-i+1} sg_i(z_i) + \dot\alpha_{i-1}$$
$$-(\delta_{i+1} + 1)sg_i(z_i) \quad (i = 3, \ldots, n) \qquad (10.17)$$

$$\bar v = -(c_n + 1)(|z_n| - \delta_n)sg_n(z_n) - \hat a^T Y - sg_n\hat D + y_r^{(n)} + \dot\alpha_{n-1}$$
$$v = \hat e \bar v \qquad (10.18)$$

Parameter Update Laws:

$$\dot{\hat e} = -\gamma\bar v(|z_n| - \delta_n)f_n sg_n(z_n) \qquad (10.19)$$

$$\dot{\hat a} = \Gamma Y(|z_n| - \delta_n)f_n sg_n(z_n) \qquad (10.20)$$

$$\dot{\hat D} = \eta(|z_n| - \delta_n)f_n \qquad (10.21)$$

- *The resulting closed loop system is globally stable.*
- *The tracking error converges to δ_1 asymptotically, i.e.,*

$$\lim_{t\to\infty} |x(t) - y_r(t)| = \delta_1, \quad |z_1| \geq \delta_1 \qquad (10.24)$$

- *The transient tracking error performance is given by*

$$\| \, |x(t) - y_r(t)| - \delta_1 \, \|_2 \leq c_1^{\frac{-1}{2n}} \left(\frac{1}{2}\tilde a(0)^T \Gamma^{-1} \tilde a(0) + \frac{\beta}{2\gamma}\tilde e(0)^2 + \frac{1}{2\eta}\tilde D(0)^2\right)^{\frac{1}{2n}}$$

$$(10.25)$$

with $z_i(0) = \delta_i, i = 1, \ldots, n$,

Remark 10.1. From the above two theorems the following conclusions can be obtained:

- The transient performance depends on the initial estimate errors $\tilde{e}(0)$, $\tilde{a}(0)$, $\tilde{D}(0)$ and the explicit design parameters. The closer the initial estimates $\hat{e}(0), \hat{a}(0)$ and $\hat{D}(0)$ to the true values e, a and D, the better the transient performance.
- The bound for $\parallel x(t) - y_r(t) \parallel_2$ is an explicit function of design parameters and thus computable. We can decrease the effects of the initial error estimates on the transient performance by increasing the adaptation gains γ, η and Γ.
- To improve the tracking error performance we can also increase the gain c_1. However, increasing c_1 will influence other performance such as $\parallel \dot{x} - \dot{y}_r \parallel_2$.

10.3 Output Feedback Control Using Backstepping and Inverse Technique

In this section, we will address the output feedback control of a class of nonlinear systems in the presence of unknown dead-zone actuator nonlinearity. We take the dead-zone into account in our controller design unlike in [74] and [140]. A new smooth inverse of the dead-zone will be introduced to compensate the effect of the dead-zone in controller design with backstepping approach as in [125]. Such a smooth inverse can avoid chattering problems that may occur in the nonsmooth inverse approach proposed in [41], [49], [126] and [142]. The specific treatment of the dead-zone may bring performance improvement. As system output feedback is employed, a state observer is required. To obtain such an observer, a new parametrization of the state observer for the plant is proposed to include two sets of parameters: one from the dead-zone nonlinearity and the other from the plant. Besides showing stability of the system, the transient performance in terms of L_2 norm of the tracking error is derived to be an explicit function of design parameters and thus our scheme allows designers to obtain the closed loop behavior by tuning design parameters in an explicit way.

10.3.1 System Model

We consider the class of uncertain nonlinear systems with unknown dead-zone nonlinearity. For completeness, the system model is given as follows:

$$x^{(n)}(t) + a_1 Y_1(x(t)) + a_2 Y_2(x(t)) + \ldots + a_r Y_r(x(t)) = bu \qquad (10.26)$$

$$y = x_1, \quad u = DZ(v) \qquad (10.27)$$

where Y_i are known continuous linear or nonlinear functions, parameters a_i and control gain b are unknown constants, $v(t)$ is the output from the controller, $u(t)$ is the input to the system and $y(t)$ is the system output. The actuator nonlinearity $DZ(v)$ is described as a dead-zone characteristic.

The control objective is to design an output feedback control law for $v(t)$ to ensure that all closed-loop signals are bounded and the plant output $y(t)$ tracks a given reference signal $y_r(t)$ under the following assumptions:

Assumption 1. The sign of b is known and the reference signal $y_r(t)$ and its first nth derivatives are known and bounded.

Assumption 2. The dead-zone parameters m_r and m_l satisfy $m_r \geq m_{r0}$ and $m_l \geq m_{l0}$, where m_{r0} and m_{l0} are two small positive constants.

10.3.2   Dead-Zone Characteristic

The parameterized model of the dead-zone characteristic $DZ(.)$ can be unified as follows.

$$u(t) = DZ(v(t)) = \begin{cases} m_r(v(t) - b_r) & v(t) \geq b_r \\ 0 & b_l < v(t) < b_r \\ m_l(v(t) - b_l) & v(t) \leq b_l \end{cases} \qquad (10.28)$$

where $b_r \geq 0, b_l \leq 0$ and $m_r > 0, m_l > 0$ are constants. In general, the break-points $|b_r| \neq |b_l|$ and the slopes $m_r \neq m_l$.

The essence of compensating dead-zone effect is to employ a dead-zone inverse as shown in [41, 49, 126, 142]. In this section, we propose a smooth inverse for the dead-zone as follows:

$$v(t) = DI(u(t)) = \frac{u(t) + m_r b_r}{m_r} \phi_r(u) + \frac{u(t) + m_l b_l}{m_l} \phi_l(u) \qquad (10.29)$$

where $\phi_r(u)$ and $\phi_l(u)$ are smooth continuous indicator functions defined as

$$\phi_r(u) = \frac{e^{u/e_0}}{e^{u/e_0} + e^{-u/e_0}} \qquad (10.30)$$

$$\phi_l(u) = \frac{e^{-u/e_0}}{e^{u/e_0} + e^{-u/e_0}} \qquad (10.31)$$

Such an inverse is shown in Figure 10.1.

As $e_0 \to 0$, $\phi_r(u)$ and $\phi_l(u)$ approaches the indicator functions defined in [49].

Remark 10.2. Note that the functions $\phi_r(u)$ and $\phi_l(u)$ are continuous and differentiable. This is different from the inverse in [41], [49] and [126], where the inverse indicator functions are nonsmooth. The latter case may cause chattering phenomenon in the recursive backstepping control.

To design adaptive controller for the system, we parameterize the dead-zone as

$$u(t) = -\theta^T \omega \qquad (10.32)$$

where

$$\theta = [m_r, m_r b_r, m_l, m_l b_l]^T \qquad (10.33)$$

$$\omega(t) = [-\sigma_r(t)v(t), \sigma_r(t), -\sigma_l(t)v(t), \sigma_l(t)]^T \qquad (10.34)$$

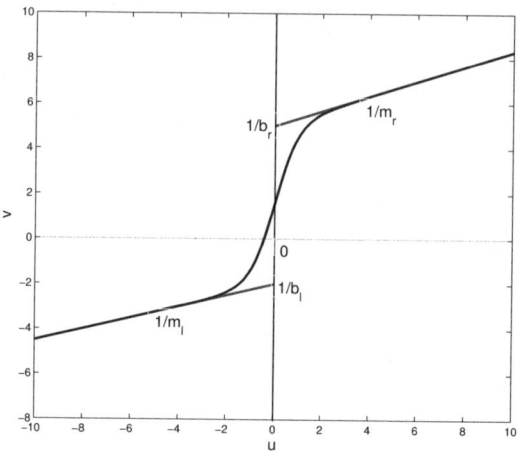

Fig. 10.1. Dead-Zone inverse

$$\sigma_r(t) = \begin{cases} 1 & if \ u(t) > 0 \\ 0 & otherwise \end{cases} \tag{10.35}$$

$$\sigma_l(t) = \begin{cases} 1 & if \ u(t) < 0 \\ 0 & otherwise \end{cases} \tag{10.36}$$

As θ is unknown and ω is unavailable, the actual control input to the plant $u_d(t)$ is designed as

$$u_d(t) = -\hat{\theta}^T \hat{\omega}(t) \tag{10.37}$$

$$\hat{\theta} = [\widehat{m_r}, \widehat{m_r b_r}, \widehat{m_l}, \widehat{m_l b_l}]^T \tag{10.38}$$

$$\hat{\omega}(t) = [-\phi_r(v)v(t), \phi_l(v), -\phi_l(v)v(t), \phi_l(v)]^T \tag{10.39}$$

where $\hat{\theta}$ is an estimate of θ. Then corresponding control output $v(t)$ is given by

$$v(t) = \widehat{DI}(u_d(t)) = \frac{u_d(t) + \widehat{m_r b_r}}{\widehat{m_r}} \phi_r(u_d) + \frac{u_d(t) + \widehat{m_l b_l}}{\widehat{m_l}} \phi_l(u_d(t)) \tag{10.40}$$

The resulting error between u and u_d is

$$u(t) - u_d(t) = (\hat{\theta} - \theta)^T \hat{\omega}(t) + d_N(t) \tag{10.41}$$

where $d_N(t) = \theta^T \hat{\omega}(t) - u(t)$. The bound of $d_N(t)$ can be obtained as

$$|d_N(t)| = |\theta^T \hat{\omega}(t) - u(t)|$$

$$\leq \begin{cases} \frac{1}{2}e^{-1}|m_r - m_l|e_0 + \frac{|m_r b_r - m_l b_l|}{e^{2b_r/e_0} + 1} & v(t) \geq b_r \\ max\{|m_r b_r|, |m_l b_l|\} & b_l < v(t) < b_r \\ \frac{1}{2}e^{-1}|m_r - m_l|e_0 + \frac{|m_r b_r - m_l b_l|}{e^{-2b_l/e_0} + 1} & v(t) \leq b_l \end{cases} \tag{10.42}$$

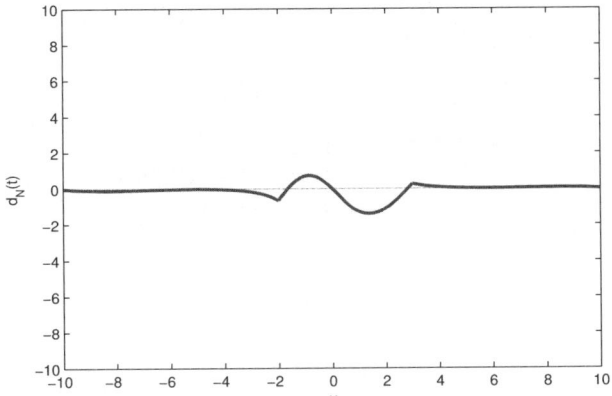

Fig. 10.2. Approximation error $d_N(t)$

where we have used that $|v|e^{-|v|} \leq e^{-1}$. It has the desired properties that $d_N(t)$ is bounded for all $t \geq 0$ and $d_N(t)$ approaches to 0 as $\hat{\theta} \rightarrow \theta$ and $e_0 \rightarrow 0$. Figure 10.2 shows the characteristic of $d_N(t)$ and it can be observed that $d_N(t)$ is bounded.

10.3.3   State Observer

As we consider output feedback, a state observer is required. To design such an observer, we re-write plant equation (10.26) as

$$\dot{x} = Ax + a^T Y e_n + bu e_n \tag{10.43}$$

$$y = cx, \quad u = DZ(v) \tag{10.44}$$

where

$$A = \begin{bmatrix} 0 \\ \vdots \ I_{n-1} \\ 0 \dots 0 \end{bmatrix}, \ a = \begin{bmatrix} -a_1 \\ \vdots \\ -a_r \end{bmatrix}, \ Y = \begin{bmatrix} Y_1 \\ \vdots \\ Y_r \end{bmatrix}, \ c = \begin{bmatrix} 1 \\ \vdots \\ 0 \end{bmatrix}^T \ e_n = \begin{bmatrix} 0 \\ \vdots \\ 1 \end{bmatrix}$$

To construct an observer for (10.43) and (10.44), we choose $k = [k_1, \dots, k_n]^T$ such that all eigenvalues of $A_0 = A - kc$ are at some desired stable locations. If the signal $u(t)$ were available we would implement the following filters

$$\hat{x}(t) = \xi_0 - \sum_{i=1}^{r} a_i \xi_i + b\eta \tag{10.45}$$

$$\dot{\xi}_0 = A_0 \xi_0 + ky + \chi \tag{10.46}$$

$$\dot{\xi}_i = A_0 \xi_i + Y_i e_n, \ i = 1, \dots, r \tag{10.47}$$

$$\dot{\eta} = A_0 \eta + e_n u \tag{10.48}$$

where χ is a design signal specified later. It can be shown that the state estimation error $\epsilon = x(t) - \hat{x}(t)$ satisfies

$$\dot{\epsilon} = A_0\epsilon - \chi \tag{10.49}$$

Hence, $lim_{t\to\infty}\epsilon(t) = 0$ exponentially if $\chi = 0$.

In our control problem, the signal $u(t)$ is not available. Thus the signal η in (10.48) needs to be re-parameterized. From (10.37) and (10.41), we know that

$$u(t) = -\theta^T\hat{\omega}(t) + d_N(t) \tag{10.50}$$

Let $p = \frac{d}{dt}$. With $\Delta(p) = det(pI - A_0)$, we express (10.48)

$$\eta(t) = [\eta_1(t), \eta_2(t), \ldots, \eta_n(t)]^T$$
$$= [q_1(p), q_2(p), \ldots, q_n(p)]^T \frac{1}{\Delta(p)}u(t) \tag{10.51}$$

for some known polynomials $q_i(p), i = 1, \ldots, n$. Using (10.50) and (10.51), we obtain

$$\eta_i(t) = -\theta^T\hat{\omega}_i(t) + d_i(t) \tag{10.52}$$

$$\hat{\omega}_i(t) = \frac{q_i(p)I_4}{\Delta(p)}\hat{\omega}(t) \tag{10.53}$$

$$d_i(t) = \frac{q_i(p)}{\Delta(p)}d_N(t) \tag{10.54}$$

Based on (10.52), $\hat{\omega}_i$ is available for controller design in place of u. Denoting the second components of ξ_0, ξ_i as $\xi_{02}, \xi_{i2}, i = 0, \ldots, r$, respectively, we have

$$\hat{x}_2 = \xi_{02} - \sum_{i=1}^{r} a_i\xi_{i2} - b\theta^T\hat{\omega}_2(t) + bd_2(t) \tag{10.55}$$

$$\hat{\omega}_2(t) = \frac{(p + k_1)I_4}{p^n + k_1p^{n-1} + \ldots + k_{n-1}p + k_n}\hat{\omega}(t) \tag{10.56}$$

The term $\theta^T\hat{\omega}_2^{(n-1)}$ renders a place to the signal u_d at the last step of the backstepping design.

10.3.4   Backstepping Design with Dead-Zone Inverse

Design Procedure

As usual in backstepping approach, the following change of coordinates is made.

$$z_1 = y - y_r \tag{10.57}$$
$$z_i = -\hat{\theta}^T\hat{\omega}_2^{(i-2)} - \hat{e}y_r^{(i-1)} - \alpha_{i-1}, \ i = 2, 3, \ldots, \rho \tag{10.58}$$

where $\hat{e}$ is an estimate of $e = 1/b$ and α_{i-1} is the virtual control at the *ith step* and will be determined in later discussion.

Firstly, we define functions $sg_i(z_i)$ and $f_i(z_i)$ as follows

$$
sg_i(z_i) = \begin{cases} \dfrac{z_i}{|z_i|} & |z_i| \geq \delta_i \\[2mm] \dfrac{z_i^{(2q+1)}}{(\delta_i^2 - z_i^2)^{n-i+2} + |z_i|^{(2q+1)}} & |z_i| < \delta_i \end{cases} \tag{10.59}
$$

$$
f_i(z_i) = \begin{cases} 1 & |z_i| \geq \delta_i \\ 0 & |z_i| < \delta_i \end{cases} \tag{10.60}
$$

where $\delta_i (i = 1, \ldots, n)$ is a positive design parameter and $q = round\{(n - i + 2)/2\}$, where $round\{x\}$ means the element of x to the nearest integer. Clearly $2q+1 \geq (n-i+2)$. It can be shown that $sg_i(z_i)$ is $(n-i+2)th$ order differentiable.

As the backstepping design procedures are similar to [120], only the first and the last steps of the design, i.e. *steps 1 and n* below, are elaborated in details.

• *Step* 1: We start with the equation for the tracking error z_1 obtained from (10.43), (10.55) and (10.57) to obtain

$$
\dot{z}_1 = \xi_{02} + a^T \xi_{(2)} + bz_2 + b\alpha_1 - b\tilde{\theta}^T \hat{\omega}_2(t) + d(t) + \epsilon_2 - b\tilde{e}\dot{y}_r \tag{10.61}
$$

where $d(t) = bd_2(t)$, $\tilde{\theta} = \theta - \hat{\theta}$, $\xi_{(2)} = [\xi_{i1}, \ldots, \xi_{r2}]$. From (10.42) and (10.54), there exists a positive constant D such that

$$
|d(t)| \leq D.
$$

Remark 10.3. The unknown bound D of $d(t)$ will be estimated online and thus it is not assumed to be known in contrast with [41], [49] and [126]. In fact, bounded external disturbance can also be treated in the same way, even though disturbance is not considered explicitly in this section.

Now select the first virtual control law α_1 as

$$
\alpha_1 = \hat{e}\bar{\alpha}_1 \tag{10.62}
$$

$$
\bar{\alpha}_1 = -(c_1 + \frac{\hat{b}^2}{4})(|z_1| - \delta_1)^n sg_1 - \xi_{02} - \hat{a}^T \xi_{(2)}
$$

$$
-\hat{D}_1 sg_1 - (\delta_2 + 1)\sqrt{\hat{b}^2 + \delta_0} \cdot sg_1 \tag{10.63}
$$

where c_1 is a positive constant, δ_0 is a small positive real number, $\hat{e}, \hat{a}$ and $\hat{b}$ are estimates of e, a and b, $\hat{D}_1$ is an estimate of D. We define a positive definite function V_1 as

$$
V_1 = \frac{1}{n+1}(|z_1| - \delta_1)^{n+1} f_1 + \frac{1}{2}|b|\tilde{\theta}^T \Gamma_\theta^{-1} \tilde{\theta} + \frac{1}{2}\tilde{a}^T \Gamma_a \tilde{a}
$$

$$
+\frac{|b|}{2\gamma_1}\tilde{e}^2 + \frac{1}{2\gamma_{d1}}\tilde{D}_1^2 + \frac{1}{2l_1}\epsilon^T P\epsilon \tag{10.64}
$$

where $\tilde{a} = a - \hat{a}, \tilde{e} = e - \hat{e}$, Γ_θ, Γ_a are positive definite matrices, γ_1, γ_{d1} are positive constants, and $P = P^T > 0$ satisfies the equation $PA_0 + A_0^T P = -2I$. We select the adaptive update laws as

$$\dot{\hat{\theta}} = Proj\{-\text{sign}(b)\Gamma_\theta \hat{\omega}_2(t)(|z_1| - \delta_1)^{2n} f_1 sg_1\} \tag{10.65}$$

$$\dot{\hat{e}} = -\text{sign}(b)\gamma_1(\bar{\alpha}_1 + \dot{y}_r)(|z_1| - \delta_1)^{2n} f_1 sg_1 \tag{10.66}$$

$$\dot{\hat{D}}_1 = \gamma_{d1}(|z_1| - \delta_1)^n f_1 \tag{10.67}$$

where $Proj(.)$ is a smooth projection operation to ensure the estimates $\hat{m}_r(t) \geq m_{r0}$ and $\hat{m}_l(t) \geq m_{l0}$. Such an operation can be found in [1]. Then form (10.64) to (10.66), we obtain the time derivative of V_1 as

$$\dot{V}_1 \leq -(c_1 + \frac{\hat{b}^2}{4})(|z_1| - \delta_1)^{2n} f_1 + \tilde{a}^T(\tau_{a1} - \Gamma_a^{-1}\dot{\hat{a}}) + \epsilon^T(\tau_{\chi 1} - \frac{1}{l_1}P\chi)$$

$$+ (|z_1| - \delta_1)^n f_1 sg_1\left(bz_2 - (\delta_2 + 1)\sqrt{\hat{b}^2 + \delta_0}\right) - \frac{1}{l_1}\epsilon^T \epsilon \tag{10.68}$$

$$\tau_{a1} = \xi_{(2)}(|z_1| - \delta_1)^n f_1 sg_1 \tag{10.69}$$

$$\tau_{\chi 1} = e_2(|z_1| - \delta_1)^n f_1 sg_1 \tag{10.70}$$

where $e_2 = [0, 1, 0, \ldots, 0]^T$.
• Step 2: Using (10.45), (10.46), (10.58)and (10.62), we write

$$\dot{z}_2 = -\frac{d}{dt}\left(\hat{\theta}^T \hat{\omega}_2 - \hat{e}\dot{y}_r\right) - \dot{\alpha}_1$$

$$= -\dot{\hat{\theta}}^T \hat{\omega}_2 + z_3 - \dot{\hat{e}}\dot{y}_r + \alpha_2 - \frac{\partial \alpha_1}{\partial y}\left(\xi_{02} + a^T \xi_{(2)} - b\theta^T \hat{\omega}_2(t) + d(t) + \epsilon_2\right)$$

$$- \frac{\partial \alpha_1}{\partial \hat{e}}\dot{\hat{e}} - \frac{\partial \alpha_1}{\partial y_r}\dot{y}_r - \frac{\partial \alpha_1}{\partial \dot{y}_r}\ddot{y}_r - \frac{\partial \alpha_1}{\partial \xi_0}(A_0\xi_0 + ky + \chi)$$

$$- \sum_{i=1}^{r}\frac{\partial \alpha_1}{\partial \xi_i}\dot{\xi}_i - \frac{\partial \alpha_1}{\partial \hat{a}}\dot{\hat{a}} - \frac{\partial \alpha_1}{\partial \hat{D}_1}\dot{\hat{D}}_1 - \frac{\partial \alpha_1}{\partial \hat{\omega}_2}\dot{\hat{\omega}}_2$$

$$= z_3 + \alpha_2 + \beta_2 - \frac{\partial \alpha_1}{\partial y}a^T \xi_{(2)} + \frac{\partial \alpha_1}{\partial y}\Theta^T \hat{\omega}_2(t) - \frac{\partial \alpha_1}{\partial \hat{a}}\dot{\hat{a}}$$

$$- \frac{\partial \alpha_1}{\partial \xi_0}\chi - \frac{\partial \alpha_1}{\partial y}d(t) - \frac{\partial \alpha_1}{\partial y}\epsilon_2 \tag{10.71}$$

where

$$\beta_2 = -\dot{\hat{e}}\dot{y}_r - \dot{\hat{\theta}}^T \hat{\omega}_2 - \frac{\partial \alpha_1}{\partial y}\xi_{02} - \frac{\partial \alpha_1}{\partial \hat{e}}\dot{\hat{e}} - \frac{\partial \alpha_1}{\partial y_r}\dot{y}_r - \frac{\partial \alpha_1}{\partial \dot{y}_r}\ddot{y}_r - \frac{\partial \alpha_1}{\partial \xi_0}(A_0\xi_0 + ky)$$

$$- \sum_{i=1}^{r}\frac{\partial \alpha_1}{\partial \xi_i}\dot{\xi}_i - \frac{\partial \alpha_1}{\partial \hat{\omega}_2}\dot{\hat{\omega}}_2 - \frac{\partial \alpha_1}{\partial \hat{D}_1}\dot{\hat{D}}_1 \tag{10.72}$$

where $\Theta = b\theta$.

Remark 10.4. Noted that in (10.72), the term $\hat{\omega}_2$ is continuous and differentiable because $\hat{\omega}$ is continuous function from the definitions of the dead-zone (10.30), (10.31), (10.39) and (10.56).

We choose the virtual control law for α_2 and the adaptive updated law for $\hat{b}$, the estimate of b, as

$$\alpha_2 = -(c_2 + 1)(|z_2| - \delta_2)^{n-1} sg_2 - \beta_2 - (\delta_3 + 1) sg_2 - \frac{\partial \alpha_1}{\partial y} \hat{\Theta}^T \hat{\omega}_2(t)$$

$$+ \frac{\partial \alpha_1}{\partial y} \hat{a}^T \xi_{(2)} + \sqrt{\| \frac{\partial \alpha_1}{\partial y} \|^2 + \delta_0} \cdot \hat{D}_2 sg_2 + \frac{\partial \alpha_1}{\partial \hat{a}} \Gamma_a \tau_{a2} + \frac{\partial \alpha_1}{\partial \xi_0} l_1 P^{-1} \tau_{\chi 2}$$

$$\tag{10.73}$$

$$\dot{\hat{b}} = \gamma_2 (|z_1| - \delta_1)^n f_1 sg_1 z_2 \tag{10.74}$$

$$\dot{\hat{D}}_2 = \gamma_{d2} \sqrt{\| \frac{\partial \alpha_1}{\partial y} \|^2 + \delta_0} \cdot (|z_2| - \delta_2)^{n-1} f_2 \tag{10.75}$$

$$\tau_{a2} = \tau_{a1} - \frac{\partial \alpha_1}{\partial y} \xi_{(2)} (|z_2| - \delta_2)^{n-1} f_2 sg_2 \tag{10.76}$$

$$\tau_{\chi 2} = \tau_{\chi 1} - \frac{\partial \alpha_1}{\partial y} (|z_2| - \delta_2)^{n-1} f_2 sg_2 e_2 \tag{10.77}$$

where c_2, γ_2 and γ_{d2} are positive constants. Defining a positive Lyapunov function V_2 as

$$V_2 = V_1 + \frac{1}{n}(|z_2| - \delta_2)^n f_2 + \frac{1}{2}\tilde{\Theta}^T \Gamma_\Theta^{-1} \tilde{\Theta} + \frac{1}{2\gamma_2}\tilde{b}^2 + \frac{1}{2\gamma_{d2}}\tilde{D}_2^2 \tag{10.78}$$

Then the derivative of V_2 is

$$\dot{V}_2 \leq -\sum_{i=1}^{2} c_i(|z_i| - \delta_i)^{2(n-i+1)} f_i + \tilde{a}^T(\tau_{a2} - \Gamma_a^{-1}\dot{\hat{a}}) + \epsilon^T(\tau_{\chi 2} - \frac{1}{l_1}P\chi)$$

$$+ [\frac{\partial \alpha_1}{\partial \hat{a}}(\Gamma_a \tau_{a2} - \dot{\hat{a}}) + \frac{\partial \alpha_1}{\partial \xi_0}(l_1 P^{-1} \tau_{\chi 2} - \chi)](|z_2| - \delta_2)^{n-1} f_2 sg_2 - \frac{1}{l_1}\epsilon^T \epsilon$$

$$+ \tilde{\Theta}^T(\tau_{\Theta 2} + \Gamma_\Theta^{-1}\dot{\hat{\Theta}}) + (|z_2| - \delta_2)^{n-1}(|z_3| - \delta_3 - 1)f_2 + M_2 \tag{10.79}$$

$$\tau_{\Theta 2} = -\frac{\partial \alpha_1}{\partial y}\hat{\omega}_2(|z_2| - \delta_2)^{n-1} f_2 sg_2 \tag{10.80}$$

where

$$M_2 = -\frac{\hat{b}^2}{4}(|z_1| - \delta_1)^{2n} f_1 + |\hat{b}|(|z_1| - \delta_1)^n(|z_2| - \delta_2 - 1)f_1 - (|z_2| - \delta_2)^{2(n-1)} f_2$$

Now we show that $M_2 < 0$. This is quite clear for $|z_2| < \delta_2 + 1$. For $|z_2| \geq \delta_2 + 1$

$$M_2 \leq -\frac{\hat{b}^2}{4}(|z_1| - \delta_1)^{2n} f_1 + \frac{\hat{b}^2}{4}(|z_1| - \delta_1)^{2n} f_1^2$$

$$+(|z_2| - \delta_2 - 1)^2 - (|z_2| - \delta_2)^{2(n-1)}$$
$$< (|z_2| - \delta_2)^2 - (|z_2| - \delta_2)^{2(n-1)}$$
$$= (|z_2| - \delta_2)^2(1 - (|z_2| - \delta_2)^{2(n-2)})$$
$$\leq 0 \tag{10.81}$$

- *Step i, $i = 3$, ..., n:* We choose

$$\alpha_i = -(c_i + 1)(|z_i| - \delta_i)^{n-i+1} sg_2 - \beta_i - (\delta_{i+1} + 1)sg_i + \frac{\partial \alpha_{i-1}}{\partial y} \hat{a}^T \xi_{(2)}$$

$$+ \sqrt{\| \frac{\partial \alpha_{i-1}}{\partial y} \|^2 + \delta_0} \cdot \hat{D}_i sg_i + \frac{\partial \alpha_{i-1}}{\partial \hat{a}} \Gamma_a \tau_{ai} + \frac{\partial \alpha_{i-1}}{\partial \xi_0} l_1 P^{-1} \tau_{\chi i}$$

$$- \sum_{k=3}^{i-1} (|z_k| - \delta_k)^{n-k+1} f_k sg_k \frac{\partial \alpha_{k-1}}{\partial \hat{\Theta}} \frac{\partial \alpha_{i-1}}{\partial y} \hat{\omega}_2 - \frac{\partial \alpha_{i-1}}{\partial y} \hat{\Theta}^T \hat{\omega}_2(t)$$

$$+ \sum_{k=2}^{i-1} (|z_k| - \delta_k)^{n-k+1} f_k sg_k \Big[- \frac{\partial \alpha_{k-1}}{\partial \hat{a}} \frac{\partial \alpha_{i-1}}{\partial y} \xi_{(2)}$$

$$- \frac{\partial \alpha_{k-1}}{\partial \xi_0} \frac{\partial \alpha_{i-1}}{\partial y} l_1 P^{-1} e_2 \Big] + \frac{\partial \alpha_{i-1}}{\partial \hat{\Theta}} \Gamma_\Theta \tau_{\Theta i} \tag{10.82}$$

and

$$\dot{\hat{D}}_i = \gamma_{di} \sqrt{\| \frac{\partial \alpha_{i-1}}{\partial y} \|^2 + \delta_0} \cdot (|z_i| - \delta_i)^{n-i+1} f_i \tag{10.83}$$

$$\tau_{ai} = \tau_{ai-1} - \frac{\partial \alpha_{i-1}}{\partial y} \xi_{(2)}(|z_i| - \delta_i)^{n-i+1} f_i sg_i \tag{10.84}$$

$$\tau_{\chi i} = \tau_{\chi i-1} - \frac{\partial \alpha_{i-1}}{\partial y} (|z_i| - \delta_i)^{n-i+1} f_i sg_i e_2 \tag{10.85}$$

$$\tau_{\Theta i} = \tau_{\Theta i-1} - \frac{\partial \alpha_{i-1}}{\partial y} (|z_i| - \delta_i)^{n-i+1} f_i sg_i \hat{\omega}_2 \tag{10.86}$$

and

$$V_i = \sum_{k=1}^{i} \left(\frac{1}{n-k+2}(|z_k| - \delta_k)^{n-k+2} f_k + \frac{1}{2\gamma_{dk}} \tilde{D}_k^2 \right) + \frac{1}{2}|b|\tilde{\theta}^T \Gamma_\theta^{-1} \tilde{\theta}$$

$$+ \frac{1}{2}\tilde{a}^T \Gamma_a \tilde{a} + \frac{|b|}{2\gamma_1} \tilde{e}^2 + \frac{1}{2}\tilde{\Theta}^T \Gamma_\Theta^{-1} \tilde{\Theta} + \frac{1}{2\gamma_2} \tilde{b}^2 + \frac{1}{2l_1} \epsilon^T P \epsilon \tag{10.87}$$

where $\hat{\Theta}, \hat{D}_k$ are estimates of $\Theta = b\theta$ and D, $\tilde{\Theta} = \Theta - \hat{\Theta}$, $\tilde{b} = b - \hat{b}$, $\tilde{D}_k = D - \hat{D}_k$, β_i contains all known terms, $c_i, \gamma_{di}, i = 1, \ldots, n$ are positive constants, Γ_Θ is a positive definite matrix.

Step n: Using (10.40), (10.50) and (10.56), we have

$$\hat{\theta}^T \hat{\omega}_2^{(n-1)} = \hat{\theta}^T \frac{(p^n + k_1 p^{n-1})I_4}{p^n + k_1 p^{n-1} + \ldots + k_{n-1}p + k_n} \hat{\omega}(t)$$

$$= -u_d(t) + \omega_0 \tag{10.88}$$

where ω_0 is given by

$$\omega_0 = -\frac{(k_2 p^{n-2} + \ldots + k_{n-1} p + k_n) I_4}{p^n + k_1 p^{n-1} + \ldots + k_{n-1} p + k_n} \hat{\omega}(t) \tag{10.89}$$

With this equation, the derivative of $z_n = -\hat{\theta}^T \hat{\omega}_2^{(n-2)} - \hat{e} y_r^{(n-1)} - \alpha_{n-1}$ is

$$\dot{z}_n = u_d + \beta_n - \frac{\partial \alpha_{n-1}}{\partial y} a^T \xi_{(2)} + \frac{\partial \alpha_{n-1}}{\partial y} \Theta^T \hat{\omega}_2(t) - \frac{\partial \alpha_{n-1}}{\partial \hat{a}} \dot{\hat{a}} - \sum_{j=1}^{n-1} \frac{\partial \alpha_{n-1}}{\partial \hat{D}_j} \dot{\hat{D}}_j$$

$$- \frac{\partial \alpha_{n-1}}{\partial \hat{\Theta}} \dot{\hat{\Theta}} - \frac{\partial \alpha_{n-1}}{\partial \xi_0} \chi - \frac{\partial \alpha_{n-1}}{\partial y} d(t) - \frac{\partial \alpha_{n-1}}{\partial y} \epsilon_2 \tag{10.90}$$

where β_n contains all known terms.

We choose the update laws for $\hat{a}$, $\hat{D}$, $\hat{\Theta}$

$$\dot{\hat{a}} = \Gamma_a \tau_{an} \tag{10.91}$$

$$\dot{\hat{D}}_n = \gamma_{dn} \sqrt{\| \frac{\partial \alpha_{n-1}}{\partial y} \|^2 + \delta_0} \cdot (|z_n| - \delta_n) f_n \tag{10.92}$$

$$\dot{\hat{\Theta}} = -\Gamma_{\Theta} \tau_{\Theta n} \tag{10.93}$$

and the design signal as

$$\chi = l_1 P^{-1} \tau_{\chi n} \tag{10.94}$$

Finally the control law is given by

$$v(t) = \frac{u_d(t) + \widehat{m_r b_r}}{\widehat{m_r}} \phi_r(u_d) + \frac{u_d(t) + \widehat{m_l b_l}}{\widehat{m_l}} \phi_l(u_d) \tag{10.95}$$

$$u_d = \alpha_n \tag{10.96}$$

10.3.5   Stability Analysis

Define a positive definite Lyapunov function V_n as

$$V_n = V_{n-1} + \frac{1}{2}(|z_n| - \delta_n)^2 f_n + \frac{1}{2\gamma_{dn}} \tilde{D}_n^2 \tag{10.97}$$

With this choice of adaptive controller and parameter update laws, the derivative of V_n becomes

$$\dot{V}_n \leq -\sum_{i=1}^{n} c_i(|z_i| - \delta_i)^{2(n-i+1)} f_i + \tilde{a}^T(\tau_{an} - \Gamma_a^{-1}\dot{\hat{a}}) + \epsilon^T(\tau_{\chi n} - \frac{1}{l_1} P \chi)$$

$$+ \tilde{\Theta}^T(\tau_{\Theta n} + \Gamma_{\Theta}^{-1}\dot{\hat{\Theta}}) - \frac{1}{l_1}\epsilon^T \epsilon + \sum_{k=3}^{n}(|z_k| - \delta_k)^{n-k+1} f_k \frac{\partial \alpha_{k-1}}{\partial \hat{\Theta}}(\Gamma_{\Theta}\tau_{d\Theta} + \dot{\hat{\Theta}})$$

$$+ \sum_{k=2}^{n}(|z_k| - \delta_k)^{n-k+1} f_k \Big[\frac{\partial \alpha_{k-1}}{\partial \hat{a}}(\Gamma_a \tau_{an} - \dot{\hat{a}}) + \frac{\partial \alpha_{k-1}}{\partial \xi_0}(l_1 P^{-1}\tau_{\chi n} - \chi) \Big]$$

$$\leq -\sum_{i=1}^{n} c_i(|z_i| - \delta_i)^{2(n-i+1)} f_i - \frac{1}{l_1}\epsilon^T \epsilon \tag{10.98}$$

From (10.98), we get the following Lemma.

Lemma 10.1. *The adaptive controller designed above ensures that* $z_1, \ldots, z_n$, $\hat{\theta}$, $\hat{e}$, $\hat{b}, \hat{a}, \hat{\Theta}, \hat{D}_i, \epsilon$ *are all bounded.*

With Lemma 10.1, all the signals in the closed-loop can be shown to be bounded and a bound can be established for the tracking error, as stated in the following theorem.

Theorem 10.3. *Consider the system consisting of the parameter estimators given by (10.65), (10.66), (10.74), (10.91)-(10.93), adaptive controllers designed using (10.95) and (10.96) with virtual control laws (10.62), (10.73) and (10.82), and plant (10.26) with a dead-zone nonlinearity (10.28). The system is stable in the sense that all signals in the closed loop are bounded. Furthermore*

- *The tracking error converges to* $[-\delta_1, \delta_1]$ *asymptotically, i.e.,*

$$\lim_{t \to \infty} |y(t) - y_r(t)| = \delta_1, \quad |z_1| \geq \delta_1 \tag{10.99}$$

- *The transient tracking error performance is given by*

$$\| \, |y(t) - y_r(t)| - \delta_1 \, \|_2 \leq \frac{1}{c_1^{2n}} \Big(\frac{1}{2} \tilde{a}(0)^T \Gamma_a^{-1} \tilde{a}(0) + \frac{|b|}{2\Gamma_\theta} \tilde{\theta}(0)^2 + \frac{1}{2\Gamma_\Theta} \tilde{\Theta}(0)^2$$

$$+ \frac{|b|}{2\gamma_1} \tilde{e}(0)^2 + \sum_{i=1}^{n} \frac{1}{2\gamma_{di}} \tilde{D}_i(0)^2 + \frac{1}{2\gamma_2} \tilde{b}(0)^2 + \frac{1}{2l_1} \epsilon(0)^2 \Big)^{1/2n} \tag{10.100}$$

with $z_i(0) = 0, i = 1, \ldots, n,$

Proof: From Lemma 10.1, we have that $z_1, \ldots, z_n, \hat{\theta}, \hat{e}, \hat{b}, \hat{a}, \hat{\Theta}, \hat{D}_i, \epsilon$ are bounded. The tracking error performance can be obtained from (10.98) following similar approaches to those in [125]. What we need to prove is the boundedness of state x, controller output v and plant input u. From state observers (10.46) and (10.47), and (10.49), we have that $\xi_0, \ldots, \xi_r$ are bounded. Re-writing plant (10.26) as

$$p^n y + \sum_{i=1}^{r} a_i Y_i \big(y, py, \ldots, p^{n-1} y \big) = bu \tag{10.101}$$

and using (10.51), we have

$$\eta_2 = \frac{q_2(p)}{\Delta(p)} u = \frac{p^n q_2(p)}{b\Delta(p)} y + \frac{q_2(p)}{b\Delta(p)} \sum_{i=1}^{r} a_i Y_i \big(y, py, \ldots, p^{n-1} y \big) \tag{10.102}$$

Since $\Delta(p) = p^n + k_1 p^{n-1} + \ldots + k_n$ is Hurwitz, so $\frac{q_2(p)}{b\Delta(p)}$ is stable. We have that η_2 is bounded because y is bounded. From (10.50) and (10.52), we have $\eta_2 = -\theta^T \hat{\omega}_2(t) + d_2(t)$. As $d_2(t) \in L^\infty$, then $\theta^T \hat{\omega}_2 \in L^\infty$.

Express (10.53) as

$$\hat{\omega}_2(t) = \big[-\frac{q_2(p)}{\Delta(p)} \phi_r(v)v, \frac{q_2(p)}{\Delta(p)} \phi_r(v), -\frac{q_2(p)}{\Delta(p)} \phi_l(v)v, \frac{q_2(p)}{\Delta(p)} \phi_l(v) \big]^T$$

$$\tag{10.103}$$

$$\theta^T \hat{\omega}_2(t) = -m_r \frac{q_2(p)}{\Delta(p)} \phi_r(v)v + m_r b_r \frac{q_2(p)}{\Delta(p)} \phi_r(v)$$

$$-m_l \frac{q_2(p)}{\Delta(p)} \phi_l(v)v + m_l b_l \frac{q_2(p)}{\Delta(p)} \phi_l(v) \qquad (10.104)$$

Because $\phi_r(v) \in L^\infty$, $\phi_l(v) \in L^\infty$ and $\frac{q_2(p)}{\Delta(p)}$ is stable, the terms $\frac{q_2(p)}{\Delta(p)} \phi_r(v)$ and $\frac{q_2(p)}{\Delta(p)} \phi_l(v)$ in (10.103) are bounded.

We now show that $\hat{\omega}_2$ is bounded in two cases:

Case 1. If $v(t)$ is bounded, $\hat{\omega}_2$ is bounded directly from (10.103).

Case 2. In case that $v(t)$ is unbounded, we divide $R^+ = [0, \infty)$ into two subsequences $R^+ = R_1 \cup R_2$, where $R_1 = \{t \, |v(t) \geq 0\}$ and $R_2 = \{t \, |v(t) < 0\}$. Then the following two situations are considered.

(i). $t \in R_1$. From (10.31) we get

$$\phi_l(v) \cdot v = \frac{e^{-v/e_0}}{e^{v/e_0} + e^{-v/e_0}} \cdot v = \frac{v}{1 + e^{2v/e_0}}$$

Thus $\phi_l(v) \cdot v \to 0$, when $v \to +\infty$ *for* $t \in R_1$.   (10.105)

So in equation (10.104), the third term $m_l \frac{q_2(p)}{\Delta(p)} \phi_l(v)v(t) \to 0$, with the boundedness of second term and fourth term and $\theta^T \hat{\omega}_2 \in L^\infty$, we see that the first term $m_r \frac{q_2(p)}{\Delta(p)} \phi_r(v)v$ is bounded for $t \in R_1$.

(ii). $t \in R_2$. Similarly from (10.30) we can show that

$$\phi_r(v) \cdot v = \frac{e^{v/e_0}}{e^{v/e_0} + e^{-v/e_0}} \cdot v = \frac{v}{1 + e^{-2v/e_0}}$$

Thus $\phi_r(v) \cdot v \to 0$, when $v \to -\infty$ *for* $t \in R_2$.   (10.106)

and the third term $m_l \frac{q_2(p)}{\Delta(p)} \phi_l(v)v$ is bounded for $t \in R_2$.

Combining (i) and (ii), we get that for all $t \in R^+$, $\frac{q_2(p)}{\Delta(p)} \phi_r(v)v$ and $\frac{q_2(p)}{\Delta(p)} \phi_l(v)v$ are bounded. Then $\hat{\omega}_2$ is bounded from (10.103). In summary, from the two cases we obtain the boundedness of $\hat{\omega}_2$.

Since $\hat{\theta}^T \hat{\omega}_2$ and z_2 are bounded, from $z_2 = -\hat{\theta}^T \hat{\omega}_2 - \hat{e}\dot{y}_r - \alpha_1$ we can obtain the boundedness of α_1. From (10.62), we have $\bar{\alpha}_1$ is bounded. From (10.82), $\alpha_2, \ldots, \alpha_n$ are bounded, and so is χ. From (10.96) we have that $u_d(t)$ is bounded, and so are $v = \widehat{DI}(u_d)$ and $u = DI(v)$. It following from (10.53) that $\hat{\omega}_i \in L^\infty, i = 1, \ldots, n$. From (10.48), we have that η is bounded. Then $\hat{x}$ is bounded from (10.45) and finally $x(t) = \hat{x}(t) + \epsilon(t)$ is bounded from (10.45-10.48).

$\triangle\triangle\triangle$

Remark 10.5. The transient performance depends on the initial estimate errors and the explicit design parameters. The closer the initial estimates to the true values, the better the transient performance. The bound for $\| y(t) - y_r(t) \|_2$ is an explicit function of design parameters and thus computable. We can decrease the effects of the initial error estimates on the transient performance by increasing the adaptation gains $\gamma_1, \gamma_2, \gamma_{di}$ and $\Gamma_a, \Gamma_\theta, \Gamma_\Theta$.

10.4   Illustrative Examples

10.4.1   Example 1: State Feedback Backstepping Control

In this section, we illustrate the above methodologies for state feedback control of a simple system which is described as:

$$\ddot{x} = a_1 \frac{1 - e^{-x(t)}}{1 + e^{-x(t)}} - a_2(\dot{x}^2 + 2x) - 0.2a_3 \sin(3t) + bu(t) \qquad (10.107)$$

where $u(t)$ represents the output of the dead-zone nonlinearity. The actual parameter values are $b = 1$ and $a_1 = a_2 = a_3 = 1$. The parameters of the dead-zone at $b_r = 0.5, b_l = -0.6, m = 1$. The objective is to control the system state x to follow a desired trajectory $y_r(t) = 2.5 \sin(t)$.

In the simulation of Scheme I, the robust adaptive control law (10.9)-(10.12) was used, taking $c_1 = 2, c_2 = 2, \gamma = 0.5, \Gamma = 0.5 I_3, \eta = 0.5$. The initial values are chosen as follows: $\hat{e}(0) = 0.25, \hat{a}(0) = [1.5\ 1\ 1]^T, \hat{D}(0) = 2, x(0) = [1\ 1.05]^T$ and $v(0) = 0$. The simulation results presented in the Figure 10.3 and Figure 10.4 are system tracking error and input. The effectiveness of adaptive Scheme I is demonstrated by the fact that the tracking error is reduced to zero after a few periods of the reference input as shown in Figure 10.3. The chattering phenomena in Figure 10.4 is caused by the *sign* function used in the controller. It can be avoided by adaptive Scheme II.

In the simulation of Scheme II by using the robust adaptive control law (10.18)-(10.21), we choose $c_1, \gamma, \eta, \Gamma$ and the initial values to be same as above and $\delta_1 = 0.05$. The simulation results presented in the Figure 10.5 and Figure 10.6 are system tracking error and input. In Figure 10.5, the tracking error is reduced to $\delta_1 = 0.05$ after a few periods of the reference input. The control input v is bounded and has no chattering problem as shown in Figure 10.6.

As a conclusion, all the results verify our theoretical findings and show the effectiveness of the control schemes.

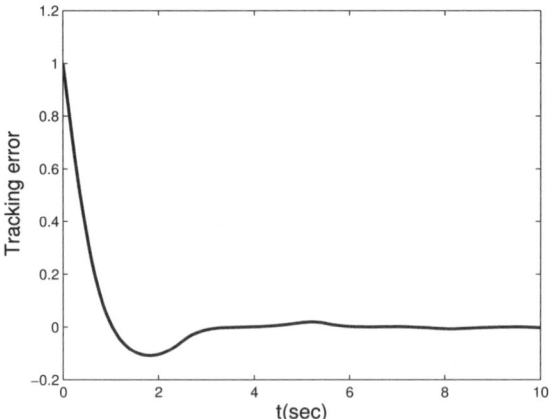

Fig. 10.3. Tracking error-Scheme I

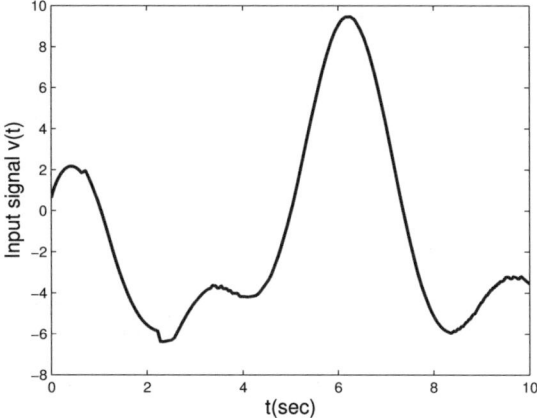

Fig. 10.4. Control signal $v(t)$-Scheme I

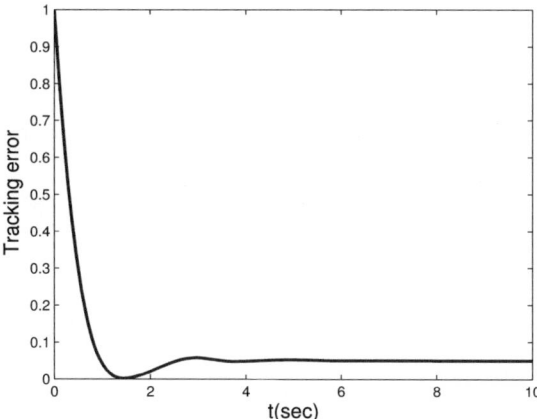

Fig. 10.5. Tracking error-Scheme II

10.4.2   Example 2: Output Feedback Inverse Control

In this section, we illustrate the above methodology for output feedback control of the same system as in [43] and [74], which is described as:

$$\dot{x} = a\frac{1 - e^{-x(t)}}{1 + e^{-x(t)}} + bu(t) \tag{10.108}$$
$$u = DZ(v)$$

where u represents the output of the dead-zone nonlinearity. The actual parameter values are $b = 1$ and $a = 1$, and the dead-zone parameter values are $m_r = 1.05, m_l = 1.05, b_r = 0.3, b_l = -0.5$. The objective is to control the system state x to follow a desired trajectory $y_r(t) = \sin(2t)$.

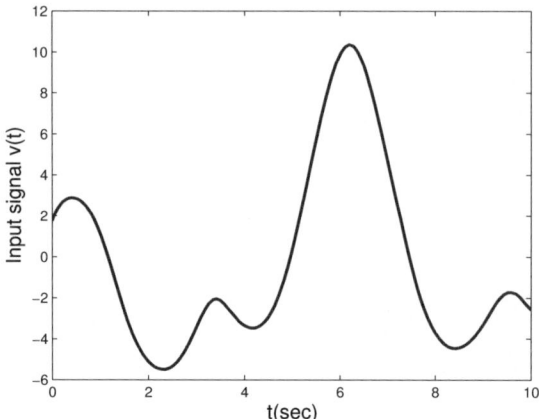

Fig. 10.6. Control signal $v(t)$-Scheme II

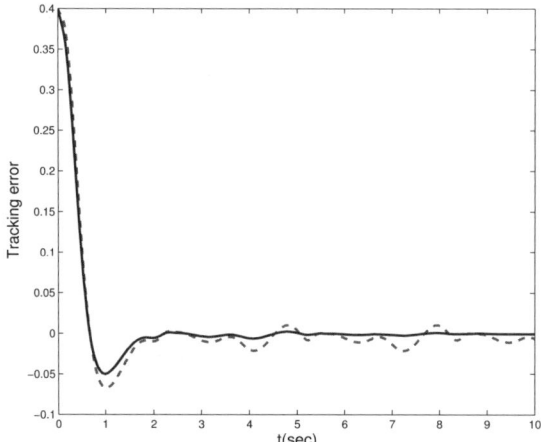

(dashed: scheme in [74], solid: proposed scheme)

Fig. 10.7. Tracking error

In the simulations, taking $c_1 = 4, \Gamma_a = 0.1, \gamma_1 = 0.3, \gamma_2 = 0.2, \Gamma_\theta = [0.1, 0.1,$ $0.1, 0.1]^T$, $e_0 = 1, \delta_1 = 0.02$ and the initial parameters $\hat{e}(0) = 0.3, \hat{a}(0) =$ $1.5, \hat{D}(0) = 0.4, \hat{\theta}(0) = [1, 1, 0.2, -0.3]^T$. The initial state is chosen as $x(0) = 0.4$. The parameters and the initial states are the same as in [74]. For comparison, the scheme in [74] and our proposed scheme are both applied to the system. The simulation results presented in the Figure 10.7 and Figure 10.8 are the tracking error and the controller output $v(t)$.

Note that the tracking error with the proposed scheme is reduced to zero after a few periods of the reference input as shown in Figure 10.7. However, for the scheme in [74], the tracking error converges to a residual. It is remarkable that

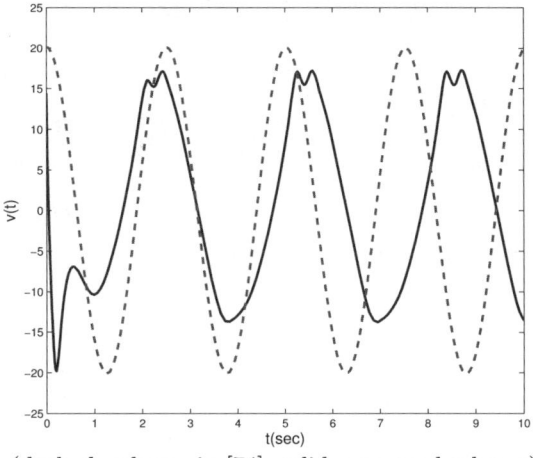

(dashed: scheme in [74], solid: proposed scheme)

Fig. 10.8. Control signal $v(t)$

the tracking performance of our scheme is better than that of other scheme, while the control effort is about same.

As a conclusion, the simulation results verify our theoretical findings and show the effectiveness of our control scheme. Also system performance is improved by our scheme.

10.4.3   Example 3: Application to Servo-Valve

In this section, we apply our proposed scheme to servo-valve as in Chapter 6. Its spool occludes the orifice with some overlap so that for a range of spool positions v there is no fluid flow u. This overlap prevents leakage losses which increase with wear and tear. Considering the spool position as the input v, and the load position y as output, the hydraulic system is represented in Figure 10.9 as a dead-zone block. It is located as the input of linear dynamics with transfer function $G(s) = \frac{K}{Ms^2+Bs}$, where $K = \frac{Ak_x}{k_p}$, $B = f + \frac{A^2}{k_p}$, $k_x = \frac{\partial g}{\partial x}$, $k_p = \frac{\partial g}{\partial P}$, $g = g(x, P) = $ flow, $A = $ area of position, $P = $ pressure, and $f = $ viscous friction. The system is modelled as

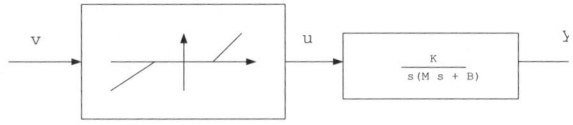

Fig. 10.9. Block diagram

$$M\ddot{y} + B\dot{y} = Ku(v) \qquad (10.109)$$

where $u(v)$ represents the dead-zone nonlinearity. The adaptive control law is designed as follows

$$v(t) = \frac{u_d(t) + \widehat{m_r b_r}}{\widehat{m_r}}\phi_r(u_d) + \frac{u_d(t) + \widehat{m_l b_l}}{\widehat{m_l}}\phi_l(u_d(t)) \qquad (10.110)$$

$$u_d = \hat{e}\bar{u} \qquad (10.111)$$

$$\bar{u} = -(c_2 + 1)(|z_2| - \delta_2)sg_2 + \dot{\hat{\alpha}}_1 + \hat{a}\dot{y} - sg_2\hat{D} \qquad (10.112)$$

$$\alpha_1 = -(c_1 + 1)(|z_1| - \delta_1)^2 sg_1 \qquad (10.113)$$

$$\dot{\hat{e}} = -\gamma\bar{u}(|z_2| - \delta_2)f_2 sg_2 \qquad (10.114)$$

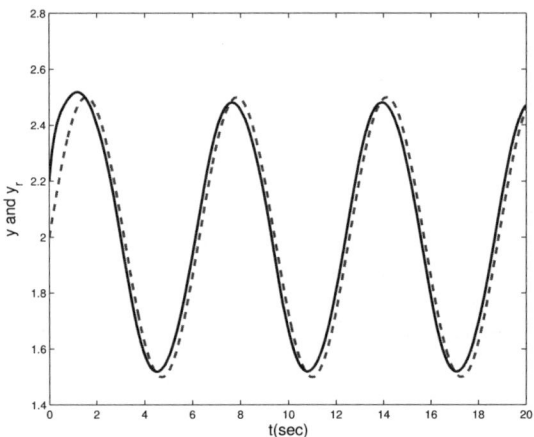

Fig. 10.10. Without considering dead-zone: Load position (y: solid line; y_r: dashed line)

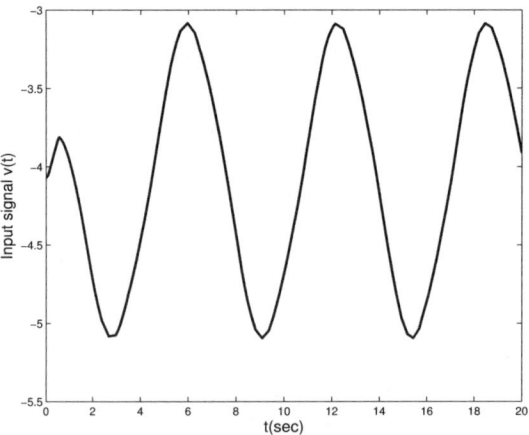

Fig. 10.11. Without considering dead-zone: Spool position v

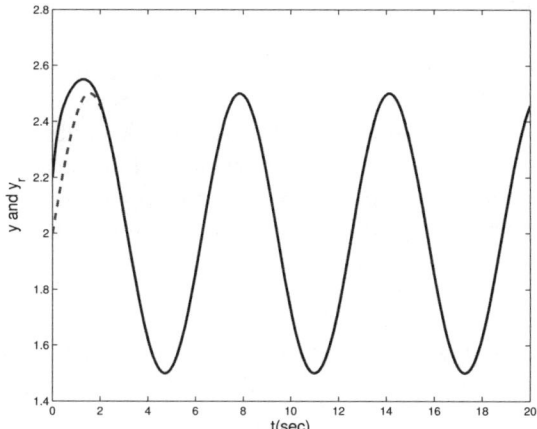

Fig. 10.12. Our proposed scheme: Load position (y: solid line; y_r: dashed line)

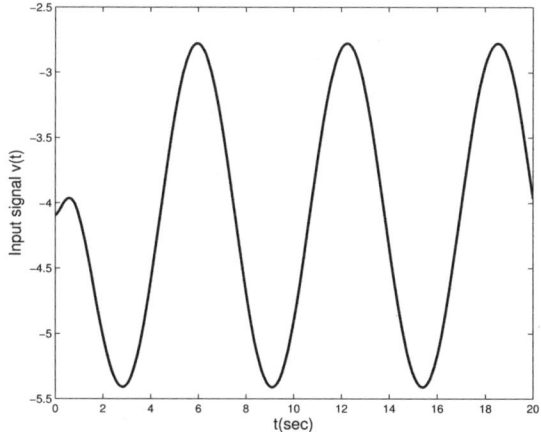

Fig. 10.13. Our proposed scheme: Spool position v

$$\dot{\hat{a}} = -\eta\dot{y}(|z_2| - \delta_2)f_2sg_2 \tag{10.115}$$

$$\dot{\hat{\theta}} = Proj\left\{ - \text{sign}(\frac{K}{M})\Gamma_\theta\hat{\omega}(|z_2| - \delta_2)f_2 \right\} \tag{10.116}$$

where $z_1 = y - y_r, z_2 = \dot{y} - \alpha_1 - \dot{y}_r, y_r = 2 + 0.5\sin(t)$, $\hat{a}$ is an estimate of $a = B/M$ and $\hat{e}$ is an estimate of $e = M/K$. In the simulation, the design parameters are chosen as $c_1 = c_2 = 1, \gamma = 0.5, \eta = 0.6, \Gamma_\theta = 0.2I_4$ and the initial value is chosen as $y(0) = 2.2, \hat{e}(0) = \hat{a}(0) = 1, \hat{\theta}(0) = [1, 1, 0.4, -0.4]^T$. The simulation results presented in Figures 10.10-10.13 show that the load position y and y_r and the spool position v using the controller without considering dead-zone and with our proposed smooth inverse. Clearly, our proposed scheme improves the system performance greatly.

10.5   Summary

In this chapter, we present two types of robust adaptive backstepping control algorithms: state feedback control and output feedback control of nonlinear systems with unknown dead-zone. For state feedback control two backstepping adaptive controller design schemes are developed. For output feedback control, we propose a new smooth adaptive inverse to compensate the effect of the unknown dead-zone. Such an inverse can avoid possible chattering phenomenon which may be caused by nonsmooth inverse. The inverse function is employed in the backstepping controller design. For the design and implementation of the controllers, no knowledge is assumed on the unknown system parameters and nonlinearity. Besides showing stability, we also give an explicit bound on the L_2 performance of the tracking error in terms of design parameters. Simulation results illustrate the effectiveness of proposed schemes.

11 Adaptive Control of Systems with Input Saturation

In this chapter, we present a new scheme to design adaptive controllers for uncertain nonlinear systems in the presence of input saturation. Similar to previous chapters, the developed controllers do not require uncertain parameters within a known compact set. Besides showing global stability, transient performance is also established and can be adjusted by tuning certain design parameters.

11.1 Introduction

In many practical dynamic systems, physical input saturation on hardware dictates that the magnitude of the control signal is always constrained. Saturation is a potential problem for actuators of control systems. It often severely limits system performance, giving rise to undesirable inaccuracy or leading instability. The development of adaptive control schemes for systems with input saturation has been a task of major practical interest as well as theoretical significance.

However, the number of available results by taking saturation into account in the design and analysis is still limited due to the difficulty of the problem. For linear stable systems with known parameters and input saturation, a few control schemes have been proposed, for examples, anti-windup schemes in [143, 144], low-gain control in [145, 146] and linear feedback regulation in [147]. When the system parameters are unknown, adaptive control schemes have been proposed, for examples, model reference adaptive control in [59, 63], predictive control in [61], discrete-time control approaches in [46], indirect adaptive regulator in [60, 62], where uncertain parameters must be inside a known compact set. An adaptive force-balancing control scheme with actuator limits for a MEMS gyroscope was also presented in [148], where the plant is a stable second-order uncertain linear system. The system tracking error is shown to approach a signal generated by an artificially constructed system.

Backstepping approach is a Lyapunov-based recursive design procedure. With this technique, transient performance can be established and improved with explicit tuning of design parameters. A great deal of attention has been paid to tackle both linear and nonlinear systems with unknown parameters. A number

J. Zhou & C. Wen: Adapt. Backstepping Ctrl. of Uncertain Systems, LNCIS 372, pp. 189–197, 2008.
springerlink.com

of results have been obtained as summarized in [1]. Some robustness issues have also been addressed, see for examples, [34, 120]. However, the effect of saturation nonlinearity has not been addressed with this approach, especially in the absence of a priori knowledge of system parameters. To solve such a problem, certain modifications of standard backstepping controllers are required.

In this chapter, we will address the problem of controlling a class of uncertain nonlinear systems in the presence of saturation as in [149]. To deal with saturation, we construct a new system with the same order as that of the plant similar to [148]. With the error between the control input and saturated input as the input of the constructed system, a number of signals are generated to compensate the effect of saturation. With the proposed adaptive backstepping controller, the system tracking error is shown to approach a signal generated by the constructed system. The tracking error is also adjustable by an explicit choice of design parameters. Thus our designed backstepping scheme allows designers to obtain the closed loop behavior by tuning design parameters in an explicit way.

11.2   System Description and Problem Statement

The system model is given as follows:

$$x^{(n)}(t) + \sum_{i=1}^{r} a_i Y_i\big(x(t), \dot{x}(t), \ldots, x^{(n-1)}(t)\big) = u(v) \qquad (11.1)$$

where $Y_i(x, \dot{x}, \ldots, x^{(n-1)})$ are known continuous linear or nonlinear functions, parameters a_i are unknown constants, v is the control input, and $u(v(t)) \in R$ denotes the plant input subject to saturation described by

$$u(v(t)) = sat(v(t)) = \begin{cases} sign(v(t))u_M & |v(t)| \geq u_M \\ v(t) & |v(t)| < u_M \end{cases} \qquad (11.2)$$

where u_M is the saturation bound of $u(t)$.

For the development of control laws, the following assumption is made.

Assumption 1. The plant is bounded input bounded output stable.

The control objectives are to design backstepping adaptive control law $v(t)$ such that

- The closed loop system is globally stable in sense that all the signals in the system are uniformly ultimately bounded;
- The tracking error $y(t) - y_r(t)$ is adjustable by an explicit choice of design parameters.

11.3   Design of Adaptive Controllers

Now equation (11.1) is rewritten in the following form

$$\dot{x}_1 = x_2$$

$$\vdots$$

$$\dot{x}_{n-1} = x_n$$

$$\dot{x}_n = -\sum_{i=1}^{r} a_i Y_i\big(x_1(t), x_2(t), \ldots, x_{(n-1)}(t)\big) + bu(v)$$

$$= a^T Y + u(v) \tag{11.3}$$

$$y = x_1 \tag{11.4}$$

where $x_1 = x, x_2 = \dot{x}, \ldots, x_n = x^{(n-1)}$, $a = [-a_1, -a_2, \ldots, -a_r]^T$ and $Y = [Y_1, Y_2, \ldots, Y_r]^T$.

In order to compensate the effect of the saturation, the following system is constructed to generate signals $\lambda(t) = [\lambda_1, \ldots, \lambda_n]^T$

$$\dot{\lambda}_1 = \lambda_2 - c_1 \lambda_1$$

$$\dot{\lambda}_i = \lambda_{i+1} - c_i \lambda_i, \qquad i = 2, 3, \ldots, n$$

$$\dot{\lambda}_n = -c_n \lambda_n + \Delta u \tag{11.5}$$

where c_i are positive constants and $\Delta u = u(v) - v$.
The following change of coordinates is made.

$$z_1 = y - y_r - \lambda_1 \tag{11.6}$$

$$z_i = x_i - \alpha_{i-1} - y_r^{(i-1)} - \lambda_i, \qquad i = 2, 3, \ldots, n \tag{11.7}$$

where α_{i-1} is the virtual control at the *ith step* to be determined.

Remark 11.1. With the error Δu as the input of the constructed system, it has no effect on z_i. Thus it will not affect the design of controllers. Then by following the standard backstepping approach, the adaptive law will ensure the boundedness of parameter estimates regardless of Δu. On the other hand, such estimates will depend on Δu when standard backstepping is used without using the transformed systems.

In the following, backstepping control scheme is proposed. To illustrate the design procedures, only the first and the last step are elaborated in details.

• *Step 1:* Starting from the equations for the tracking error obtained from (11.3) to (11.7), we get

$$\dot{z}_1 = x_2 - \lambda_2 + c_1 \lambda_1 - \dot{y}_r$$

$$= z_2 + \alpha_1 + c_1 \lambda_1 \tag{11.8}$$

We design the virtual control law α_1 as

$$\alpha_1 = -c_1(x_1 - y_r) \tag{11.9}$$

where $c_1 > 1/2$ is a positive design parameter. A positive Lyapunov function V_1 is defined as

$$V_1 = \frac{1}{2} z_1^2 \tag{11.10}$$

Then the derivative of V_1 along with (11.8) and (11.9) is given as

$$\begin{aligned} \dot{V}_1 &= -c_1 z_1^2 + z_1 z_2 \\ &\le -c_1 z_1^2 + \frac{1}{2} z_1^2 + \frac{1}{2} z_2^2 \\ &= -\bar{c}_1 z_1^2 + \frac{1}{2} z_2^2 \end{aligned} \tag{11.11}$$

where $\bar{c}_1 = c_1 - \frac{1}{2} > 0$.

• *Step i* ($i = 2, \ldots, n-1$): For $z_i = x_i - \alpha_{i-1} - y_r^{(i-1)} - \lambda_i$, we choose virtual control law α_i as

$$\alpha_i = -c_i(x_i - \alpha_{i-1} - y_r^{(i-1)}) + \dot{\alpha}_{i-1}(x_1, \ldots, x_{i-1}) \tag{11.12}$$

where $c_i, i = 2, \ldots, n-1$ are positive design parameters satisfying $c_i > 1$. From (11.7) and (11.12) we obtain

$$z_i \dot{z}_i = -c_i z_i^2 + z_i z_{i+1} \tag{11.13}$$

We choose Lyapunov function as

$$V_i = \sum_{k=1}^{i} \frac{1}{2} z_k^2 \tag{11.14}$$

Then the derivative of V_i along with (11.12) and (11.13) is given by

$$\begin{aligned} \dot{V}_i &\le -c_i z_i^2 + z_i z_{i+1} + \frac{1}{2} z_i^2 \\ &\le -\sum_{i=1}^{i} \bar{c}_i z_i^2 + \frac{1}{2} z_{i+1}^2 \end{aligned} \tag{11.15}$$

where $\bar{c}_i = c_i - 1 > 0$.

• *Step n:* From (11.3) and (11.7) for $i = n$, we obtain

$$\dot{z}_n = v + a^T Y - \dot{\alpha}_{n-1} + c_n \lambda_n - y_r^{(n)} \tag{11.16}$$

We design the adaptive control law $v(t)$ as follows

$$v = -c_n(x_n - \alpha_{n-1} - y_r^{(n-1)}) - \hat{a}^T Y + \dot{\alpha}_{n-1}(x_1, \ldots, x_{n-1}) + y_r^{(n)} \tag{11.17}$$

where c_n is a positive design parameter satisfying $c_n > \frac{1}{2}$, $\hat{a}$ is an estimate of a.

The parameter update law is designed as

$$\dot{\hat{a}} = \Gamma Y z_n \tag{11.18}$$

where Γ is a positive definite matrix. We define a positive Lyapunov function V_n as

$$V = \sum_{i=1}^{n} \frac{1}{2}z_i^2 + \frac{1}{2}\tilde{a}^T\Gamma^{-1}\tilde{a} \tag{11.19}$$

where $\tilde{a} = a - \hat{a}$. Then the derivative of V along with (11.16) to (11.18) is given by

$$\dot{V} = \sum_{i=1}^{n} z_i\dot{z}_i + \tilde{a}^T\Gamma^{-1}\dot{\tilde{a}}$$

$$\leq -\sum_{i=1}^{n} \bar{c}_i z_i^2 + \tilde{a}^T\Gamma^{-1}(\Gamma Y z_n - \dot{\hat{a}})$$

$$= -\sum_{i=1}^{n} \bar{c}_i z_i^2 \tag{11.20}$$

where $\bar{c}_n = c_n - \frac{1}{2}$.

This shows that V is uniformly bounded. Thus $z_i, i = 1, \ldots, n$ and $\hat{a}$ are bounded. From Assumption 1, we have that $x_i, i = 1, \ldots, n$ are bounded as the plant is stable and its input is bounded. So that the boundedness of $\alpha_1, \ldots, \alpha_{n-1}$ and control signal $v(t)$ can be obtained from (11.9), (11.12) and (11.17). Thus $\Delta u = u(v) - v$ is also bounded. Therefore boundedness of all signals in the closed loop system is ensured as stated in the following theorem.

Theorem 11.1. *Consider the uncertain nonlinear system (11.1) in the presence of input saturation satisfying Assumption 1. With the application of controller (11.17) and the parameter update law (11.18), the following statements hold:*

- *The steady state tracking error satisfies*

$$\lim_{t\to\infty} [y(t) - y_r(t) - \lambda_1(t)] = 0 \tag{11.21}$$

- *A bound of the transient tracking error will be given by*

$$\| y(t) - y_r(t) \|_2 \leq \frac{1}{\sqrt{\bar{c}_1}}(\frac{1}{2}\tilde{a}(0)^T\Gamma^{-1}\tilde{a}(0))^{1/2} + \frac{1}{\sqrt{c_0}} \| \Delta u \|_2 \tag{11.22}$$

Proof: From (11.20) we established that V is non increasing. Hence, $z_i, i = 1, \ldots, n$, $\hat{a}$ are bounded. By applying the LaSalle-Yoshizawa theorem to (11.20), it further follows that $z_i(t) \to 0, i = 1, \ldots, n$ as $t \to \infty$, which implies that $\lim_{t\to\infty}[y(t) - y_r(t) - \lambda_1] = 0$.

From (11.20) we also have that

$$\| z_1 \|_2^2 = \| y - y_r - \lambda_1 \|_2^2 = \int_0^\infty |z_1(\tau)|^2 d\tau$$

$$\leq \frac{1}{\bar{c}_1}(V(0) - V(\infty)) \leq \frac{1}{\bar{c}_1}V(0) \tag{11.23}$$

Thus, by setting $z_i(0) = 0, i = 1, \ldots, n$, we obtain

$$V(0) = \frac{1}{2}\tilde{a}(0)^T \Gamma^{-1} \tilde{a}(0) \qquad (11.24)$$

a decreasing function of Γ, independent of $\bar{c}_1$. This means that the bound resulting from (11.23) and (11.24) is

$$\| y(t) - y_r(t) - \lambda_1(t) \|_2 \le \frac{1}{\sqrt{\bar{c}_1}} \left(\frac{1}{2}\tilde{a}(0)^T \Gamma^{-1} \tilde{a}(0)\right)^{1/2} \qquad (11.25)$$

Now we derive the bound of λ_1.

We construct the positive Lyapunov function $V_\lambda = \sum_{i=1}^{n} \frac{1}{2}\lambda_i^2$. Then the derivative of V_λ is given as

$$\dot{V}_\lambda = -c_1\lambda_1^2 + \lambda_1\lambda_2 - c_2\lambda_2^2 + \lambda_2\lambda_3 + \ldots + \lambda_{n-1}\lambda_n - c_n\lambda_n^2 + \lambda_n \Delta u$$

$$\le \sum_{i=1}^{n} -\bar{c}_i\lambda_i^2 + \Delta u^2$$

$$\le -c_0 \| \lambda \|^2 + \Delta u^2 \qquad (11.26)$$

where $\bar{c}_1 = c_1 - \frac{1}{2}$, $\bar{c}_i = c_i - 1 (i = 2, \ldots, n-1)$, $\bar{c}_n = c_n - \frac{3}{4}$, $c_0 = min_{1 \le i \le n}\bar{c}_i$. Integrating both sides of (11.26), we have

$$\| \lambda \|_2^2 = \int_0^{\infty} \| \lambda \|^2 \, d\tau$$

$$\le \frac{1}{c_0}\left[(V_\lambda(0) - V_\lambda(\infty)) + \int_0^{\infty} (\Delta u)^2 d\tau\right] \qquad (11.27)$$

By setting $\lambda_i(0) = 0$, the initial value of the Lyapunov function is $V_\lambda(0) = 0$. Then a bound on the state $\| \lambda \|_2$ is established as follows

$$\| \lambda \|_2 \le \frac{1}{\sqrt{c_0}} \| \Delta u \|_2 \qquad (11.28)$$

Thus from (11.25) and (11.28), it is obtained

$$\| y - y_r \|_2 \le \frac{1}{\sqrt{\bar{c}_1}} \left(\frac{1}{2}\tilde{a}(0)^T \Gamma^{-1} \tilde{a}(0)\right)^{1/2} + \frac{1}{\sqrt{c_0}} \| \Delta u \|_2 \qquad (11.29)$$

$$\triangle\triangle\triangle$$

From Theorem 11.1 the following conclusions can be obtained.

Remark 11.2. The transient performance depends on the initial estimate error $\tilde{a}(0)$ and the explicit design parameters. The closer the initial estimate $\hat{a}(0)$ to the true value a, the better the transient performance.

Remark 11.3. The bound for $\| y(t) - y_r(t) \|_2$ is an explicit function of design parameters and thus computable. We can decrease the effects of the initial error estimate on the transient performance by increasing the adaptation gain Γ and parameter c_1.

Remark 11.4. The bound of $\| y(t) - y_r(t) \|_2$ depends on the bound of Δu, the effects of which on system performance can be decreased by increasing parameter c_0. If $\Delta u \to 0$ as $t \to \infty$, we have $\lambda_1 \to 0$. Then $\lim_{t \to \infty}[y(t) - y_r(t)] = 0$. This implies that if the system has no saturation or the control signal is not saturated as $t \to \infty$, then perfect tracking is ensured.

11.4 Simulation Study

In this section, we illustrate the above methodology on the following example. We consider a second-order system depicted in Figure 11.1 which is modelled by

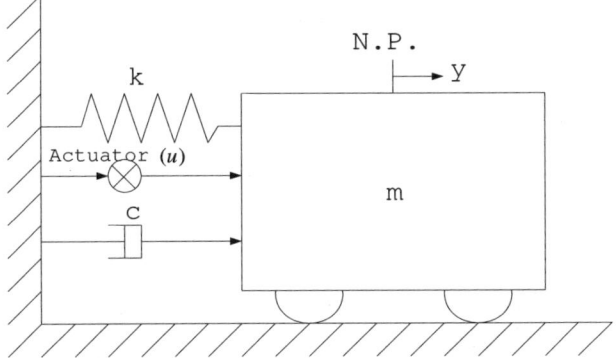

Fig. 11.1. Spring, mass and damper system

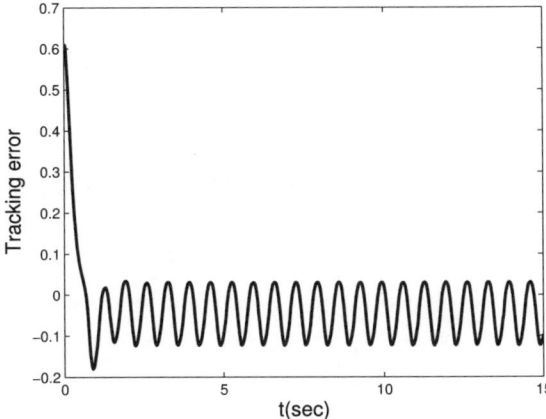

Fig. 11.2. Tracking error with the controller designed using standard backstepping approach

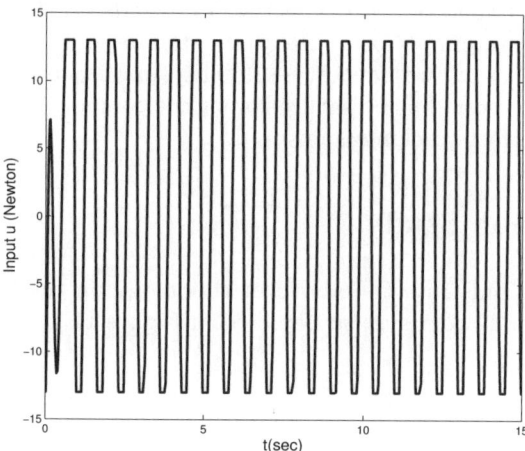

Fig. 11.3. Control signal with the controller designed using standard backstepping approach

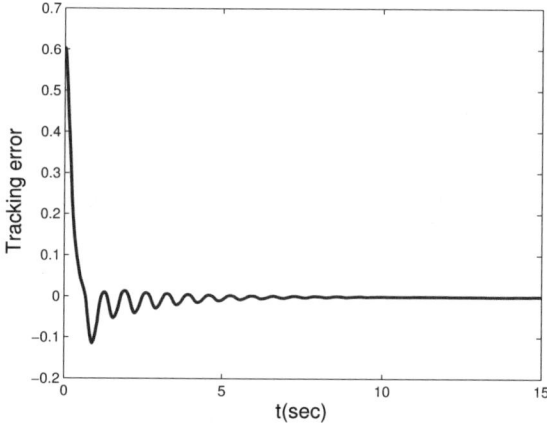

Fig. 11.4. Tracking error with proposed scheme

$$\dot{x}_1 = x_2$$
$$\dot{x}_2 = -\frac{k}{m}x_1 - \frac{c}{m}x_2 + \frac{1}{m}sat(v) \qquad (11.30)$$
$$y = x_1$$

where x_1 and x_2 are the position and velocity, m is the mass of the object, k is the stiffness constant of the spring and c is the damping. The input saturation limit is $13N$. The true parameters are set as $m = 1.25kg$, $c = 2N{\cdot}s/m$, $k = 8N/m$, which are not needed to be known in our controller design. The design parameters are chosen as $c_1 = c_2 = 5$ and $\Gamma = 0.5I_2$. The adaptive control law and parameter

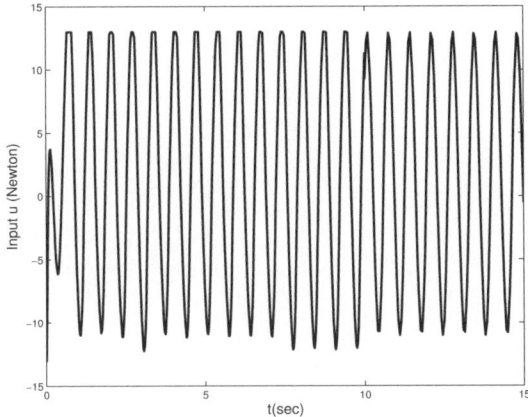

Fig. 11.5. Control signal with proposed scheme

update laws are followed by (11.17) and (11.18), where $a = [\frac{-k}{m}, \frac{-c}{m}]^T, b = \frac{1}{m}$. The initial parameter values are selected as $\hat{a}(0) = [-6, -1]^T$ and $\hat{e}(0) = 1$. The desired trajectory is given as $r(t) = -0.2\cos(2\pi \times 1.5t) + 0.2$ $[m]$ and the initial conditions are $x_1(0) = 0.6m, x_2(0) = 0.8$.

Two controllers, designed using the standard backstepping approach [1] and the proposed scheme respectively, are applied to system (11.30). Simulation results on system tracking error and control signal are presented in Figures 11.2 to 11.5. Significantly improved performance is clearly seen with the proposed scheme. It is also observed that the input signal is not saturated for $t > 8$ second and the perfect tracking is obtained. These results indicate that the proposed backstepping adaptive controller is effective and practically useful.

11.5 Conclusion

This chapter presents a new scheme to design adaptive backstepping controller for a class of uncertain nonlinear systems in the presence of input saturation. We propose a new control law to compensate the effect of the saturation nonlinearity using backstepping technique. The developed backstepping control does not require the model parameters within known intervals. Besides showing global stability, we also give an explicit bound on the performance of the tracking error in terms of design parameters. Simulation results illustrate the effectiveness of our proposed scheme. Also improvement of system performance over a backstepping adaptive controller designed without considering saturation is observed.

12 Control of a Hysteretic Structural System in Base Isolation Scheme

In this chapter, we present two adaptive backstepping control algorithms for a second-order uncertain hysteretic structural system found in base isolation scheme for seismic active protection of building structures. The hysteretic nonlinear behavior is described by a Bouc-Wen model. It is shown that not only stability is guaranteed by the proposed controller, but also both transient and asymptotic performances are quantified as explicit functions of the design parameters so that designers can tune the design parameters in an explicit way to obtain the required closed loop behavior.

12.1 Introduction

The modelling and identification of nonlinear hysteretic systems is a problem widely encountered in the structural dynamics field. Nonlinear hysteretic behavior is seen commonly in structures experiencing strong ground earthquake excitation. Because of the hysteretic nature of the restoring force in such situation, the nonlinear force cannot be expressed in the form of an algebraic function involving the instantaneous values of the state variables of the system. Studies of this problem have been reported in the works of [150, 151, 152, 153, 154, 155]. In [150], an adaptive controller was designed for a class of state-feedback nonlinear systems with unknown hysteresis to counteract the effect of an earthquake excitation. To represent the behavior of a seismic base isolation scheme which has a nonlinear hysteretic behavior, the Bouc-Wen model in connection to a second-order structural system is used. This behavior is described in ([156, 157, 158, 159, 160]). The system considered arises from a class of nonlinear oscillators, which are common in structural engineering models [150, 161]. The proposed controller is designed to counteract the effect of an earthquake excitation and mitigate the seismic displacement response of the system. In the controller design, the true hysteretic behavior is not required to be known. However, the system uncertain parameters must be within some known intervals and the effect of the hysteresis is treated as a bounded disturbance. The bound of the effect is also required for the design. Certain structural information in the model is not exploited.

In this chapter, we develop two backstepping adaptive control design schemes for a second-order uncertain hysteretic structural system as in [162]. In the

J. Zhou & C. Wen: Adapt. Backstepping Ctrl. of Uncertain Systems, LNCIS 372, pp. 199–213, 2008.
springerlink.com   © Springer-Verlag Berlin Heidelberg 2008

design, no knowledge is assumed on the term multiplying the control and other uncertain parameters. In the first scheme, we use some available structure information in the design and the residual effect of the hysteresis is treated as a bounded disturbance. An update law is used to estimate the bound involving this partial hysteresis effect and external disturbance. In the second scheme, we further take the structure of the Bouc-Wen model describing the hysteresis into account in the controller design, if apriori knowledge on some parameters of the model is available. It is shown that the proposed controller can guarantee stability and achieve tracking performance. Also with the proposed scheme, both transient and asymptotic performances are quantified as explicit functions of the design parameters so that designers can tune the design parameters in an explicit way to obtain the required closed loop behavior. Compared with the scheme in [150], system performance with the first scheme applied is still improved even though we need much less knowledge from the system. When the second scheme is applied, the performance has been significantly improved compared with the first scheme and the scheme in [150]. Simulation results verify the effectiveness of our adaptive controllers.

12.2   Problem Formulation

A hysteresis friction model will be developed to simply and appropriately describe the dynamics of Hysteretic Structural System in Base Isolation Scheme. The hysteresis model of Bouc as modified by Wen [156] possesses an appealing mathematical simplicity, and is able to represent a large class of hysteretic behavior, from inelastic stress-strain relationships found in structures to magnetoelectrical behavior.

Given the states $x(t), z(t) : T \to R$, the Bouc-Wen model is described by

$$\dot{z} = A\dot{x} - \beta|\dot{x}||z|^{n-1}z - \lambda\dot{x}|z|^n \qquad (12.1)$$

Nominally, x and z denote the position of an oscillator and a restoring force acting on the oscillator, respectively. A hysteretic relation is observed between x and z. The parameters A, β and λ control the scale and shape of the hysteresis curve, n is an integer that governs the smoothness of the transition from elastic to plastic response. Illustrations of their effect are shown in Figures 12.1-12.2. Note that for the model to exhibit positive energy dissipation through each cycle, thermodynamic laws require that $\lambda > 0$. Figures 12.1 and 12.2 show the effect of n and the loop shapes with $n = 1$ and different A, β and λ.

In the limit as $n \to \infty$, $\dot{z} = \dot{x}[sgn(z + A) - sgn(z - A)]/2$.

Consider now the following second-order uncertain nonlinear system illustrated in Figure 12.3 modelled as

$$m\ddot{x} + \bar{c}\dot{x} + \Phi(x, t) = f(t) + u(t) \qquad (12.2)$$

where m is an unknown positive parameter, $\bar{c}$ is an uncertain parameter and Φ represents a nonlinear component, $f(t)$ is an external disturbance with unknown

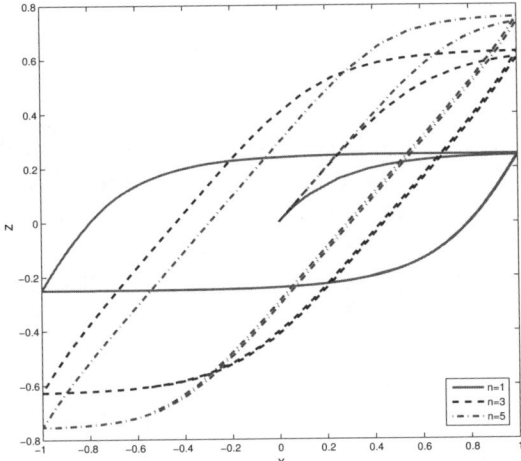

Fig. 12.1. The effect of increasing n on the hysteretic characteristic for $A = 1, \beta = 1$, $\lambda = 2$ and $n = 1, 3, 5$

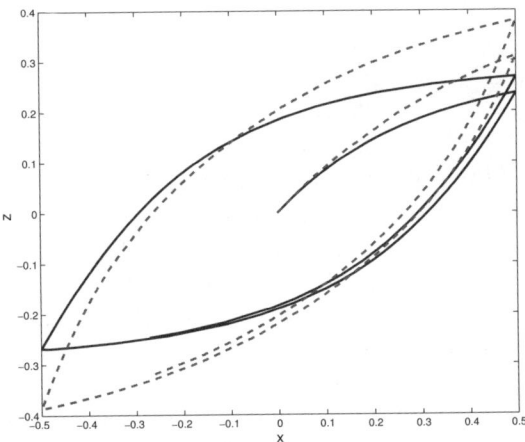

Fig. 12.2. Hysteretic loop shapes for $n = 1, A = 1, \beta = \lambda = 0.5$ (solid) and $n = 1, A = 1, \beta = -0.9\lambda = 0.5$ (dashed)

bound, and $u(t)$ is control input. In the structural system, m and $\bar{c}$ are the mass and the damping coefficients, respectively, and restoring force Φ characterizes a hysteretic behavior of isolator material, which is usually made with inelastic rubber bearings, x is the position, $u(t)$ is an active control force supplied by appropriate actuators, $f(t)$ is an exciting unknown force, which is described as $f(t) = -ma(t)$, where $a(t)$ is earthquake ground acceleration. The hysteresis force Φ is described in the following form ([156, 157, 158, 160, 163]).

$$\Phi(x,t) = \alpha k x(t) + (1 - \alpha)Dkz(t) \tag{12.3}$$

$$\dot{z} = D^{-1}[A\dot{x} - \beta|\dot{x}||z|^{n-1}z - \lambda\dot{x}|z|^n] \qquad (12.4)$$

This model represents the restoring force $\Phi(x, t)$ by superposition of elastic component $\alpha k x$ and a hysteretic component $(1 - \alpha)Dkz$, in which $D > 0$ is the yield constant displacement and α is the post to pre-yielding stiffness ratio. The hysteretic part involves an auxiliary variable z which is the solution of the nonlinear first order differential equation (12.4). The Bouc-Wen model in (12.1) is able to capture, in an analytic form, a range of shapes of hysteretic cycles which match the behavior of a wide class of nonlinear structures. It is widely used in structure dynamics, particularly to describe rubber bearing isolation schemes. Now we prove the boundedness of $z(t)$. From dynamic system (12.4), we have

$$\dot{z} = -D^{-1}|z|^{n-1}[\ \beta|\dot{x}|z + \lambda\dot{x}|z|] + D^{-1}A\dot{x}$$
$$= -D^{-1}|z|^{n-1}|\dot{x}|[\beta + \lambda\text{sign}(z)\text{sign}(\dot{x})]z + D^{-1}A\dot{x} \qquad (12.5)$$

We construct a positive Lyapunov function $V_z = z(t)^2/2$. Its derivative takes different forms depending on the signs of $\dot{x}$ and z. The following analysis determines the condition on the Bouc-Wen model parameters, such that $z(t)$ is globally bounded as in [164].

Consider the case $A > 0$. There are three possibilities.

* $P1 : \beta + \lambda > 0$ and $\beta - \lambda \geq 0$;
* $P2 : \beta + \lambda > 0$ and $\beta - \lambda < 0$;
* $P3 : \beta + \lambda \leq 0$.

Let us now focus on the case $P1$. Indeed, setting $Q1 := \{\dot{x} \geq 0$ and $z \geq 0\}$, and denoting $\dot{V}_{Q_1}$ as the expression of the derivative of the Lyapunov function V_z over the set $Q1$, we have $\dot{V}_{Q1} = z\dot{x}D^{-1}(A - (\beta+\lambda)z^n)$. Thus $\dot{V}_{Q1} \leq 0$. Similarly, $\dot{V}_{Q2} \leq 0$ for $|z| \geq z^0$, where $Q2 = \{\dot{x} \leq 0$, and $z \leq 0\}$.

Also setting $Q3 = \{\dot{x} \geq 0$, and $z \leq 0\}$, we have $\dot{V}_{Q3} = z\dot{x}D^{-1}(A+(\beta-\lambda)|z|^n)$. In this case, $\dot{V}_{Q3} \leq 0$ for all values of z. The same conclusion is drawn in the case that $Q4 = \{\dot{x} \leq 0$, and $z \geq 0\}$.

We then conclude that for all possible signs of $\dot{x}$ and z, we have $\dot{V}_z \leq 0$ for all $|z| \geq z^0$. By theorem 4.10 in [164] we conclude that $z(t)$ is bounded for every

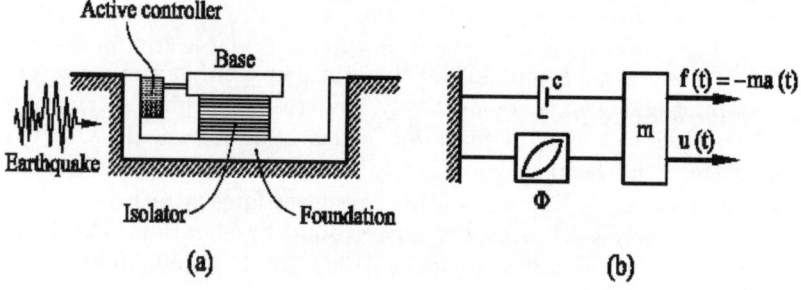

Fig. 12.3. Base isolation system (a) and physical model (b)

piecewise function $\dot{x}$ and every initial condition $z(0)$. The bounds on $z(t)$ can be derived as follows: If the initial condition of z is such that $|z(0)| \leq z^0$, then $|z| \leq z^0$ for all $t \geq 0$. If the initial condition of z is such that $z(0) \geq z^0$, then $|z| \leq z(0)$ for all $t \geq 0$.

We now turn to the case $P2$ by considering $\dot{V}_z$ in the regions $\{\dot{x} \geq 0, \ and \ z \geq z^0\}$, $\{\dot{x} \geq 0, \ and \ -z^1 \leq z \leq 0\}$, $\{\dot{x} \leq 0, \ and \ 0 \leq z \leq z^1\}$, and $\{\dot{x} \leq 0, \ and \ -z^0 \leq z \leq 0\}$, where

$$z^1 = \sqrt[n]{A/(\lambda - \beta)}$$

Following the similar argument earlier, we can show that $\dot{V}_z \leq 0$ for the initial state $z(0)$ satisfying that $|z(0)| \leq z^1$.

Following the same analysis for the case $P3$, we can see that z may be unbounded for some functions $\dot{x}$. This implies that the region of boundedness of $z(t)$ is empty in this case.

A similar analysis can be carried out for the case $A < 0$ and $A = 0$ and a conclusion draw from the analysis is summarized in the following lemma.

Lemma 12.1. *Consider the nonlinear dynamic system (12.4). Then for any piecewise continuous signal x and $\dot{x}$, the output $z(t)$ is globally bounded if and only if the parameters of system (12.4) satisfies the inequality $\beta > |\lambda|$.*

The control objective is to design a backstepping adaptive control law such that

- The closed loop is bounded.
- The tracking error $x(t) - y_r(t)$ is made arbitrarily small both in the transient period and steady state by an explicit choice of the design parameters, where $y_r(t)$ is a known bounded reference signal.

12.3 Control Design and Main Results

In this section, we develop two adaptive backstepping design schemes. In Scheme I, we use some available structure information in the design and the residual effect of the hysteresis is treated as a bounded disturbance with unknown bound. An update law is used to estimate the bound involving the effect of the hysteresis and the external disturbance. In Scheme II, we assume certain apriori information of the parameters in the Bouc-Wen model (12.4) is available and construct a variable $\bar{z}(t)$ to approximate $z(t)$ in (12.4) in our controller design. To illustrate the backstepping procedures, only the first scheme is elaborated in details.

12.3.1 Control Scheme I

The Bouc-Wen nonlinear restoring force $\Phi(x, t)$ in (12.3) can be parameterized as follows.

$$\Phi(t) = \theta_1 x(t) + R(t) \tag{12.6}$$

where $\theta_1 = \alpha k$ is uncertain parameter and $R(t) = (1 - \alpha)Dkz(t)$. Note that $x(t)$ is an available signal.

For the residual term R we have the following inequality:

$$|R(t)| \leq (1 - \alpha_{min})D_{max}k_{max}max_{t\geq0}|z(t)| \tag{12.7}$$

Then we rewrite equations (12.2) and (12.6) in the following form

$$\dot{x}_1 = x_2 \tag{12.8}$$

$$\dot{x}_2 = \frac{1}{m}\left(u(t) - \bar{c}x_2 - \theta_1 x(t) - R(t) + f(t)\right)$$

$$= \theta^T \varphi(t) + \frac{1}{m}\left(u(t) + d(t)\right) \tag{12.9}$$

where $x_1 = x$, $x_2 = \dot{x}$, $\theta = [\frac{\bar{c}}{m}, \frac{\theta_1}{m}]^T$ is a constant vector of uncertain parameters, $\varphi = [-x_2, -x(t)]^T$ and $d(t) = f(t) - R(t)$. Note that $R(t)$ is bounded as $z(t)$ has been shown bounded in Lemma 12.1. So $d(t)$ is bounded with unknown bound F.

Before presenting the adaptive control design using the backstepping technique to achieve the desired control objectives, the following change of coordinates is made.

$$\varsigma_1 = x_1 - y_r \tag{12.10}$$

$$\varsigma_2 = x_2 - \dot{y}_r - \alpha_1 \tag{12.11}$$

where α_1 is the virtual control and will be determined in later discussion.

• *Step 1:* We design the virtual control law α_1 as

$$\alpha_1 = -c_1\varsigma_1 \tag{12.12}$$

where c_1 is a positive design parameter. From (12.8) and (12.12) we have

$$\varsigma_1\dot{\varsigma}_1 = -c_1\varsigma_1^2 + \varsigma_1\varsigma_2 \tag{12.13}$$

• *Step 2:* From (12.9) and (12.11), we have

$$\dot{\varsigma}_2 = \theta^T\varphi + \frac{1}{m}\left(d(t) + u(t)\right) - \ddot{y}_r - \dot{\alpha}_1 \tag{12.14}$$

Then the control law and parameter update laws are given below.

$$u = -\hat{F}\text{sign}(\varsigma_2) + \hat{m}\bar{u} \tag{12.15}$$

$$\bar{u} = -c_2\varsigma_2 - \varsigma_1 - \hat{\theta}\varphi + \ddot{y}_r + \dot{\alpha}_1 \tag{12.16}$$

$$\dot{\hat{\theta}} = \Gamma\varphi\varsigma_2 \tag{12.17}$$

$$\dot{\hat{m}} = -\gamma\bar{u}\varsigma_2 \tag{12.18}$$

$$\dot{\hat{F}} = \gamma_f|\varsigma_2| \tag{12.19}$$

where c_2, γ and γ_f are designed positive parameters, Γ is a positive definite design matrix. $\hat{\theta}$, $\hat{m}$ and $\hat{F}$ are estimates of θ, m and F.

Remark 12.1. Note that a parameter update law is used to estimate the bound F of the disturbance $d(t)$, so there is no need to know this bound.

We define a positive Lyapunov function as

$$V = \frac{1}{2}\varsigma_1^2 + \frac{1}{2}\varsigma_2^2 + \frac{1}{2m\gamma}\tilde{m}^2 + \frac{1}{2}\tilde{\theta}^T \Gamma^{-1}\tilde{\theta} + \frac{1}{2m\gamma_f}\tilde{F}^2 \qquad (12.20)$$

where $\tilde{m} = m - \hat{m}$, $\tilde{\theta} = \theta - \hat{\theta}$ and $\tilde{F} = F - \hat{F}$.

Note that $\frac{1}{m}u$ in (12.14) can be expressed as

$$\frac{1}{m}u = \frac{1}{m}\hat{m}\bar{u} - \frac{1}{m}\hat{F}\text{sign}(\varsigma_2)$$

$$= \bar{u} - \frac{1}{m}\tilde{m}\bar{u} - \frac{1}{m}\hat{F}\text{sign}(\varsigma_2) \qquad (12.21)$$

Then the derivative of V along with (12.15-12.18) is given by

$$\begin{aligned}
\dot{V} &= \varsigma_1\dot{\varsigma}_1 + \varsigma_2\dot{\varsigma}_2 + \tilde{\theta}^T \Gamma^{-1}\dot{\tilde{\theta}} + \frac{1}{m\gamma}\tilde{m}\dot{\tilde{m}} + \frac{1}{m\gamma_f}\tilde{F}\dot{\tilde{F}} \\
&\leq -c_1\varsigma_1^2 - c_2\varsigma_2^2 + \tilde{\theta}^T \Gamma^{-1}\left(\Gamma\varphi\varsigma_2 - \dot{\hat{\theta}}\right) \\
&\quad - \frac{1}{m\gamma}\tilde{m}\left(\gamma\bar{u}\varsigma_2 + \dot{\hat{m}}\right) + \frac{1}{m\gamma_f}\tilde{F}\left(\gamma_f|\varsigma_2| - \dot{\hat{F}}\right) \\
&= -c_1\varsigma_1^2 - c_2\varsigma_2^2 \qquad (12.22)
\end{aligned}$$

Based on (12.22), we can obtain the result on system stability and performance as stated below.

Theorem 12.1. *Consider the uncertain nonlinear system (12.2). With the application of the controller (12.15) and the parameter update laws (12.17), (12.18) and (12.19), the following statements hold:*

- *The resulting closed loop system is global uniform ultimate bounded.*
- *The asymptotic tracking is achieved, i.e.,*

$$\lim_{t\to\infty}\left[x(t) - y_r(t)\right] = 0 \qquad (12.23)$$

- *The transient displacement tracking error performance is given by*

$$\| x(t) - y_r(t) \|_2 \leq \frac{1}{\sqrt{c_1}} \times \left(\frac{1}{2}\tilde{\theta}^T(0)\Gamma^{-1}\tilde{\theta}(0) + \frac{1}{2m\gamma}\tilde{m}(0)^2 + \frac{1}{2m\gamma_f}\tilde{F}(0)^2\right)^{1/2} \qquad (12.24)$$

- *The transient velocity tracking error performance is given by*

$$\| \dot{x} - \dot{y}_r \|_2 \leq \left(\frac{1}{\sqrt{c_2}} + \sqrt{c_1}\right)\left(\frac{1}{2}\tilde{\theta}^T(0)\Gamma^{-1}\tilde{\theta}(0) + \frac{1}{2m\gamma}\tilde{m}(0)^2 + \frac{1}{2m\gamma_f}\tilde{F}(0)^2\right)^{1/2} \qquad (12.25)$$

Proof: Equation (12.22) shows that $V(t)$ is globally uniformly bounded. This implies that $\varsigma_1, \varsigma_2, \tilde{\theta}, \tilde{m}, \tilde{F}$ are bounded. The state variables x_1, x_2 and the parameter estimates $\hat{\theta}, \hat{m}, \hat{F}$ are also bounded. Thus u is bounded from (12.15) because of the boundedness of $\varsigma_1, \varsigma_2, \hat{\theta}, \hat{m}, \hat{F}$.

Since V is non increasing from (12.22), we have

$$\| \varsigma_1 \|_2^2 = \int_0^\infty |\varsigma_1(\tau)|^2 d\tau \leq \frac{1}{c_1}(V(0) - V(\infty)) \leq \frac{1}{c_1}V(0) \qquad (12.26)$$

Thus, by setting $\varsigma_1(0) = \varsigma_2(0) = 0$, we obtain

$$V(0) = \frac{1}{2}\tilde{\theta}^T(0)\Gamma^{-1}\tilde{\theta}(0) + \frac{1}{2m\gamma}\tilde{m}(0)^2 + \frac{1}{2m\gamma_f}\tilde{F}(0)^2 \qquad (12.27)$$

a decreasing function of γ, γ_f and Γ, independent of c_1. This means that the bounds resulting from (12.26) and (12.27)

$$\| \varsigma_1 \|_2 \leq \frac{1}{\sqrt{c_1}}\left(\frac{1}{2}\tilde{\theta}^T(0)\Gamma^{-1}\tilde{\theta}(0) + \frac{1}{2m\gamma}\tilde{m}(0)^2 + \frac{1}{2m\gamma_f}\tilde{F}(0)^2\right)^{1/2} \qquad (12.28)$$

can be asymptotically reduced either by increasing c_1 or by simultaneously increasing γ, γ_f and Γ. The bound for $\| \varsigma_1 \|_2$ is explicit.

From equations (12.9) to (12.12), we get

$$\| \dot{x} - \dot{y}_r \|_2 = \| \varsigma_2 - c_1\varsigma_1 \|_2 \leq \| \varsigma_2 \|_2 + c_1 \| \varsigma_1 \|_2 \qquad (12.29)$$

Similarly, we can get $\| \varsigma_2 \|_2 \leq \frac{1}{\sqrt{c_2}}\sqrt{V(0)}$. Along with (12.28) we get

$$\| \dot{x} - \dot{y}_r \|_2 \leq \left(\frac{1}{\sqrt{c_2}} + \sqrt{c_1}\right)\left(\frac{1}{2}\tilde{\theta}^T(0)\Gamma^{-1}\tilde{\theta}(0) + \frac{1}{2m\gamma}\tilde{m}(0)^2 + \frac{1}{2m\gamma_f}\tilde{F}(0)^2\right)^{1/2} \qquad (12.30)$$

Remark 12.2. From Theorem 12.1 the following conclusions can be obtained:

- Boundedness of the adaptive system is guaranteed to be global, uniform and ultimate for any positive values of the design parameters $c_1, c_2, \gamma, \gamma_f$ and Γ. No a priori information is required about the parameter uncertainty.
- We can decrease the effects of the initial error estimates on the transient performance by increasing the adaptation gains γ, γ_f and Γ. And thus the bound for $\| x - y_r \|$ is an explicit function of desired parameters.
- The transient performance depends on the initial estimate errors $\tilde{\theta}(0)$ and $\tilde{m}(0)$. The closer the initial estimates $\hat{\theta}(0)$, $\hat{F}(0)$ and $\hat{m}(0)$ to the true values θ, F and m, the better the transient performance. The asymptotic behavior is not affected by the initial estimate errors.

- To improve the displacement tracking error performance we can also increase the gain c_1. However, increasing the gain c_1 will also increase the velocity tracking error as shown above. Improving the closed loop displacement behavior may be done at the expense of the increase in the control signal amplitude. This suggests to fix the gain c_1 to some acceptable value and adjust the other gains. By fixing the gain c_1, increasing the gain c_2 or by simultaneously increasing γ, γ_f and Γ, we can achieve a velocity tracking error as small as desired.

12.3.2   Control Scheme II

In this section, we assume that certain apriori information of the Bouc-Wen model parameters is available. Thus we further exploit the structure of the model in our controller design to improve system performance.

The Bouc-Wen nonlinear restoring force $\Phi(x,t)$ in (12.3) can be parameterized as follows.

$$\Phi(t) = \theta_1 x(t) + \theta_2 z(t) \tag{12.31}$$

where $\theta_1 = \alpha k$ and $\theta_2 = (1-\alpha)Dk$ are uncertain parameters.

Assumption. Parameters A, β, D, λ are inside some known intervals.

With the above assumption, a signal $\bar{z}(t)$ can be generated using an equation

$$\dot{\bar{z}} = D_0^{-1}[A_0\dot{x} - \beta_0|\dot{x}||\bar{z}|^{n-1}\bar{z} - \lambda_0\dot{x}|\bar{z}|^n] \tag{12.32}$$

where $A_0, \beta_0, D_0, \lambda_0$ are inside the known intervals. With this $\bar{z}(t)$, we approximate $\Phi(x,t)$ by $\bar{\Phi}(x,t)$ as $\bar{\Phi}(x,t) = \theta_1 x(t) + \theta_2 \bar{z}(t)$.

Remark 12.3. From Lemma 12.1, $\bar{z}(t) - z(t)$ is bounded. Note that $\Phi - \bar{\Phi} = \theta_2(z - \bar{z})$. So $\Phi - \bar{\Phi}$ is also bounded. This bounded error can then be combined with external disturbance to get f with its combined bound F estimated as in Scheme I. It is expected that $\parallel \Phi - \bar{\Phi} \parallel \leq \parallel \Phi \parallel$.

Then we rewrite equations (12.2) and (12.6) in the following form

$$\dot{x}_1 = x_2 \tag{12.33}$$

$$\dot{x}_2 = \theta^T \varphi(t) + \frac{1}{m}\left(u(t) + f(t)\right) \tag{12.34}$$

where $x_1 = x, x_2 = \dot{x}, \theta = [\frac{\bar{c}}{m}, \frac{\theta_1}{m}, \frac{\theta_2}{m}]^T$ is a constant vector of uncertain parameters, and $\varphi = [-x_2, -x(t), -\bar{z}(t)]^T$.

The controller design is similar to the Scheme I. We only give the resulting control laws.

$$u = -\hat{F}\mathrm{sign}(\varsigma_2) + \hat{m}\bar{u} \tag{12.35}$$

$$\bar{u} = -c_2\varsigma_2 - \varsigma_1 - \hat{\theta}\varphi + \ddot{y}_r + \dot{\alpha}_1 \tag{12.36}$$

$$\alpha_1 = -c_1\varsigma_1 \tag{12.37}$$

$$\dot{\hat{\theta}} = \Gamma\varphi\varsigma_2 \tag{12.38}$$

$$\dot{\hat{m}} = -\gamma\bar{u}\varsigma_2 \tag{12.39}$$

$$\dot{\hat{F}} = \gamma_f|\varsigma_2| \tag{12.40}$$

where c_1, c_2, γ and γ_f are designed positive parameters, Γ is a positive definite design matrix, $\hat{\theta}$, $\hat{m}$ and $\hat{F}$ are estimates of θ, m and F.

Remark 12.4. Certain information of the hysteretic structure is used in our controller design, unlike the Scheme I and the scheme in [150], where the effect of hysteresis is treated as a bounded disturbance. This is reflected in $\hat{\theta}$ and φ of the designed controller in (12.35)-(12.40).

Following the similar analysis to Scheme I, we can establish that $x, \dot{x}, \hat{\theta}, \hat{m}, \hat{F}$ are all bounded. Then from Lemma 12.1 and (12.35), (12.36), u is also bounded. Thus the result on system stability and performance can be established and now stated in the following theorem.

Theorem 12.2. *Consider the uncertain nonlinear system (12.2). With the application of the controller (12.35) and the parameter update laws (12.38), (12.39) and (12.40), the following statements hold:*

- *The resulting closed loop system is global uniform ultimate bounded.*
- *The asymptotic tracking is achieved, i.e.,*

$$\lim_{t \to \infty} [x(t) - y_r(t)] = 0 \tag{12.41}$$

- *The transient displacement tracking error performance is given by*

$$\| x(t) - y_r(t) \|_2 \leq \frac{1}{\sqrt{c_1}} \left(\frac{1}{2} \tilde{\theta}^T(0) \Gamma^{-1} \tilde{\theta}(0) + \frac{1}{2m\gamma} \tilde{m}(0)^2 + \frac{1}{2m\gamma_f} \tilde{F}(0)^2 \right)^{1/2} \tag{12.42}$$

- *The transient velocity tracking error performance is given by*

$$\| \dot{x} - \dot{y}_r \|_2 \leq \left(\frac{1}{\sqrt{c_2}} + \sqrt{c_1} \right) \left(\frac{1}{2} \tilde{\theta}^T(0) \Gamma^{-1} \tilde{\theta}(0) + \frac{1}{2m\gamma} \tilde{m}(0)^2 + \frac{1}{2m\gamma_f} \tilde{F}(0)^2 \right)^{1/2} \tag{12.43}$$

12.4   Simulation Results

In this section we test our proposed backstepping controller. For simulation studies, the following values are selected as "true" parameters for the system and the hysteresis model: $m = 156 \times 10^3 [Kg]$, $k = 6 \times 10^6 [N/m]$, $c = 2 \times 10^4 [Ns/m]$, $\alpha = 0.6, D = 0.6[m], A = 1, \beta = 0.5, \lambda = 0.4, n = 3$. In fact, it is not required to know the exact values of these parameters to implement the controller.

The control objective is to mitigate the seismic displacement response of the system, so the reference trajectory $y_r(t)$ is set to 0.

When Scheme I is used, we take the following set of design parameters: $\gamma = 0.1, \Gamma = I_3, \gamma_f = 0.5, c_1 = 0.3, c_2 = 3, \hat{m}(0) = 300 \times 10^3, \hat{D}(0) = 200 \times 10^3$, and $\hat{\theta}(0) = [0.15, 12, 5]^T$.

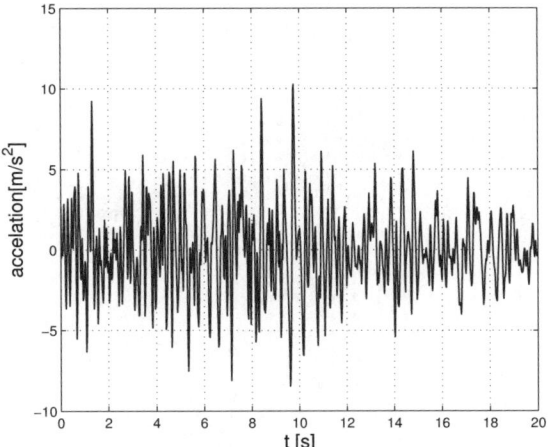

Fig. 12.4. Earthquake ground acceleration

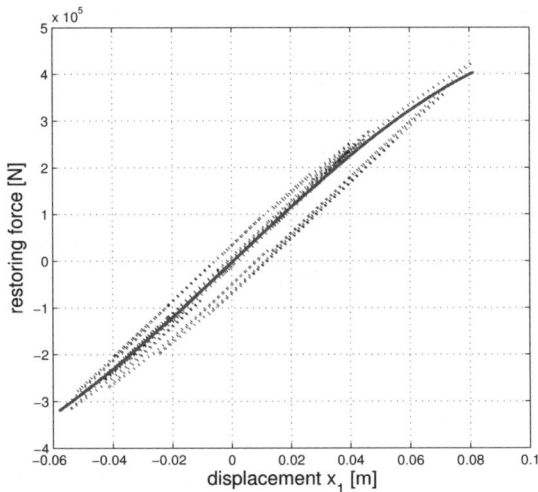

Fig. 12.5. Hysteresis identification

When Scheme II is used, we take the following set of design parameters: $\gamma = 0.1, \Gamma = I_2, \gamma_f = 0.5, c_1 = 0.3, c_2 = 3, \hat{m}(0) = 300 \times 10^3, \hat{F}(0) = 200 \times 10^3, \hat{\theta}(0) = [0.15, 12]^T, A_0 = 1.5, \beta_0 = 0.6, \lambda_0 = 0.3, D = 0.5$. Note that the uncertainties of these parameters are $50\%, 20\%, 25\%, 16.7\%$, respectively.

Figure 12.4 shows the earthquake ground acceleration. Figure 12.5 displays the results of the hysteresis behavior.

The simulation results with the proposed Scheme I are presented in Figures 12.6 to 12.8. Figures 12.6 and 12.7 show the time histories of the displacement

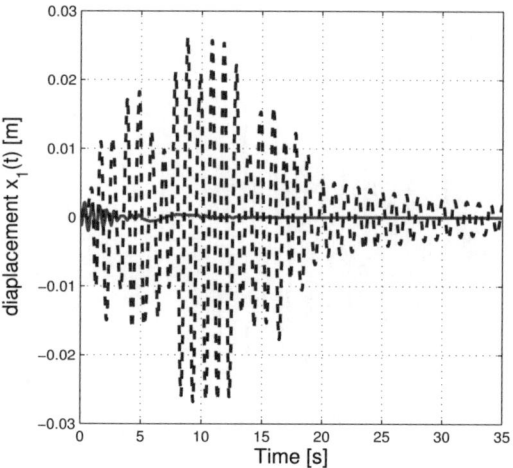

Fig. 12.6. Displacement x_1 (Without Control: dashed; With Scheme I: solid)

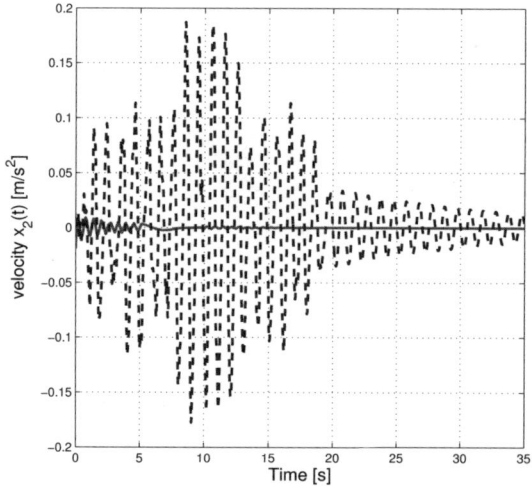

Fig. 12.7. Velocity x_2 (Without Control: dashed; With Scheme I: solid)

x_1 and the velocity x_2 without control and with control using Scheme I. After t = 20 seconds, the excitation stops and the uncontrolled case corresponds to free vibration response. The open loop system exhibits a low damping behavior. On the contrary, the proposed control drive the response towards zero rapidly, thus introducing a significant damping effect into the system. Figure 12.8 shows the time history of the control signal with Scheme I. Clearly, system performance is improved by Scheme I, even though we need much less apriori knowledge from the system.

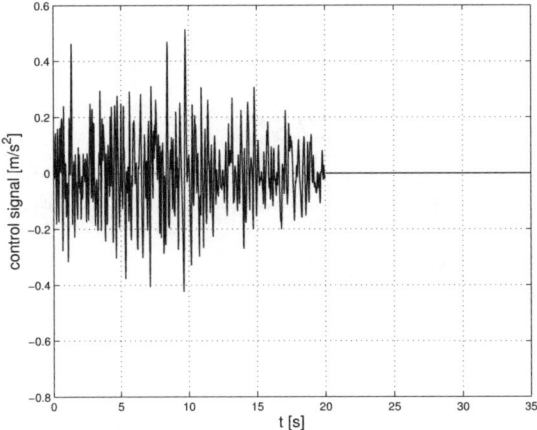

Fig. 12.8. Control Signal $u(t)/m$ with Scheme I

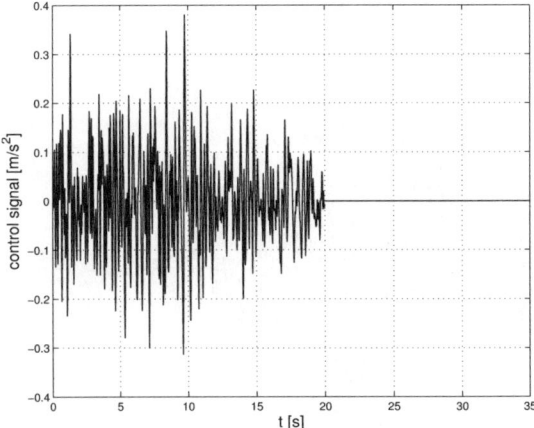

Fig. 12.9. Control Signal $u(t)/m$ with Scheme II

The simulation results with the proposed Scheme II are also presented in Figures 12.9 to 12.13. Figures 12.9-12.11 show the time history of the control signal, the displacement x_1 and the velocity x_2. Figure 12.12 displays system hysteresis behavior $\Phi(t)$ and the approximated hysteresis behavior $\bar{\Phi}(t)$ with Scheme II. Figure 12.13 displays the behavior of $\Phi - \bar{\Phi}$. As noted, $|\Phi - \bar{\Phi}|$ is smaller than $|\Phi(t)|$. A significant reduction in the magnitude of x and $\dot{x}$ can be observed with Scheme II.

As a conclusion, the simulation results verify our theoretical findings and show the effectiveness of our control schemes. System performance is improved by our proposed schemes. Also Scheme II is better than Scheme I in improving system performance, but requires more apriori knowledge.

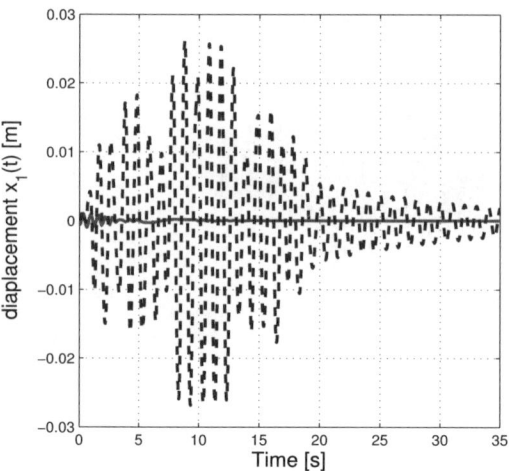

Fig. 12.10. Displacement x_1 (Without Control: dashed; With Scheme II: solid)

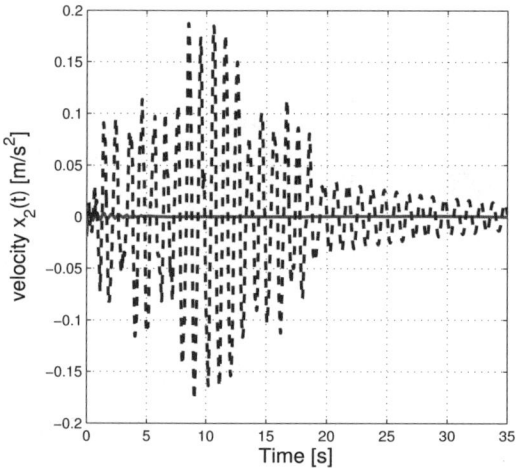

Fig. 12.11. Velocity x_2 (Without Control: dashed; With Scheme II: solid)

12.5 Summary

This chapter has presented two backstepping adaptive controllers for a second-order uncertain building structural system involving hysteretic phenomena. The hysteretic nonlinear behavior is described by the so-called Bouc-Wen model. The control strategies have been applied to a system found in base isolation schemes for seismic active protection of building structures. In the first scheme, the partial effect of the hysteresis is treated as a bounded disturbance. In the second scheme, we further take the structure of the hysteresis into account in our

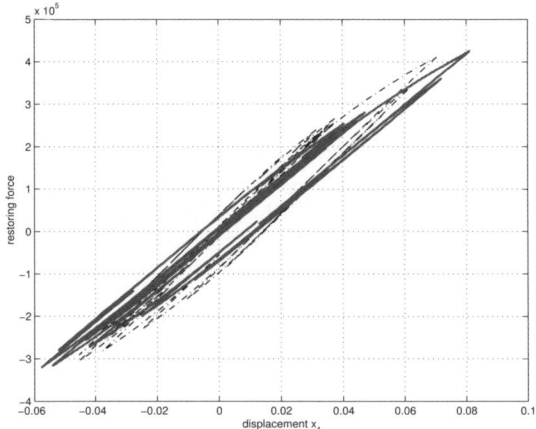

Fig. 12.12. True system hysteresis Φ: solid; approximated hysteresis $\bar{\Phi}$ with Scheme II: dashed

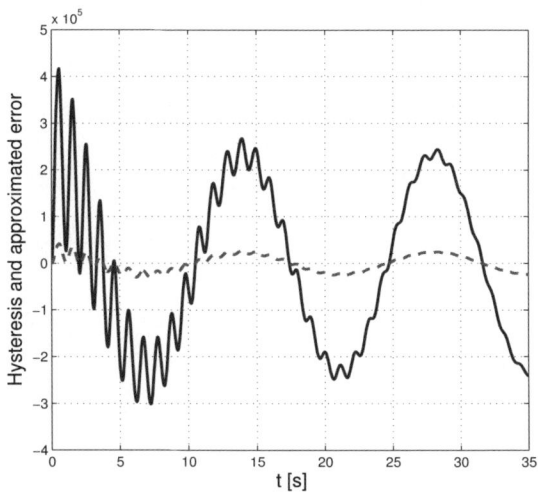

Fig. 12.13. True hysteresis Φ: solid; $\Phi - \bar{\Phi}$: dashed

controller design. It is shown that the proposed controllers can guarantee global boundedness of all signals and achieve tracking to a desired precision. Numerical results show that the adaptive control law is working satisfactorily in that the response induced by seismic action is significant reduced.

13 Control of a Piezo-Positioning Mechanism with Hysteresis

Piezo-positioning mechanisms are often used in high-precision positioning applications. Due to their materials, nonlinear hysteretic behavior is commonly observed in such mechanisms and can be described by a LuGre model. In this chapter, we develop two robust adaptive backstepping control algorithms for piezo-positioning mechanisms. In the first scheme, we take the structure of the LuGre model into account in the controller design, if the parameters of the model are known. A nonlinear observer is designed to estimate the hysteresis force. In the second scheme, there is no apriori information required from these parameters and thus they can be allowed totally uncertain. In this case, the LuGre model is divided into two parts. While the unknown parameters of one part is incorporated with unknown system parameters for estimation, the effect of the other part is treated as a bounded disturbance. An update law is used to estimate the bound involving this partial hysteresis effect and the external load. For both schemes, it is shown that not only stability is guaranteed by the proposed controller, but also both transient and asymptotic performances are quantified as explicit functions of the design parameters so that designers can tune the design parameters in an explicit way to obtain the required closed loop behavior.

13.1 Introduction

Piezo-positioning mechanisms are often used in high-precision positioning applications, such as nanometer. Since the materials of the piezo-positioning mechanisms are ferroelectric, nonlinear hysteretic behavior is commonly observed in such mechanisms in response to an applied electric field. This leads to problems of severe inaccuracy, instability, and restricted system performance due to the hysteresis nonlinearity. Moreover, the hysteresis characteristic is usually unknown and the states to represent hysteresis dynamics are often unavailable. These usually cause the increasing difficulties in servo control design with high performance requirement for piezo-positioning mechanisms. Studies of this problem have been reported in the works of [165, 166]. In [166], a feed-forward model-reference control was designed to improve scanning accuracy of PZT piezoelectric actuator. Studies of controlling hysteresis nonlinearity have also been reported

J. Zhou & C. Wen: Adapt. Backstepping Ctrl. of Uncertain Systems, LNCIS 372, pp. 215–226, 2008.
springerlink.com   © Springer-Verlag Berlin Heidelberg 2008

in the works of [43, 45, 56, 55, 126, 120]. In [55, 56, 126], an adaptive hysteresis inverse cascaded with the plant was employed to cancel the effects of hysteresis. In [43, 120] a dynamic hysteresis model was defined to pattern a hysteresis rather than constructing an inverse model to mitigate the effects of the hysteresis. In [167], a reinforcement discrete neuro-adaptive control was proposed for unknown piezoelectric actuator systems with dominant hysteresis. In [168], an adaptive wavelet neural network control was proposed to control a piezo-positioning mechanism with hysteresis estimation. However, it is assumed that system uncertain parameters must be within some known intervals. It is also assumed that system states must be inside compact sets in order to ensure the error due to neural network approximation bounded. This implies that the closed-loop system is bounded-input bounded-output stable even before the controller is designed.

In this chapter, in order to consider the hysteresis, the Lugre model presented in [169] is used and the proposed hysteresis model with parametrization is integrated into a mechanical motion dynamics with lumped external load to completely represent the overall dynamics of a piezo-positioning mechanism. We develop two simple adaptive backstepping control schemes for the piezo-positioning mechanism. In the first scheme, we take the structure of the Lugre model into account in the controller design, if parameters of the model are known. A nonlinear observer is designed by using Lyapunov technique to estimate the unavailable state. In the second scheme, there is no apriori information required from these parameters and thus they can be allowed totally uncertain. In this case, the LuGre model is divided into two parts. While the unknown parameters of one part is incorporated with unknown system parameters for estimation, the effect of the other part is treated as a bounded disturbance. An update law is used to estimate the bound involving this partial hysteresis effect and the external load. Besides showing stability of the system for both schemes, the transient performance in terms of L_2 norm of the tracking error is derived to be an explicit function of design parameters and thus our scheme allows designers to obtain the closed loop behavior by tuning design parameters in an explicit way.

13.2 System Description

It is well known that there exist two difficulties in the modelling of the hysteresis nonlinearity of piezo-actuators: (i) the nonlocal memory phenomenon; and (ii) the asymmetric loop between descending and ascending paths. A hysteresis friction model will be developed to simply and appropriately describe the dynamics of the piezo-positioning mechanism with a hysteresis effect. In this chapter, the following second-order uncertain nonlinear system in [168] is used to model the dynamics of a piezo-positioning mechanism.

$$M\ddot{x} + D\dot{x} + F_H + F_L = u \tag{13.1}$$

where M denotes the equivalent mass of the controlled piezo-positioning mechanism, which is positive; x is the displacement of the mechanism; $\dot{x}$ denotes the relative velocity; $\ddot{x}$ denotes the acceleration; D is the linear friction coefficient of the piezo-positioning mechanism; F_L is the external load; F_H denotes the hysteresis friction force function; u denotes the applied voltage to piezo-positioning mechanism. A block diagram representing system (13.1) is shown in Figure 13.1.

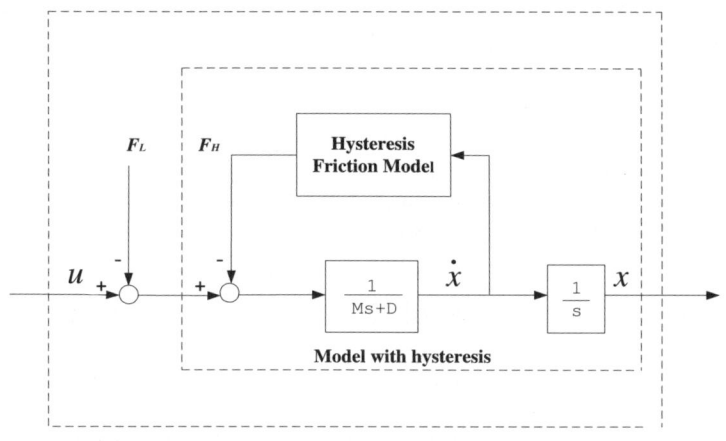

Model of Piezo-Positioning Mechanism with Hysteresis

Fig. 13.1. Block diagram of modelling piezo-positioning mechanism

The hysteresis friction force F_H is described by the so-called LuGre model [169] in the following form

$$F_H = \sigma_0 z + \sigma_1 \frac{dz}{dt} + \sigma_2 \dot{x} \tag{13.2}$$

$$\dot{z} = \dot{x} - \frac{|\dot{x}|}{g(\dot{x})} z \tag{13.3}$$

where z is an un-measurable state and represents the average deflection of the contact force, $\dot{x}$ denotes the relative velocity between the two contact surfaces, σ_0, σ_1 and σ_2 are positive constants and can be equivalently interpreted as the bristle stiffness, bristle damping and viscous damping coefficient, respectively. Moreover, the function $g(\dot{x})$ denotes the Stribeck effect curve given by

$$\sigma_0 g(\dot{x}) = f_C + (f_S - f_C)e^{-(\dot{x}/\dot{x}_S)^2} \tag{13.4}$$

where f_C is the coulomb friction which is independent of the velocity, f_S is the stiction force which is the critical force that makes an object move from the static mode, and $\dot{x}_S$ is the Stribeck velocity, see [170]. The function $g(\dot{x})$ is positive and bounded. As shown in [169], the following lemma holds.

Lemma 13.1. *Consider nonlinear dynamic system (13.3). For any piecewise continuous signal x and $\dot{x}$, the output $z(t)$ is bounded.*

Substituting (13.3) into (13.2), the hysteresis friction model can be rewritten as:

$$F_H = \sigma_0 z - \sigma_1 \frac{1}{g(\dot{x})} z|\dot{x}| + (\sigma_1 + \sigma_2)\dot{x} \tag{13.5}$$

The hysteresis friction expressed in (13.5), which is functionally related to the system velocity, is used to represent the behavior of the hysteresis introduced by the piezo-actuator. Furthermore, if the hysteresis friction is integrated into the mechanical motion dynamics, the overall dynamics including the hysteresis effect can be effectively and completely modelled.

The control objective is to design a backstepping adaptive control law for system (13.1) so that the displacement x of the piezo-positioning mechanism can track any desired bounded reference trajectory x_m.

13.3 Backstepping Control and Stability Analysis

In this section, we develop two adaptive backstepping design schemes. In Scheme I, we assume that the hysteresis parameters $\sigma_0, \sigma_1, \sigma_2$ and the function g are known. The state $z(t)$ is not measurable and hence has to be observed to estimate the hysteresis force F_H. For this we design an observer $\hat{z}(t)$ to estimate $z(t)$. In Scheme II, we assume that the parameters $\sigma_0, \sigma_1, \sigma_2, f_S, f_C, \dot{x}_S$ in the hysteresis model (13.3) are all uncertain. The residual effect of the hysteresis is treated as a bounded disturbance with unknown bound. An update law is used to estimate the bound involving the effect of the hysteresis and the external load. To illustrate the backstepping procedures, only the first scheme is elaborated in details.

Control Scheme I

When the hysteresis parameters $\sigma_0, \sigma_1, \sigma_2$ and the function g are known, we can exploit the structure of the model in our controller design to improve system performance.

We rewrite equations (13.1) in the following form

$$\dot{x}_1 = x_2$$
$$\dot{x}_2 = \frac{1}{M}u - \theta x_2 - \frac{1}{M}F_H - \frac{1}{M}F_L \tag{13.6}$$

where $x_1 = x, x_2 = \dot{x}$ and $\theta = \frac{D}{M}$ is uncertain parameter. Note that F_L is bounded with unknown bound F_o. Before presenting the adaptive control design using the backstepping technique to achieve the desired control objectives, the following change of coordinates is made.

$$\varsigma_1 = x_1 - x_m \tag{13.7}$$
$$\varsigma_2 = x_2 - \dot{x}_m - \alpha_1 \tag{13.8}$$

where α_1 is a virtual control and will be determined in later discussion.

The variable $z(t)$ is not measurable and hence has to be observed to estimate the hysteresis force F_H. For this, we design a nonlinear observer to estimate the variable z as follows:

$$\dot{\hat{z}} = \dot{x} - \frac{|x_2|}{g(x_2)}\hat{z} + \phi(t) \tag{13.9}$$

where $\phi(t)$ is a nonlinear function derived later. Note that $\frac{|x_2|}{g(x_2)}$ is positive and bounded.

- *Step 1:* From (13.6) to (13.8), we obtain that

$$\dot{\varsigma}_1 = \varsigma_2 + \alpha_1 \tag{13.10}$$

We design the virtual control law α_1 as

$$\alpha_1 = -c_1\varsigma_1 \tag{13.11}$$

where c_1 is a positive design parameter. From (13.10) and (13.11) we have

$$\varsigma_1\dot{\varsigma}_1 = -c_1\varsigma_1^2 + \varsigma_1\varsigma_2 \tag{13.12}$$

- *Step 2:* From (13.6) and (13.8), we have

$$\dot{\varsigma}_2 = \frac{1}{M}u - \theta x_2 - \frac{1}{M}F_H - \frac{1}{M}F_L - \ddot{x}_m - \dot{\alpha}_1 \tag{13.13}$$

Then function $\phi(t)$ in (13.9), the control law and parameter update laws are obtained as follows by considering a Lyapunov function in (13.21) with details given later.

$$\phi(t) = -\sigma_0\varsigma_2 + \sigma_1\frac{|x_2|}{g(x_2)}\varsigma_2 \tag{13.14}$$

$$u(t) = \hat{M}\bar{u} + \hat{F}_H - \hat{F}_o sign(\varsigma_2) \tag{13.15}$$

$$\bar{u}(t) = -c_2\varsigma_2 - \varsigma_1 + \hat{\theta}x_2 + \ddot{x}_m + \dot{\alpha}_1 \tag{13.16}$$

$$\dot{\hat{\theta}} = -\gamma_\theta x_2\varsigma_2 \tag{13.17}$$

$$\dot{\hat{M}} = -\gamma_M \bar{u}\varsigma_2 \tag{13.18}$$

$$\dot{\hat{F}}_o = \gamma_F|\varsigma_2| \tag{13.19}$$

$$\hat{F}_H = \sigma_0\hat{z} - \sigma_1\frac{|x_2|}{g(x_2)}\hat{z} + (\sigma_1 + \sigma_2)x_2 \tag{13.20}$$

where c_2, γ_θ, γ_M and γ_F are positive design parameters, $\hat{\theta}$, $\hat{M}$ and $\hat{F}_o$ are estimates of θ, M and F_o, respectively, F_o is the unknown bound of the external load F_L.

Remark 13.1. Note that a parameter update law is used to estimate the bound F_o of the external load F_L, so there is no need to know this bound.

We now show how (13.14)-(13.20) are derived by the following function.

$$V = \frac{1}{2}\varsigma_1^2 + \frac{1}{2}\varsigma_2^2 + \frac{1}{2M\gamma_M}\tilde{M}^2 + \frac{1}{2\gamma_\theta}\tilde{\theta}^2 + \frac{1}{2M\gamma_F}\tilde{F}_o^2 + \frac{1}{2M}\tilde{z}^2 \quad (13.21)$$

where $\tilde{M} = M - \hat{M}$, $\tilde{\theta} = \theta - \hat{\theta}$, $\tilde{F}_o = F_o - \hat{F}_o$ and $\tilde{z} = z - \hat{z}$.

Note that $\frac{1}{M}u$ in (13.13) can be expressed as

$$\begin{aligned}
\frac{1}{M}u &= \frac{1}{M}\hat{M}\bar{u} + \frac{1}{M}\hat{F}_H - \frac{1}{M}\hat{F}_o sign(\varsigma_2) \\
&= \bar{u} - \frac{1}{M}\tilde{M}\bar{u} + \frac{1}{M}\hat{F}_H - \frac{1}{M}\hat{F}_o sign(\varsigma_2)
\end{aligned} \quad (13.22)$$

Then the derivative of V along with (13.9), (13.13) and (13.22) is given by

$$\begin{aligned}
\dot{V} &= \varsigma_1\dot{\varsigma}_1 + \varsigma_2\dot{\varsigma}_2 + \frac{1}{\gamma_\theta}\tilde{\theta}\dot{\tilde{\theta}} + \frac{1}{M\gamma_M}\tilde{M}\dot{\tilde{M}} + \frac{1}{M\gamma_F}\tilde{F}_o\dot{\tilde{F}}_o + \frac{1}{M}\tilde{z}\dot{\tilde{z}} \\
&= -c_1\varsigma_1^2 + \varsigma_1\varsigma_2 + \varsigma_2(\frac{1}{M}u - \theta x_2 - \frac{1}{M}F_H - \frac{1}{M}F_L - \ddot{x}_m - \dot{\alpha}_1) \\
&\quad + \frac{1}{\gamma_\theta}\tilde{\theta}\dot{\tilde{\theta}} + \frac{1}{M\gamma_M}\tilde{M}\dot{\tilde{M}} + \frac{1}{M\gamma_F}\tilde{F}_o\dot{\tilde{F}}_o + \frac{1}{M}\tilde{z}(\dot{z} - \dot{\hat{z}}) \\
&= -c_1\varsigma_1^2 + \varsigma_2(\bar{u} + \varsigma_1 - \theta x_2 - \ddot{x}_m - \dot{\alpha}_1) - \frac{1}{M}\varsigma_2(F_H - \hat{F}_H) + \frac{1}{M}\tilde{z}(\dot{z} - \dot{\hat{z}}) \\
&\quad - \frac{1}{\gamma_\theta}\tilde{\theta}^T\dot{\hat{\theta}} - \frac{1}{M\gamma_M}\tilde{M}(\gamma_M\bar{u}\varsigma_2 + \dot{\hat{M}}) + \frac{1}{M\gamma_F}\tilde{F}_o(\gamma_F|\varsigma_2| - \dot{\hat{F}}_o) \quad (13.23)
\end{aligned}$$

So $\bar{u}$ given in (13.16) is designed based on the second term of (13.23). Then the derivative of V is given by

$$\begin{aligned}
\dot{V} &\le -c_1\varsigma_1^2 - c_2\varsigma_2^2 - \frac{1}{M}\frac{|x_2|}{g(x_2)}\tilde{z}^2 + \frac{1}{M}\tilde{z}\left(-\sigma_0\varsigma_2 + \frac{\sigma_1|x_2|}{g(x_2)}\varsigma_2 - \phi(t)\right) \\
&\quad - \frac{1}{\gamma_\theta}\tilde{\theta}^T\left(\gamma_\theta x_2\varsigma_2 + \dot{\hat{\theta}}\right) - \frac{1}{M\gamma_M}\tilde{M}\left(\gamma_M\bar{u}\varsigma_2 + \dot{\hat{M}}\right) \\
&\quad + \frac{1}{M\gamma_F}\tilde{F}_o\left(\gamma_F|\varsigma_2| - \dot{\hat{F}}_o\right) \quad (13.24)
\end{aligned}$$

Then $\phi(t)$, $\dot{\hat{\theta}}$, $\dot{\hat{M}}$ and $\dot{\hat{F}}_o$ in (13.14), (13.17), (13.18) and (13.19) are obtained by making the last four terms of (13.24) zero, respectively. Thus,

$$\dot{V} \le -c_1\varsigma_1^2 - c_2\varsigma_2^2 - \frac{1}{M}\frac{|x_2|}{g(x_2)}\tilde{z}^2 \quad (13.25)$$

Based on (13.25), we can obtain the result on system stability and performance as stated below.

Theorem 13.1. *Consider the uncertain nonlinear system (13.1). With the application of the controller (13.15), the observer (13.9) and the parameter update laws (13.17), (13.18) and (13.19), the following statements hold:*

- *The resulting closed loop system is globally bounded input bounded output (BIBO) stable.*
- *The asymptotic tracking is achieved, i.e.,*

$$\lim_{t \to \infty} [x(t) - x_m(t)] = 0 \qquad (13.26)$$

- *The transient displacement tracking error performance is given by*

$$\| x(t) - x_m(t) \|_2$$
$$\leq \frac{1}{\sqrt{c_1}} \left(\frac{1}{2\gamma_\theta} \tilde{\theta}(0)^2 + \frac{1}{2M\gamma_M} \tilde{M}(0)^2 + \frac{1}{2M\gamma_F} \tilde{F}_o(0)^2 + \frac{1}{2M} \tilde{z}(0)^2 \right)^{1/2}$$

$$(13.27)$$

- *The transient velocity tracking error performance is given by*

$$\| \dot{x} - \dot{x}_m \|_2$$
$$\leq \left(\frac{1}{\sqrt{c_2}} + \sqrt{c_1} \right) \left(\frac{1}{2\gamma_\theta} \tilde{\theta}(0)^2 + \frac{1}{2M\gamma_M} \tilde{M}(0)^2 + \frac{1}{2M\gamma_F} \tilde{F}_o(0)^2 + \frac{1}{2M} \tilde{z}(0)^2 \right)^{1/2}$$

$$(13.28)$$

Proof: Equation (13.25) shows that $V(t)$ is globally uniformly bounded. This implies that $\varsigma_1, \varsigma_2, \tilde{\theta}, \tilde{M}, \tilde{F}_o, \tilde{z}$ are bounded. The state variables x_1, x_2 and the parameter estimates $\hat{\theta}, \hat{M}, \hat{F}_o, \hat{z}$ are also bounded. Thus u is bounded from (13.15) because of the boundedness of $\varsigma_1, \varsigma_2, \hat{\theta}, \hat{M}, \hat{F}_o, \hat{z}$. By applying the LaSalle-Yoshizawa theorem to (13.25), it further follows that $z_i(t) \to 0, i = 1, 2$ as $t \to \infty$, which implies that $\lim_{t \to \infty} [x(t) - x_m(t)] = 0$.

Since V is non increasing from (13.25), we have

$$\| \varsigma_1 \|_2^2 = \int_0^\infty |\varsigma_1(\tau)|^2 d\tau \leq \frac{1}{c_1} (V(0) - V(\infty)) \leq \frac{1}{c_1} V(0) \qquad (13.29)$$

The initial value of the Lyapunov function is

$$V(0) = \frac{1}{2} \varsigma_1^2(0) + \frac{1}{2} \varsigma_2^2(0) + \frac{1}{2\gamma_\theta} \tilde{\theta}(0)^2 + \frac{1}{2M\gamma_M} \tilde{M}(0)^2$$
$$+ \frac{1}{2M\gamma_F} \tilde{F}_o(0)^2 + \frac{1}{2M} \tilde{z}(0)^2 \qquad (13.30)$$

Note that $\tilde{\theta}(0), \tilde{M}(0), \tilde{F}_o(0)$ and $\varsigma_1(0) = x_1(0) - x_m(0)$ are clearly independent of $c_1, \gamma_M, \gamma_\theta$ and γ_F. We can set $\varsigma_1(0)$ and $\varsigma_2(0)$ to zero by appropriately initializing the reference trajectory $x_m(0)$ and $\dot{x}_m(0)$ as follows

$$x_m(0) = x_1(0) \qquad (13.31)$$
$$\dot{x}_m(0) = x_2(0) \qquad (13.32)$$

Thus, by setting $\varsigma_1(0) = \varsigma_2(0) = 0$, we obtain

$$V(0) = \frac{1}{2\gamma_\theta} \tilde{\theta}(0)^2 + \frac{1}{2M\gamma_M} \tilde{M}(0)^2 + \frac{1}{2M\gamma_F} \tilde{F}_o(0)^2 + \frac{1}{2M} \tilde{z}(0)^2 \quad (13.33)$$

a decreasing function of γ_θ, γ_F and γ_M, independent of c_1. This means that the bound resulting from (13.29) and (13.33) satisfies

$$\| \varsigma_1 \|_2 \leq \frac{1}{\sqrt{c_1}} \Big(\frac{1}{2\gamma_\theta} \tilde{\theta}(0)^2 + \frac{1}{2M\gamma_M} \tilde{M}(0)^2 + \frac{1}{2M\gamma_F} \tilde{F}_o(0)^2 + \frac{1}{2M} \tilde{z}(0)^2 \Big)^{1/2}$$

(13.34)

and can be asymptotically reduced either by increasing c_1 or by simultaneously increasing γ_θ, γ_M and γ_F. Thus the bound for $\| \varsigma_1 \|_2$ is an explicit function of design parameters.

From equations (13.10) to (13.11), we get

$$\| \dot{x} - \dot{y}_r \|_2 = \| \varsigma_2 - c_1 \varsigma_1 \|_2 \leq \| \varsigma_2 \|_2 + c_1 \| \varsigma_1 \|_2$$

(13.35)

Similarly, we can get $\| \varsigma_2 \|_2 \leq \frac{1}{\sqrt{c_2}} \sqrt{V(0)}$. Along with (13.34) we get

$$\| \dot{x} - \dot{y}_r \|_2$$
$$\leq \Big(\frac{1}{\sqrt{c_2}} + \sqrt{c_1} \Big) \Big(\frac{1}{2\gamma_\theta} \tilde{\theta}(0)^2 + \frac{1}{2M\gamma_M} \tilde{M}(0)^2 + \frac{1}{2M\gamma_F} \tilde{F}_o(0)^2 + \frac{1}{2M} \tilde{z}(0)^2 \Big)^{1/2}$$

(13.36)

Remark 13.2. From Theorem 13.1 the following conclusions can be obtained:

- Boundedness of signals in the adaptive system is guaranteed to be global, uniform and ultimate for any positive values of the design parameters c_1, c_2, γ_θ, γ_M and γ_F.
- The transient performance depends on the initial estimate errors $\tilde{\theta}(0)$, $\tilde{M}(0)$, $\tilde{F}_o(0)$ and $\tilde{z}(0)$. The closer the initial estimates to the true values, the better the transient performance. The asymptotic behavior is not affected by the initial estimate errors. We can decrease the effects of the initial error estimates on the transient performance by increasing the adaptation gains γ_θ, γ_M and γ_F.
- To improve the displacement tracking error performance we can also increase the gain c_1. However, increasing the gain c_1 will also increase the velocity tracking error as shown above. Improving the closed loop displacement behavior may be done at the expense of the increase in the control signal amplitude. This suggests to fix the gain c_1 to some acceptable value and adjust the other gains. By fixing the gain c_1, increasing the gain c_2 or by simultaneously increasing γ_θ, γ_M and γ_F, we can achieve a velocity tracking error as small as desired.

Control Scheme II

In this section, there is no apriori information required from parameters σ_0, σ_1, σ_2, f_C, $f_S \dot{x}_S$ and thus they can be allowed totally uncertain. The LuGre hysteresis friction force F_H in (13.5) can be divided into two parts as follows.

$$F_H = (\sigma_1 + \sigma_2)\dot{x} + R(t)$$

(13.37)

$$R(t) = \sigma_0 z - \sigma_1 \frac{|\dot{x}|}{g(\dot{x})} z$$

(13.38)

From Lemma 13.1, we have that $R(t)$ is bounded. Then we combine $(\sigma_1 + \sigma_2)\dot{x}$ with $D\dot{x}$ in (1) and rewrite equations (13.1) and (13.37) in the following form

$$\dot{x}_1 = x_2$$
$$\dot{x}_2 = \frac{1}{M}u - \theta x_2 - \frac{1}{M}d(t) \tag{13.39}$$

where $x_1 = x, x_2 = \dot{x}$, $\theta = \frac{1}{M}(D + \sigma_1 + \sigma_2)$, and $d(t) = R + F_L$. So $d(t)$ is bounded with unknown bound F_o.

Now $d(t)$ can be handled in the same way as F_L in Scheme I. Thus the controller design in this case is similar to the Scheme I and we only give the resulting control laws.

$$u = \hat{M}\bar{u} - \hat{F}_o sign(\varsigma_2) \tag{13.40}$$
$$\bar{u} = -c_2\varsigma_2 - \varsigma_1 + \hat{\theta}x_2 + \ddot{x}_m + \dot{\alpha}_1 \tag{13.41}$$
$$\alpha_1 = -c_1\varsigma_1 \tag{13.42}$$
$$\dot{\hat{\theta}} = -\gamma_\theta x_2\varsigma_2 \tag{13.43}$$
$$\dot{\hat{M}} = -\gamma_M \bar{u}\varsigma_2 \tag{13.44}$$
$$\dot{\hat{F}}_o = \gamma_F|\varsigma_2| \tag{13.45}$$
$$\varsigma_1 = x_1 - x_m \tag{13.46}$$
$$\varsigma_2 = x_2 - \dot{x}_m - \alpha_1 \tag{13.47}$$

where c_1, c_2, γ_θ, γ_M and γ_F are designed positive parameters, $\hat{\theta}$, $\hat{M}$ and $\hat{F}_o$ are estimates of θ, M and F_o, respectively.

Following the similar analysis to Scheme I, we can establish that $\varsigma_1, \varsigma_2, \hat{\theta}, \hat{M}, \hat{F}_o, u$ are all bounded. Thus similar to Theorem 13.1, the results on system stability and performance can be established and now stated in the following theorem.

Theorem 13.2. *Consider the uncertain nonlinear system (13.1). With the application of the controller (13.40) and the parameter update laws (13.43), (13.44) and (13.45), the following statements hold:*
- *The resulting closed loop system is globally bounded input bounded output (BIBO) stable.*
- *The asymptotic tracking is achieved, i.e.,*

$$\lim_{t \to \infty} [x(t) - x_m(t)] = 0 \tag{13.48}$$

- *The transient displacement tracking error performance is given by*

$$\| x(t) - x_m(t) \|_2 \leq \frac{1}{\sqrt{c_1}} \times \left(\frac{1}{2\gamma_\theta}\tilde{\theta}(0)^2 + \frac{1}{2M\gamma_M}\tilde{M}(0)^2 + \frac{1}{2M\gamma_F}\tilde{F}_o(0)^2\right)^{1/2} \tag{13.49}$$

• *The transient velocity tracking error performance is given by*

$$\| \dot{x} - \dot{x}_m \|_2$$
$$\leq \left(\frac{1}{\sqrt{c_2}} + \sqrt{c_1} \right) \left(\frac{1}{2\gamma_\theta} \tilde{\theta}(0)^2 + \frac{1}{2M\gamma_M} \tilde{M}(0)^2 + \frac{1}{2M\gamma_F} \tilde{F}_o(0)^2 \right)^{1/2}$$

$$(13.50)$$

13.4 Simulation Results

In this section we test our proposed backstepping controllers on model (13.1). For simulation studies, the following values are selected as "true" parameters

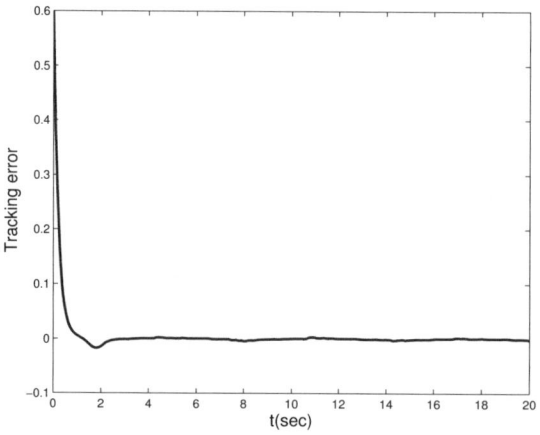

Fig. 13.2. Time history of tracking error $x - x_m$ with Scheme I

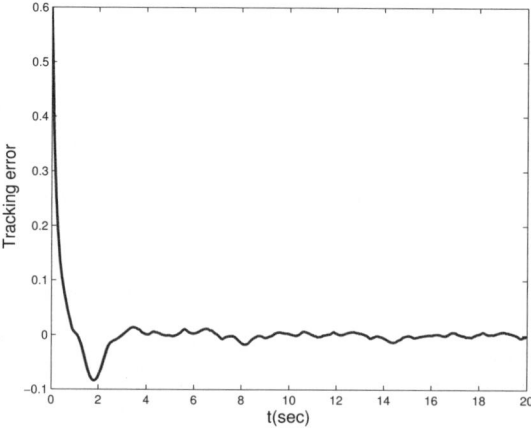

Fig. 13.3. Time history of tracking error $x - x_m$ with Scheme II

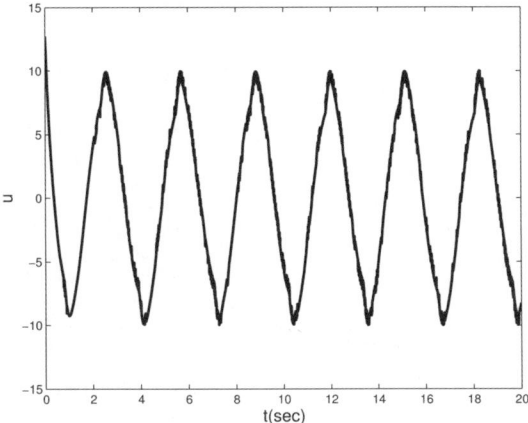

Fig. 13.4. Time history of control input $u(t)$ with Scheme I

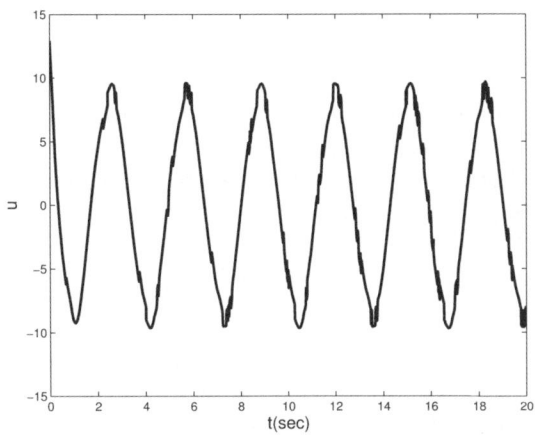

Fig. 13.5. Time history of control input $u(t)$ with Scheme II

for the system and the hysteresis model: $\sigma_0 = 5N/m, \sigma_1 = 0.9Ns/m, \sigma_2 = 0.4Ns/m$, $f_C = 1N, f_S = 1.5N, \dot{x}_S = 0.01m/s$, $M = 1kg, D = 0.15Ns/m$ and $F_L = 0.3\sin(t)$.

The design objective is to drive the displacement x of the piezo-positioning mechanism to track the reference trajectory $x_m(t) = 2\sin(2t)$.

For both schemes, we take the following set of design parameters: $\gamma_\theta = 0.4, \gamma_M = 0.2, \gamma_F = 0.4, c_1 = c_2 = 4$. The initials are set as $x(0) = 0.6, \dot{x}(0) = 2, z(0) = 0, \hat{z}(0) = 0, \hat{M}(0) = 1.2, \hat{F}_O(0) = 0.8$, and $\hat{\theta}(0) = 0.2$, respectively.

When Scheme II is used, we take the following set of design parameters: $\gamma_\theta = 0.4, \gamma_M = 0.2, \gamma_F = 0.4, c_1 = c_2 = 4$. The initials are same with Scheme I.

The simulation results with the proposed two schemes are presented in Figures 13.2 to 13.6, respectively. Figures 13.2 and 13.3 show the time history of the

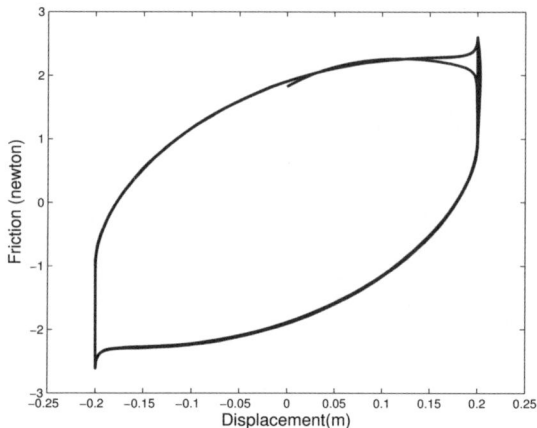

Fig. 13.6. Hysteresis identification

displacement x and the trajectory x_m, while figures 13.4 and 13.5 show the time history of the control input u. Figure 13.6 displays the results of the hysteresis behavior.

It is observed that Scheme I is better than Scheme II in improving system performance. This is expected as more apriori knowledge is used. Overall, the simulation results verify our theoretical findings and show the effectiveness of our control schemes.

13.5 Conclusion

In this chapter, two backstepping adaptive controllers have been presented for a piezo-positioning mechanism involving hysteretic phenomena. The hysteretic nonlinear behavior is described by the LuGre model. In the first scheme, we take the structure of the hysteresis into account in our controller design, if the parameters of the Lugre model are all known. A nonlinear observer is designed by considering a Lyapunov function. In the second scheme, there is no apriori information required from these parameters and part of the hysteresis effect is treated as a bounded disturbance. It is shown that the proposed controllers can guarantee global boundedness of signals and achieve asymptotic tracking. Numerical results show that the designed adaptive control laws work satisfactorily.

A Appendices

Appendix A

Lyapunov Stability [1]

For all control systems and adaptive control systems in particular, stability is
the primary requirement. Consider the time-varying system

$$\dot{x} = f(x, t) \tag{A.1}$$

where $x \in R^n$, and $f : R^n \times R_+ \to R^n$ is piecewise continuous in t and locally
Lipschiz in x. The solution of (A.1) which starts from the point x_0 at time
$t_0 \geq 0$ is denoted as $x(t; x_0, t_0)$ with $x(t_0; x_0, t_0) = x_0$. If the initial condition x_0
is perturbed to $\tilde{x}_0$, then, for stability, the resulting perturbed solution $x(t; \tilde{x}_0, t_0)$
is required to stay close to $x(t; x_0, t_0)$ for all $t \geq t_0$. In addition, for asymptotic
stability, the error $x(t; \tilde{x}_0, t_0) \to x(t; x_0, t_0)$ is required to vanish as $t \to \infty$. So
the solution $x(t; x_0, t_0)$ of (A.1) is

- *bounded*, if there exists a constant $B(x_0, t_0) > 0$ such that

$$|x(t; x_0, t_0)| < B(x_0, t_0), \quad \forall\, t \geq t_0;$$

- *stable*, if for each $\epsilon > 0$ there exists a $\delta(\epsilon, t_0) > 0$ such that

$$|\tilde{x}_0 - x_0| < \delta, \quad |x(t; \tilde{x}_0, t_0) - x(t; x_0, t_0)| < \epsilon, \quad \forall\, t \geq t_0;$$

- *attractive*, if there exists a $r(t_0) > 0$ and, for each $\epsilon > 0$, a $T(\epsilon, t_0) > 0$ such
  that

$$|\tilde{x}_0 - x_0| < r, \quad |x(t; \tilde{x}_0, t_0) - x(t; x_0, t_0)| < \epsilon, \quad \forall\, t \geq t_0 + T;$$

- *asymptotically stable*, if it is stable and attractive; and
- *unstable*, if it is not stable.

Theorem A.1 (Uniform Stability). Let $x = 0$ be an equilibrium point of (A.1) and $D = \{x \in R^n \mid |x| < r\}$. Let $V : D \times R^n \to R_+$ be a continuously differentiable function such that $\forall t \geq 0$, $\forall x \in D$, such that

$$\gamma_1(|x|) \leq V(x, t) \leq \gamma_2(|x|)$$
$$\frac{\partial V}{\partial t} + \frac{\partial V}{\partial x} f(x, t) \leq -\gamma_3(|x|)$$

Then the equilibrium $x = 0$ is

- uniformly stable, if γ_1 and γ_2 are class κ functions on $[0, r)$ and $\gamma_3(.) \geq 0$ on $[0, r)$;
- uniformly asymptotically stable, if γ_1, γ_2 and γ_3 are class κ functions on $[0, r)$;
- exponentially stable, if $\gamma_i(\rho) = k_i \rho^\alpha$ on $[0, r), k_i > 0, \alpha > 0, i = 1, 2, 3$;
- globally uniformly stable, if $D = R^n$, γ_1 and γ_2 are class κ_∞ functions, and $\gamma_3(.) \geq 0$ on R_+;
- globally unifromly asymptotically stable, if $D = R^n$, γ_1 and γ_2 are class κ_∞ functions, and γ_3 is a class of κ function on R_+; and
- globally exponentially stable, if $D = R^n$ and $\gamma_i(\rho) = k_i \rho^\alpha$ on $R_+, k_i > 0, \alpha > 0, i = 1, 2, 3$.

Appendix B

LaSalle-Yoshizawa Theorem [1]

Theorem B.1 (LaSalle-Yoshizawa). Let $x = 0$ be an equilibrium point of (A.1) and suppose f is locally Lipschitz in x uniformly in t. Let $V : R^n \times R_+ \to R_+$ be a continuously differentiable function such that

$$\gamma_1(|x|) \leq V(x, t) \leq \gamma_2(|x|) \tag{B.1}$$
$$\dot{V} = \frac{\partial V}{\partial t} + \frac{\partial V}{\partial x} f(x, t) \leq -W(x) \leq 0 \tag{B.2}$$

$\forall t \geq 0$, $\forall x \in R^n$, where γ_1 and γ_2 are class k_∞ functions and W is a continuous function. Then, all solutions of (A.1) are globally uniformly bounded and satisfy

$$lim_{t \to \infty} W(x(t)) = 0 \tag{B.3}$$

In addition, if $W(x)$ is positive definite, then the equilibrium $x = 0$ is globally uniformly asymptotically stable.

Appendix C

Parameter Projection [1]

Defining the following convex set

$$II_\epsilon = \{\hat{\theta} \in IR^p | P(\hat{\theta}) \leq \epsilon\}, \quad II = \{\hat{\theta} \in IR^p | P(\hat{\theta}) \leq 0\} \tag{C.1}$$

which is a union of the set II and an $O(\epsilon)$-boundary layer around it. Let us denote the interior of II_ϵ by II° and observe that $\nabla_{\hat{\theta}} P$ represents an outward normal vector at $\hat{\theta} \in \partial II_\epsilon$. The standard projection operator is

$$Proj\{\tau\} = \begin{cases} \tau & \hat{\theta} \in II^o \ or \ \nabla_{\hat{\theta}} P^t \tau \leq 0 \\ (I - c(\hat{\theta}) \Gamma \frac{\nabla_{\hat{\theta}} P \nabla_{\hat{\theta}} P^T}{\nabla_{\hat{\theta}} P^T \Gamma \nabla_{\hat{\theta}} P}) \tau & \hat{\theta} \in II_\epsilon / II^o \ and \ \nabla_{\hat{\theta}} P^T \tau > 0 \end{cases}$$

(C.2)

$$c(\hat{\theta}) = min\{1, \frac{P(\hat{\theta})}{\epsilon}\}$$

(C.3)

where Γ belongs to the set G of all positive definite symmetric $p \times p$ matrices. It is helpful to note that $c(\partial II_\epsilon) = 1$.

Theorem C.1 (Projection Operator). The following are the properties of the projection operator (C.2):

(i). The mapping $Proj: IR^p \times II_\epsilon \times G \to IR^p$ is locally Lipschiz in its arguments $\tau, \hat{\theta}, \Gamma$.

(ii). $Proj\{\tau\}^T \Gamma^{-1} Proj\{\tau\} \leq \tau^T \Gamma^{-1} \tau, \quad \forall \ \hat{\theta} \in II_\epsilon$.

(iii). Let $\Gamma(t), \tau(t)$ be continuously differentiable and $\dot{\hat{\theta}} = Proj\{\tau\}$, $\hat{\theta}(0) \in II_\epsilon$. Then, on its domain of definition, the solution $\hat{\theta}(t)$ remains in II_ϵ.

(iv). $-\tilde{\theta}^T \Gamma^{-1} Proj\{\tau\} \leq -\tilde{\theta}^T \Gamma^{-1} \tau, \forall \ \hat{\theta} \in II_\epsilon, \ \theta \in II$.

Appendix D

Internal Model Principle

Consider w generated by an exosystem

$$\dot{w} = Sw$$

(D.1)

where S is an unknown matrix having distinct eigenvalues with zero real parts. Such as

$$S = \begin{bmatrix} S_1 & \cdots & 0 \\ \cdot & \cdots & \cdot \\ 0 & \cdots & S_m \end{bmatrix}, \quad S_1 = \begin{bmatrix} 0 & \beta_1 \\ -\beta_1 & 0 \end{bmatrix} \quad \cdots \quad S_m = \begin{bmatrix} 0 & \beta_m \\ -\beta_m & 0 \end{bmatrix}$$

(D.2)

where $w = col(w_{11}, w_{12}, \ldots, w_{m1}, w_{m2})$, $\beta_1, \ldots, \beta_m$ are constants.

Lemma D.1. Let A be a $n \times n$ matrix having all eigenvalues with nonzero real part and S be a matrix which the eigenvalues are zero real parts and distinct as in (D.2). Let $\mathcal{P}$ denote the set of all homogeneous polynomials of degree p in $w_{11}, w_{12}, \ldots, w_{m1}, w_{m2}$ with coefficients in $\mathcal{R}$. For any $q(w) \in \mathcal{P}^n$, the equation

$$\frac{\partial \pi(w)}{\partial w} Sw = A\pi(w) + q(w) \tag{D.3}$$

has a unique solution $\pi(w)$, which is an element of $\mathcal{P}^n$.

Proof. Follows the proof as in [171]. $\mathcal{P}$ is indeed a vector space over $\mathcal{R}$, of finite dimension $d(p, m)$. Set

$$X_i = w_{i1} - jw_{i2}, \quad \bar{X}_i = w_{i1} + jw_{i2} \tag{D.4}$$

and note that any $b(w) \in \mathcal{P}$ can be written as

$$b(w) = \sum_{i_1 + j_1 + \ldots + i_m + j_m = p} b_{i_1 j_1 \ldots i_m j_m} X_i^{i_1} \bar{X}_1^{j_1} \ldots X_m^{i_m} \bar{X}_m^{j_m} \tag{D.5}$$

where $b_{i_1 j_1 \ldots i_m j_m}$ are unique determined and

$$b_{i_1 j_1 \ldots i_m j_m} = \bar{b}_{j_1 i_1 \ldots j_m i_m} \tag{D.6}$$

because the coefficients of $b(w)$ are real numbers. Choose any order for the set of indices $i_1 j_1 \ldots i_m j_m$ and write $b(w)$ in the form

$$b(w) = BW \tag{D.7}$$

where W is $d(p, m) \times 1$ vector consisting of all products of the form the $X_i^{i_1} \bar{X}_1^{j_1} \ldots X_m^{i_m} \bar{X}_m^{j_m}$, while B is a $1 \times d(p, m)$ vector consisting of the corresponding $b_{j_1 i_1 \ldots j_m i_m}$'s. In the notation thus established, elements $q(w)$ and $\pi(w)$ of $\mathcal{P}^n$ can be expressed in the form

$$q(w) = QW, \quad \pi(w) = \Pi W, \tag{D.8}$$

where Q and Π are $n \times d(p, m)$ matrices.
  Note that

$$\frac{\partial X_i^{i_1} \bar{X}_1^{j_1} \ldots X_m^{i_m} \bar{X}_m^{j_m}}{\partial w} Sw = \lambda_{i_1 j_1 \ldots i_m j_m} X_i^{i_1} \bar{X}_1^{j_1} \ldots X_m^{i_m} \bar{X}_m^{j_m}, \tag{D.9}$$

where

$$\lambda_{i_1 j_1 \ldots i_m j_m} = j((i_1 - j_1)\beta_1 + \ldots + (i_m - j_m)\beta_m). \tag{D.10}$$

Thus,

$$\frac{\partial W}{\partial w} Sw = \tilde{S}W \tag{D.11}$$

where $\tilde{S}$ is a $d(p, m) \times d(p, m)$ diagonal matrix having all the eigenvalues on the imaginary axis.

In the notation introduced above, the equation (D.3) becomes

$$\Pi \tilde{S} = A\Pi W + QW \tag{D.12}$$

and this in turn reduces to the Sylvester equation

$$\Pi \tilde{S} = A\Pi + Q \tag{D.13}$$

Since the spectra of $\tilde{S}$ and A are disjoint, this equation has a unique solution Π.

◁ ◁ ◁

Using this property it is possible to prove the following result.

Proposition D.2. Let $F(x, u, w) = Ax + Bu + Dw$ and S as in (D.2). Assume that all matrices A_i have eigenvalues with negative real part. The the equation

$$\frac{\partial \pi(w)}{\partial w} Sw = F(\pi(w), \alpha(w), w), \quad \pi(0) = 0 \tag{D.14}$$

having a globally defined solution $\pi(w)$, whose entries are polynomials, in the components of w.

Proof. Set $\pi(w) = \Pi w$, $\alpha(w) = \Lambda w$, where Π and Λ are matrices of appropriate dimensions. Then observe that the equation

$$\frac{\partial \pi(w)}{\partial w} Sw = A\pi(w) + B\Lambda w + Dw \tag{D.15}$$

reduces to a Sylvester equation of the form

$$\Pi S = A\Pi + B\Lambda + D \tag{D.16}$$

which indeed has a unique solution Π because the spectra of S and A are disjoint.

Thus according to Lemma D.1, It is easy to show the existence and uniqueness of the solution $\pi(w)$ of (D.14), whose entries are homogeneous polynomials.

◁ ◁ ◁

References

1. Krstic, M., Kanellakopoulos, I., Kokotovic, P.V.: Nonlinear and Adaptive Control Design. Wiley, New York (1995)
2. Ioannou, P., Kokotovic, P.: An asymptotic error analysis of identifiers and adaptive observers in the presence of parasitics. IEEE Transactions on Automatic Control 27, 921–927 (1982)
3. Praly, L.: Towards a globally stable direct adaptive control scheme for not necessarily minimum phase systems. IEEE Transactions on Automatic Control 29, 946–949 (1984)
4. Praly, L.: Lyapunov design of stabilizing controllers for cascaded systems. IEEE Transactions on Automatic Control 36, 1177–1181 (1991)
5. Praly, L., d'Andrea Novel, B., Coron, J.M.: Lyapunov design of stabilizing controllers. In: Proceedings of the 28th IEEE Conference on Decision and Control, vol. 2, pp. 1047–1052 (1989)
6. Praly, L.: Towards an adaptive regulator: Lyapunov design with a growth condition. In: Proceedings of the 30th IEEE Conference on Decision and Control, vol. 2, pp. 1094–1099 (1991)
7. Pomet, J.B., Praly, L.: Adaptive nonlinear regulation: estimation from the lyapunov equation. IEEE Transactions on Automatic Control 37, 729–740 (1992)
8. Tao, G., Ioannou, P.A.: Model reference adaptive control for plants with unknown relative degree. IEEE Transactions on Automatic Control 38, 976–982 (1993)
9. Ioannou, P., Tsakalis, K.: A robust direct adaptive controller. IEEE Transactions on Automatic Control 31, 1033–1043 (1986)
10. Ioannou, P., Kokotovic, P.: Robust redesign of adaptive control. IEEE Transactions on Automatic Control 29, 202–211 (1984)
11. Middleton, R.H., Goodwin, G.C.: Adaptive control of time-varying linear systems. IEEE Transactions on Automatic Control 33, 150–155 (1988)
12. Middleton, R.H., et al.: Design issues in adaptive control. IEEE Transactions on Automatic Control 33, 50–58 (1988)
13. Middleton, R.H., Goodwin, G.C.: Indirect adaptive output-feedback control of a class on nonlinear systems. In: Proceedings of the 29th IEEE Conference on Decision and Control, vol. 5, pp. 2714–2719 (1990)
14. Wen, C.: A robust adaptive controller with minimal modifications for discrete time varying systems. In: Proceedings of the 31st IEEE Conference on Decision and Control, vol. 2, pp. 2132–2136 (1992)

15. Wen, C., Hill, D.J.: Adaptive linear control of nonlinear systems, vol. 35, pp. 1253–1257 (1990)
16. Wen, C.: Robustness of a simple indirect continuous time adaptive controller in the presence of bounded disturbances. In: Proceedings of the 31st IEEE Conference on Decision and Control, vol. 3, pp. 2762–2766 (1992)
17. Wen, C., Hill, D.J.: Global boundedness of discrete-time adaptive control just using estimator projection. Automatica 28, 1143–1157 (1992)
18. Wen, C., Hill, D.J.: Decentralized adaptive control of lineartime varying systems. In: Proceedings of 11th IFAC World Cengress Automatica control, Tallinn, U.S.S.R. (1990)
19. Wen, C., Hill, D.J.: Globally stable discrete time indirect decentralized adaptive control systems. In: Proceedings of the 31st IEEE Conference on Decision and Control, vol. 1, pp. 522–526 (1992)
20. Hill, D.J., Wen, C., Goodwin, G.C.: Stability analysis of decentralized robust adaptive control. System and Control Letters 11, 277–284 (1988)
21. Wen, C., Hill, D.J.: Robustness of adaptive control without deadzones, data normalization or persistence of excitation. Automatica 25, 943–947 (1989)
22. Ioannou, P.: Decentralized adaptive control of interconnected systems. IEEE Transactions on Automatic Control 31, 291–298 (1986)
23. Ioannou, P., Kokotovic, P.: Decentralized adaptive control of interconnected systems with reduced-order models. Automatica 21, 401–412 (1985)
24. Datta, A., Ioannou, P.: Decentralized adaptive control. In: Leondes, P.C.T. (ed.) Advances in Control and Dynamic systems, Academic, San Diego (1992)
25. Datta, A., Ioannou, P.: Decentralized indirect adaptive control of interconnected systems. International Journal of Adaptive Control and Signal Processing 5, 259–281 (1991)
26. Tao, G., Kokotovic, P.V.: Adaptive control of plants with unknown dead-zone. IEEE Transactions on Automatic Control 39, 59–68 (1994)
27. Tang, X.D., Tao, G., Suresh, M.J.: Adaptive actuator failure compensation for parametric strict feedback systems and an aircraft application. Automatica 39, 1975–1982 (2003)
28. Tao, G., Kokotovic, P.V.: Adaptive control of systems with unknown output backlash. IEEE Transactions on Automatic Control 40, 326–330 (1995)
29. Tao, G., Kokotovic, P.V.: Contonuous-time adaptive control of systems with unknown backlash. IEEE Transactions on Automatic Control 40, 1083–1087 (1995)
30. Lozano, R., Brogliato, B.: Adaptive control of a simple nonlinear system without a priori information on the plant parameters. IEEE Transactions on Automatic Control (1992)
31. Wen, C.: Decentralized adaptive regulation. IEEE Transactions on Automatic Control 39, 2163–2166 (1994)
32. Wen, C., Soh, Y.C.: Decentralized adaptive control using integrator backstepping. Automatica 33, 1719–1724 (1997)
33. Zhang, Y., Wen, C., Soh, Y.C.: Robust decentralized adaptive stabilization of interconnected systems with guaranteed transient performance. Automatica 36, 907–915 (2000)
34. Zhang, Y., Wen, C., Soh, Y.C.: Adaptive backstepping control design for systems with unknown high-frequency gain. IEEE Transactions on Automatic Control 45, 2350–2354 (2000)
35. Ding, Z.: Adaptive asymptotic tracking of nonlinear output feedback systems under unknown bounded disturbances. System and Science 24, 47–59 (1998)

36. Zhang, Y., Wen, C., Soh, Y.C.: Robust adaptive control of uncertain discrete-time systems. Automatica 35, 321–329 (1999)
37. Wen, C., Zhang, Y., Soh, Y.C.: Robustness of an adaptive backstepping controller without modification. Systems & Control Letters 36, 87–100 (1999)
38. Zhang, Y., Wen, C., Soh, Y.C.: Discrete-time robust adaptive control for nonlinear time-varying systems. IEEE Transactions on Automatic Control 45, 1749–1755 (2000)
39. Zhang, Y., Wen, C., Soh, Y.C.: Robust adaptive control of nonlinear discrete-time systems by backstepping without overparameterization. Automatica 37, 551–558 (2001)
40. Zhou, J., Wen, C., Zhang, Y.: Adaptive output control of a class of time-varying uncertain nonlinear systems. Journal of Nonlinear Dynamics and System Theory 5, 285–298 (2005)
41. Lewis, F.L., et al.: Dead-zone compensation in motion control systems using adaptive fuzzy logic control. IEEE Transactions on Control System Technology 7, 731–741 (1999)
42. Selmis, R.R., Lewis, F.L.: Dead-zone compensation in motion control systems using neural networks. IEEE Transactions on Automatic Control 45, 602–613 (2000)
43. Su, C.Y., et al.: Robust adaptive control of a class of nonlinear systems with unknown backlash-like hysteresis. IEEE Transactions on Automatic Control 45(12), 2427–2432 (2000)
44. Ahmad, N.J., Khorrami, F.: Adaptive control of systems with backlash hysteresis at the input. In: Proceedings of the American Control Conference, pp. 3018–3022 (1999)
45. Pare, T.E., How, J.P.: Robust stability and performance analysis of systems with hysteresis nonlinearities. In: Proceedings of the American Control Conference, pp. 1904–1908 (1998)
46. Feng, G.: Robust adaptive control of input rate constrained discrete time systems. In: Adaptive Control of Nonsmooth Dynamic Systems, pp. 333–348 (2001)
47. Zhang, C.: Adaptive control with input saturation constraints. In: Adaptive Control of Nonsmooth Dynamic Systems, pp. 361–381 (2001)
48. Tao, G., Lewis, F.L.: Adaptive Control of Nonsmooth Dynamic Systems. Springer, London (2001)
49. Tao, G., Kokotovic, P.V.: Adaptive Control of Systems with Actuator and Sensor Nonlinearities. John Willey & Sons, New York (1996)
50. Recker, D., Kokotovic, P.V., Rhode, D., Winkelman, J.: Adaptive nonlinear control of systems containing a dead-zone. In: Proceedings of 30th IEEE Conference on Decision and Control, pp. 2111–2115 (1991)
51. Tian, M., Tao, G.: Adaptive control of a class of nonlinear systems with unknown dead-zones. In: Proceedings of the 13th World Congress of IFAC, vol. E, pp. 209–213 (1996)
52. Corradini, M.L., Orlando, G.: Robust stabilization of nonlinear uncertain plants with backlash or dead zone in the actuator. IEEE Transactions on Control Systems Technology 10, 158–166 (2002)
53. Cho, H., Bai, E.W.: Convergence results for an adaptive dead zone inverse. International Journal of Adaptive Control and Signal Process 12, 451–466 (1998)
54. Senjyu, T., et al.: Position control of ultrasonic motors using adaptive backstepping control and dead-zone compensation with fuzzy inference. In: Proceeding of the IEEE ICIT, Bangkok, Thailand, pp. 560–565 (2002)

55. Sun, X., Zhang, W., Jin, Y.: Stable adaptive control of backlash nonlinear systems with bounded disturbance. In: Proceedings of the IEEE Conference on Decision and Control, pp. 274–275 (1992)
56. Tao, G., Kokotovic, P.V.: Adaptive control of plants with unknown hysteresis. IEEE Transactions on Automatic Control 40, 200–212 (1995)
57. Tao, G.: Adaptive control design and analysis. John Willey & Sons, New York (2003)
58. Shieh, H.J., et al.: Adaptive tracking control solely using displacement feedback for a piezo-positioning mechanism. In: IEE Proceedings on Control Theory and Applications, vol. 151, pp. 653–660 (2004)
59. Karason, S.P., Annaswamy, A.M.: Adaptive control in the presence of input constraints. IEEE Transactions on Automatic Control 39, 2325–2330 (1994)
60. Chaoui, F.Z., et al.: Adaptive tracking with saturating input and controller integration action. IEEE Transactions on Automatic Control 43, 1638–1643 (1998)
61. Nicolao, G.D., Scattolini, R., Sala, G.: An adaptive predictive regulator with input saturations*. Automatica 32, 597–601 (1996)
62. Chaoui, F.Z., Giri, F., M'Saad, M.: Adaptive control of input-constrained tyoe-1 plants stabilization and tracking. Automatica 37, 197–203 (2001)
63. Annaswamy, A.M., et al.: Adaptive control of a class of time-delay systems in the presence of saturation. In: Adaptive Control of Nonsmooth Dynamic Systems, pp. 289–310 (2001)
64. Kim, Y.H., Lewis, F.L.: Reinforcement adaptive control of a class of neural-net-based friction compensation control for high speed and precision. IEEE Transactions on Control System Technology 8, 118–126 (2000)
65. Polycarpou, M., Farrell, J., Sharma, M.: On-line approximation control of uncertain nonliner systems: Issues with control input saturation. In: Proceedings of the American Control Conference, Denver, Colorado, pp. 543–548 (2003)
66. Seidl, D., et al.: Neural network compensation of gear backlash hysteresis in position-controlled mechanisms. IEEE Transactions on Industrial Application 31, 1475–1483 (1995)
67. Selmic, R., Lewis, F.: Backlash compensation in nonlinear systems using dynamic inversion by neural networks. In: International Conference on Control Applications, vol. 1, pp. 1163–1168 (1999)
68. Selmic, R., Lewis, F.: Deadzone compensation in nonlinear systems using neural networks. In: Proceeding of IEEE Conference on Decision and Control, vol. 1, pp. 513–519 (1998)
69. Kim, J., et al.: A two layered fuzzy logic controller for systems with deadzones. IEEE Control Systems 41, 155–162 (1994)
70. Lewis, F., et al.: Adaptive fuzzy logic compensation of actuator deadzones. Journal of Robotic Systems 14, 501–511 (1997)
71. Woo, K., et al.: A fuzzy system compensator for backlash. In: Proceedings of IEEE International Conference on Robotics Automation, vol. 1, pp. 181–186 (1998)
72. Jang, J.O.: A deadzone compensator of a dc motor system using fuzzy logic control. IEEE Transactions on Systems, Man and Cybernetics, Part C 31, 42–48 (2001)
73. Wang, X.S., Hong, H., Su, C.Y.: Model reference adaptive control of continuous-time systems with an unknown input dead-zone. In: IEE Proceedings on Control Theory Applications, vol. 150, pp. 261–266 (2003)
74. Wang, X.S., Su, C.Y., Hong, H.: Robust adaptive control of a class of nonlinear system with unknown dead zone. In: Proceedings of the 40th IEEE Conference on Decision and Control, Orlando, Florida USA, pp. 1627–1632 (2001)

75. Azenha, A., Machado, J.: Variable structure control of robots with nonlinear friction and backlash at the joints. In: Proceedings of IEEE International Conference on Robotics Automation, vol. 1, pp. 366–371 (1996)
76. Corradini, M.L., Orlando, G.: Robust practical stabilization of nonlinear uncertain plants with input and output nonsmooth nonlinearities. IEEE Transactions on Control System Technology 11, 196–203 (2003)
77. Hwang, C.L., Chen, Y.M., Jan, C.: Trajectory tracking of large-displacement piezoelectric actuators using a nonlinear observer-based variable structure control. IEEE Transactions on Control Systems Technology 13, 56–66 (2005)
78. Corradini, M.L., Orlando, G., Parlangeli, G.: A vsc approach for the robust stabilization of nonlinear plants with uncertain nonsmooth actuator nonlinearities - a unified framework. IEEE Transactions on Automatic Control 49, 807–813 (2004)
79. Wigren, T., Nordsjo, A.: Compensation of the rls algorithm for output nonlinearities. IEEE Transactions on Automatic Control 44, 1913–1918 (1999)
80. Marino, R., Tomei, P.: Adaptive control of linear time-varying systems. Automatica 39, 651–659 (2003)
81. Ding, Z.: Global adaptive output feedback stabilization of nonlinear system of any relative degree with unknown high frequence gains. IEEE Transactions on Automatic Control 43, 1442–1446 (1998)
82. Ding, Z.: A flat-zone modification for robust adaptive control of nonlinear output feedback systems with unknown high-frequency gains. IEEE Transactions on Automatic Control 47, 358–363 (2002)
83. Ge, S.S., Wang, J.: Robust adaptive stabilization for time-varying uncertain nonlinear systems with unknown control coefficients. In: Proceedings of the IEEE Conference on Decision and Control, pp. 3952–3957 (2002)
84. Marino, R., Tomei, P.: Nonlinear Control Design: Geometric, Adaptive and Robust. Prentice Hall, New York (1995)
85. Ling, Y., Tao, G.: Adaptive backstepping control design for linear multivariable plants. In: Proceedings of the IEEE Conference on Decision and Control, pp. 2438–2443 (1996)
86. Costa, R.R., et al.: Adaptive backstepping control design for mimo plants using factorization. In: Proceedings of the American Control Conference, pp. 4601–4606 (2002)
87. Costa, R.R., et al.: Lyapunov-based adaptive control of mimo systems. Automatica 39, 1251–1257 (2003)
88. Xu, H., Ioannou, P.A.: Robust adaptive control for a class of mimo nonlinear systems with guaranteed error bounds. IEEE Transactions on Automatic Control 48, 728–742 (2003)
89. Cavallo, A., Natale, C.: Output feedback control based on high-order sliding manifold approach. IEEE Transactions on Automatic Control 48, 469–472 (2003)
90. Zhou, J., Wen, C.Y., Zhang, Y.: Adaptive backstepping control of a class of mimo systems. In: Proceedings of IEEE International Symposium on Intelligent Control, Taiwan, pp. 204–209 (2004)
91. Nikiforov, V.O.: Adaptive nonlinear tracking with complete compensation of unknown disturbances. European Journal of Control 4, 132–139 (1998)
92. Serrani, A., Isidori, A.: Semiglobal nonlinear output regulation with adaptive internal model. In: Proceedings of 39th IEEE Conference on Decision and Control, Sydney, Australia, pp. 1649–1654 (2000)
93. Hautus, M.: Linear matrix equations with applications to regulator problem. In: Landau, I.D. (ed.) Outils and Modeles Mathematique pourl Automatique, Paris (1983)

94. Nikiforov, V.O.: Adaptive nonlinear tracking with complete compensation of unknown disturbances. European Journal of Control 4, 132–139 (1998)
95. Serrani, A., Isidori, A.: Semiglobal nonlinear output regulation with adaptive internal model. In: Proceedings of the IEEE Conference on Decision and Control, Orlando, FL, pp. 1649–1854 (2000)
96. Gavel, D.T., Siljak, D.D.: Decentralized adaptive control: structural conditions for stability. IEEE Transactions on Automatic Control 34, 413–426 (1989)
97. Huseyin, O., Sezer, M.E., Siljak, D.D.: Robust decentralized control using output feedback. In: IEE Proceedings on Control Theory & Applications, pp. 310–314 (1982)
98. Shi, L., Singh, S.K.: Decentralized adaptive controller design for large-scale systems with higher order interconnections. IEEE Transactions on Automatic Control 37, 1106–1118 (1992)
99. Wen, C., Hill, D.J.: Global boundedness of discrete-time adaptive control just using estimator projection. Automatica 28, 1143–1157 (1992)
100. Wen, C., Hill, D.J.: Decentralized adaptive control of linear time-varying systems. In: Proceedings of IFAC 11th World Congress on Automatic Control, U.S.S.R. (1990)
101. Wen, C.: Indirect robust totally decentralized adaptive control of continuous-time interconnected systems. IEEE Transactions on Automatic Control, 1122–1126 (1995)
102. Ortega, R., Herrera, A.: A solution to the decentralized stabilization problem. Systems and Control Letters 20, 299–306 (1993)
103. Ortega, R.: An energy amplification condition for decentralized adaptive stabilization. IEEE Transactions on Automatic Control 41, 285–288 (1996)
104. Wen, C., Soh, Y.C.: Decentralized model reference adaptive control without restriction on subsystem relative degree. IEEE Transactions on Automatic Control 44, 1464–1469 (1999)
105. Wen, C.: Decentralized adaptive regulation. IEEE Transactions on Automatic Control 39, 2163–2166 (1994)
106. Huseyin, O., Sezer, M.E., Siljak, D.D.: Robust decentralized control using output feedback. In: IEE Proceedings on Control Theory Applications, vol. 37, pp. 310–314 (1982)
107. Hatwell, M.S., et al.: The development of a model reference adaptive controller to control the knee joint of paralegics. IEEE Transactions on Automatic Control 36, 683–691 (1991)
108. Kurdila, A.J., Webb, G.: Compensation for distributed hysteresis operators in active structural systems. Journal of Guidance Control and Dynamics 20, 1133–1140 (1997)
109. Recker, D., et al.: Adaptive nonlinear control of systems containing a dead-zone. In: Proseedings of the 30th IEEE Conference on Decision and Control, Brighton, England, pp. 2111–2155 (1991)
110. Seidl, D.R., et al.: Neural network compensation of gear backlash hysteresis in position-controlled mechanisms. IEEE Transactions on Industry applications 31, 1475–1483 (1995)
111. Tsang, K., Li, G.: Robust nonlinear nominal model following control to overcome deadzone nonlinearities. IEEE Transactions on Control Systems 48, 177–184 (2001)
112. Stepanenko, Y., Su, C.Y.: Intelligent control of piezoelectric actuators. In: Proceeding of the 37th IEEE Conference on Decision and Control, USA, pp. 4234–4239 (1998)

113. Allin, A., Inbar, G.F.: Fns control schemes for upper limb. IEEE Transactions on Biomedical Engineering 33, 818–827 (1986)
114. Bernotas, L.A., Crago, P.E., Chizeck, H.J.: Adaptive control of electrically stimulated muscle. IEEE Transactions on Biomedical Engineering 34, 140–147 (1987)
115. Senjyu, T., Kashiwagi, T., Uezato, K.: Position control of ultrasonic motors using mrac and dead-zone compensation with fuzzy inference. Automatica 17, 265–272 (2002)
116. Grigoriadis, K.M., Fialho, I.J., Zhang, F.: Linear parameter-varying anti-windup control for active microgravity isolation, ISSO-Annual Report, University of Houston, pp. 52–57 (2003)
117. Favez, J.Y., Mullhaupt, P., Bonvin, D.: Enhancing tokamak control given power supply voltage saturation. In: Proceedings of ICALEPCS, Gyeongju, Korea, pp. 40–42 (2003)
118. Pachter, M., Miller, R.B.: Manual flight control with saturating actuators. IEEE Control Systems Magazine 18, 10–20 (1998)
119. Hong, S.K., et al.: Torque calculation of hysteresis motor using vector hysteresis model. IEEE Transactions on Magnetics 36, 1932–1935 (2000)
120. Zhou, J., Wen, C., Zhang, Y.: Adaptive backstepping control of a class of uncertain nonlinear systems with unknown backlash-like hysteresis. IEEE Transactions on Automatic Control 49, 1751–1757 (2004)
121. Zhou, J., Wen, C.Y., Zhang, Y.: Adaptive output feedback control of linear systems preceded by unknown backlash-like hysteresis. In: Proceedings of IEEE Conference on Cybernetics and Intelligent Systems, Singapore, pp. 12–17 (2004)
122. Tao, G., Ma, X., Ling, Y.: Optimal and nonlinear decoupling control of systems with sandwiched backlash. Automatica 37, 165–176 (2001)
123. Zhou, J., Zhang, C.J., Wen, C.: Robust adaptive output control of uncertain nonlinear plants with unknown backlash nonlinearity. IEEE Transactions on Automatic Control 52, 503–509 (2007)
124. Zhou, J., Wen, C., Zhang, C.: Robust output control of nonlinear systems with unknown backlash. In: Proceedings of American Control Conference, NewYork, USA, pp. 498–503 (2007)
125. Zhou, J., Wen, C., Zhang, Y.: Adaptive output control of nonlinear systems with uncertain dead-zone nonlinearity. IEEE Transactions on Automatic Control 51, 504–511 (2006)
126. Taware, A., Tao, G., Teolis, C.: Design and analysis of a hybrid control scheme for sandwich nonsmooth nonlinear systems. Automatica 40, 145–150 (2002)
127. Mittal, S., Menq, C.-H.: Hysteresis compensation in electromagnetic actuators through preisach model inversion. IEEE/ASME Transactions on Mechatronics 5, 394–409 (2000)
128. Lamba, H., et al.: Subharmonic ferroresonance in an lcr circuit with hysteresis. IEEE Transactions on Magnetics 33, 2495–2500 (1997)
129. Adly, A.A.: Performance simulation of hysteresis motors using accurate rotor media models. IEEE Transactions on Magnetics 31, 3542–3544 (1995)
130. Brokate, M., Sprekels, J.: Hysteresis and Phase Transition. Springer, Heidelberg (1996)
131. Tan, X., Baras, J.S.: Adaptive identification and control of hysteresis in smart materials. IEEE Transactions on Automatic Control 50, 1469–1480 (2005)
132. Moheimani, S.O.R., Goodwin, G.C.: Guest editorial introduction to the special issue on dynamics and control of smart structures. IEEE Transactions on Control Systems Technology 9, 3–4 (2001)

133. Wen, C., Zhou, J.: Decentralized adaptive stabilization in the presence of unknown backlash-like hysteresis. Automatica 43, 426–440 (2007)
134. Macki, J.W., Nistri, P., Zecca, P.: Mathematical models for hysteresis. SIAM Review 35, 94–123 (1993)
135. Coleman, B.D., Hodgdon, M.L.: A constitutive relation for rate-independent hysteresis in ferromagnetically soft materials. International Journal of Engineering Science 24, 897–919 (1986)
136. Hodgdon, M.L.: Applications of a theory fferromagnetic hysteresis. IEEE Transactions on Magnetics 24, 218–221 (1988)
137. Hodgdon, M.L.: Mathematical theory and calculations of magnetic hysteresis curves. IEEE Transactions on Magnetics 24, 3120–3122 (1988)
138. Sezer, M.E., Siljak, D.D.: On decentralized stabilization and structure of linear large scale systems. Automatica 17, 641–644 (1981)
139. Sezer, M.E., Siljak, D.D.: Robustness of suboptimal control: Gain and phase margin. IEEE Transactions on Automatic Control 26, 907–911 (1981)
140. Wang, X.S., Su, C.Y., Hong, H.: Robust adaptive control of a class of nonlinear system with unknown dead zone. Automatica 40, 407–413 (2003)
141. Zhou, J., Wen, C.Y., Zhang, Y.: Adaptive backstepping control of a class of uncertain nonlinear systems with unknown dead-zone. In: Proceedings of IEEE Conference on Robotics, Automation and Mechatronics, Singapore, pp. 513–518 (2004)
142. Cho, H.Y., Bai, E.W.: Convergence results for an adaptive dead zone inverse. International Journal of Adaptive Control and Signal Processing 12, 451–466 (1998)
143. Kapoor, N., Teel, A.R., Daoutidis, P.: An anti-windup design for linear systems with input saturation. Automatica 34, 559–574 (1998)
144. Bemporad, A., Teel, A.R., Zaccarian, L.: Anti-windup synthesis via sampled-data piecewise affine optimal control. Automatica 40, 549–562 (2004)
145. Fliegner, T., Logemann, H., Ryan, E.P.: Low-gain integral control of continuous-time linear systems subject to input and output nonlinearities. Automatica 39, 455–462 (2003)
146. Grognard, F., Sepulchre, R., Bastin, G.: Improving the performance of low-gain designs for bounded control of linear systems. Automatica 38, 1777–1782 (2002)
147. Chaoui, F.Z., Giri, F., M'Saad, M.: Asymptotic stabilization of linear plants in the presence of input and output saturations. Automatica 37, 37–42 (2001)
148. Jagannathan, S., Hameed, M.: Adaptive force-banlancing control of mems gyroscope with actuator limits. In: Proceedings of American Control Conference, Boston, Massachusetts, pp. 1862–1867 (2004)
149. Zhou, J., Wen, C.: Robust adaptive control of uncertain nonlinear systems in the presence of input saturation. In: Proceedings of 14th IFAC Symposium on System Identification, Newcastle, Australia (2006)
150. Ikhouane, F., Manosa, V., Rodellar, J.: Adaptive control of a hysteretic structural systems. Automatica 41, 225–231 (2005)
151. Smyth, A.W., et al.: On-line parametric identification of mdof nonlinear hysteretic systems. Journal of Engineering Mechanics, 133–142 (1999)
152. Sato, T., Qi, K.: Adaptive h_∞ filter: Its applications to structural identification. Journal of Engineering Mechanics, 1233–1240 (1998)
153. Benedettini, F., Capecchi, D., Vestroni, F.: Identification of hysteretic oscillators under earthquake loading by nonparametric models. Journal of Engineering Mechanics, ASCE 121, 606–612 (1995)
154. Chassiakos, A.G., et al.: Adaptive methods for identification of hysteretic structure. In: Proceedings of the American Control Conference, pp. 2349–2353 (1995)

155. Chassiakos, A.G., et al.: On-line identification of hysteretic systems. Journal of Applied Mechanics 65, 194–203 (1998)
156. Wen, Y.: Method for random vibration of hysteretic systems. Journal of Engineering Mechanics, ASCE 35, 249–263 (1976)
157. Spencer, B.D., Sain, S.M., Carlson, F.: Phenomenological model for magnetorheological dampers. Journal of Engineering Mechanics, ASCE 123, 230–238 (1997)
158. Battaini, M., Casciati, F.: Chaotic behavior of hysteretic oscillators. Journal of Structural Control 3, 4, 7–19 (1996)
159. Barbat, A.H., Bozzo, L.: Seismic analysis of base isolated buildings. Archiv. Comput. Methods Engineering 4, 153–192 (1997)
160. Sain, P.M.: Models for hysteresis and application to structural control. In: Proceedings of the American Control Conference, pp. 16–20 (1997)
161. Soong, T.T., Dargush, G.F.: Passive energy dissipation systems in structural engineering. Wiley, Chichester (1997)
162. Zhou, J., Wen, C., Cai, W.: Adaptive control of a base isolated system for protection of building structures. Journal of Vibration and Acoustics-Transactions of the ASME 128, 261–268 (2006)
163. Kelly, J.M., Leitmann, G., Soldatos, A.: Robust control of base-isolated structures under earthquake excitation. Journal of Optimization Theory and Applications 53, 159–181 (1987)
164. Khalil, H.: Nonlinear systems. Upper Saddle River, MacMillan, NJ (1992)
165. Ge, P., Jouaneh, M.: Tracking control of a piezoceramic actuator. IEEE Transactions on Control System Technology 4, 209–216 (1996)
166. Jung, S.B., Kim, S.W.: Improvement of scanning accuracy of pzt piezoelectric actuator by feed-forward model-reference control. Precision Engineering 16, 49–55 (1994)
167. Hwang, C.L., Jan, C.: A reinforcement discrete neuro-adaptive control for unknown piezoelectric actuator systems with dominant hysteresis 14, 66–78 (2003)
168. Lin, F.-J., Shieh, H.-J., Huang, P.-K.: Adaptive wavelet neural network control with hysteresis estimation for piezo-positioning mechanism 17, 432–444 (2006)
169. Canudas-de-Wit, C., et al.: A new model for control of systems with friction. IEEE Transactions on Automatic Control 40, 419–425 (1995)
170. Armstrong-Htlouvry, B.: Control of Machines with Friction. Kluwer, Boston, MA (1991)
171. Byrnes, C.I., Priscoli, F.D., Isidori, A.: Output Regulation of Uncertain Nonlinear Systems. Birkhauser, Boston (1999)

Lecture Notes in Control and Information Sciences

Edited by M. Thoma, M. Morari

Further volumes of this series can be found on our homepage:
springer.com

Vol. 372: Zhou J.; Wen C.
Adaptive Backstepping Control of Uncertain
Systems
241 p. 2008 [978-3-540-77806-6]

Vol. 371: Blondel V.D.; Boyd S.P.;
Kimura H. (Eds.)
Recent Advances in Learning and Control
279 p. 2008 [978-1-84800-154-1]

Vol. 370: Lee S.; Suh I.H.;
Kim M.S. (Eds.)
Recent Progress in Robotics:
Viable Robotic Service to Human
410 p. 2008 [978-3-540-76728-2]

Vol. 369: Hirsch M.J.; Pardalos P.M.;
Murphey R.; Grundel D.
Advances in Cooperative Control and
Optimization
423 p. 2007 [978-3-540-74354-5]

Vol. 368: Chee F.; Fernando T.
Closed-Loop Control of Blood Glucose
157 p. 2007 [978-3-540-74030-8]

Vol. 367: Turner M.C.; Bates D.G. (Eds.)
Mathematical Methods for Robust and Nonlinear
Control
444 p. 2007 [978-1-84800-024-7]

Vol. 366: Bullo F.; Fujimoto K. (Eds.)
Lagrangian and Hamiltonian Methods for
Nonlinear Control 2006
398 p. 2007 [978-3-540-73889-3]

Vol. 365: Bates D.; Hagström M. (Eds.)
Nonlinear Analysis and Synthesis Techniques for
Aircraft Control
360 p. 2007 [978-3-540-73718-6]

Vol. 364: Chiuso A.; Ferrante A.;
Pinzoni S. (Eds.)
Modeling, Estimation and Control
356 p. 2007 [978-3-540-73569-4]

Vol. 363: Besançon G. (Ed.)
Nonlinear Observers and Applications
224 p. 2007 [978-3-540-73502-1]

Vol. 362: Tarn T.-J.; Chen S.-B.;
Zhou C. (Eds.)
Robotic Welding, Intelligence and Automation
562 p. 2007 [978-3-540-73373-7]

Vol. 361: Méndez-Acosta H.O.; Femat R.;
González-Álvarez V. (Eds.):
Selected Topics in Dynamics and Control of
Chemical and Biological Processes
320 p. 2007 [978-3-540-73187-0]

Vol. 360: Kozlowski K. (Ed.)
Robot Motion and Control 2007
452 p. 2007 [978-1-84628-973-6]

Vol. 359: Christophersen F.J.
Optimal Control of Constrained
Piecewise Affine Systems
190 p. 2007 [978-3-540-72700-2]

Vol. 358: Findeisen R.; Allgöwer
F.; Biegler L.T. (Eds.): Assessment and Future Directions of Nonlinear
Model Predictive Control
642 p. 2007 [978-3-540-72698-2]

Vol. 357: Queinnec I.; Tarbouriech
S.; Garcia G.; Niculescu S.-I. (Eds.):
Biology and Control Theory: Current Challenges
589 p. 2007 [978-3-540-71987-8]

Vol. 356: Karatkevich A.:
Dynamic Analysis of Petri Net-Based Discrete
Systems
166 p. 2007 [978-3-540-71464-4]

Vol. 355: Zhang H.; Xie L.:
Control and Estimation of Systems with
Input/Output Delays
213 p. 2007 [978-3-540-71118-6]

Vol. 354: Witczak M.:
Modelling and Estimation Strategies for Fault
Diagnosis of Non-Linear Systems
215 p. 2007 [978-3-540-71114-8]

Vol. 353: Bonivento C.; Isidori A.; Marconi L.;
Rossi C. (Eds.)
Advances in Control Theory and Applications
305 p. 2007 [978-3-540-70700-4]

Vol. 352: Chiasson, J.; Loiseau, J.J. (Eds.)
Applications of Time Delay Systems
358 p. 2007 [978-3-540-49555-0]

Vol. 351: Lin, C.; Wang, Q.-G.; Lee, T.H., He, Y.
LMI Approach to Analysis and Control of
Takagi-Sugeno Fuzzy Systems with Time Delay
204 p. 2007 [978-3-540-49552-9]

Vol. 350: Bandyopadhyay, B.; Manjunath, T.C.;
Umapathy, M.
Modeling, Control and Implementation of Smart
Structures 250 p. 2007 [978-3-540-48393-9]

Vol. 349: Rogers, E.T.A.; Galkowski, K.;
Owens, D.H.
Control Systems Theory
and Applications for Linear
Repetitive Processes
482 p. 2007 [978-3-540-42663-9]

Vol. 347: Assawinchaichote, W.; Nguang, K.S.;
Shi P.
Fuzzy Control and Filter Design
for Uncertain Fuzzy Systems
188 p. 2006 [978-3-540-37011-6]

Vol. 346: Tarbouriech, S.; Garcia, G.; Glattfelder,
A.H. (Eds.)
Advanced Strategies in Control Systems
with Input and Output Constraints
480 p. 2006 [978-3-540-37009-3]

Vol. 345: Huang, D.-S.; Li, K.; Irwin, G.W. (Eds.)
Intelligent Computing in Signal Processing
and Pattern Recognition
1179 p. 2006 [978-3-540-37257-8]

Vol. 344: Huang, D.-S.; Li, K.; Irwin, G.W. (Eds.)
Intelligent Control and Automation
1121 p. 2006 [978-3-540-37255-4]

Vol. 341: Commault, C.; Marchand, N. (Eds.)
Positive Systems
448 p. 2006 [978-3-540-34771-2]

Vol. 340: Diehl, M.; Mombaur, K. (Eds.)
Fast Motions in Biomechanics and Robotics
500 p. 2006 [978-3-540-36118-3]

Vol. 339: Alamir, M.
Stabilization of Nonlinear Systems Using
Receding-horizon Control Schemes
325 p. 2006 [978-1-84628-470-0]

Vol. 338: Tokarzewski, J.
Finite Zeros in Discrete Time Control Systems
325 p. 2006 [978-3-540-33464-4]

Vol. 337: Blom, H.; Lygeros, J. (Eds.)
Stochastic Hybrid Systems
395 p. 2006 [978-3-540-33466-8]

Vol. 336: Pettersen, K.Y.; Gravdahl, J.T.;
Nijmeijer, H. (Eds.)
Group Coordination and Cooperative Control
310 p. 2006 [978-3-540-33468-2]

Vol. 335: Kozłowski, K. (Ed.)
Robot Motion and Control
424 p. 2006 [978-1-84628-404-5]

Vol. 334: Edwards, C.; Fossas Colet, E.;
Fridman, L. (Eds.)
Advances in Variable Structure and Sliding Mode
Control
504 p. 2006 [978-3-540-32800-1]

Vol. 333: Banavar, R.N.; Sankaranarayanan, V.
Switched Finite Time Control of a Class of
Underactuated Systems
99 p. 2006 [978-3-540-32799-8]

Vol. 332: Xu, S.; Lam, J.
Robust Control and Filtering of Singular Systems
234 p. 2006 [978-3-540-32797-4]

Vol. 331: Antsaklis, P.J.; Tabuada, P. (Eds.)
Networked Embedded Sensing and Control
367 p. 2006 [978-3-540-32794-3]

Vol. 330: Koumoutsakos, P.; Mezic, I. (Eds.)
Control of Fluid Flow
200 p. 2006 [978-3-540-25140-8]

Vol. 329: Francis, B.A.; Smith, M.C.; Willems,
J.C. (Eds.)
Control of Uncertain Systems: Modelling,
Approximation, and Design
429 p. 2006 [978-3-540-31754-8]

Vol. 328: Loría, A.; Lamnabhi-Lagarrigue, F.;
Panteley, E. (Eds.)
Advanced Topics in Control Systems Theory
305 p. 2006 [978-1-84628-313-0]

Vol. 327: Fournier, J.-D.; Grimm, J.; Leblond, J.;
Partington, J.R. (Eds.)
Harmonic Analysis and Rational Approximation
301 p. 2006 [978-3-540-30922-2]

Vol. 326: Wang, H.-S.; Yung, C.-F.; Chang, F.-R.
H_∞ Control for Nonlinear Descriptor Systems
164 p. 2006 [978-1-84628-289-8]

Vol. 325: Amato, F.
Robust Control of Linear Systems Subject to
Uncertain
Time-Varying Parameters
180 p. 2006 [978-3-540-23950-5]

Vol. 324: Christofides, P.; El-Farra, N.
Control of Nonlinear and Hybrid Process Systems
446 p. 2005 [978-3-540-28456-7]

Vol. 323: Bandyopadhyay, B.; Janardhanan, S.
Discrete-time Sliding Mode Control
147 p. 2005 [978-3-540-28140-5]

Vol. 322: Meurer, T.; Graichen, K.; Gilles, E.D.
(Eds.)
Control and Observer Design for Nonlinear Finite
and Infinite Dimensional Systems
422 p. 2005 [978-3-540-27938-9]

Vol. 321: Dayawansa, W.P.; Lindquist, A.;
Zhou, Y. (Eds.)
New Directions and Applications in Control
Theory
400 p. 2005 [978-3-540-23953-6]

Vol. 320: Steffen, T.
Control Reconfiguration of Dynamical Systems
290 p. 2005 [978-3-540-25730-1]

Printing: Krips bv, Meppel, The Netherlands
Binding: Stürtz, Würzburg, Germany